Production Operations

Well Completions, Workover,
and Stimulation
Volume 2

Second Edition

Second Edition

Volume 2

Production Operations

Well Completions, Workover, and Stimulation

Thomas O. Allen
and
Alan P. Roberts

Oil & Gas Consultants International, Inc.
Tulsa

Copyright © 1978, 1982
Oil & Gas Consultants International, Inc.
4554 South Harvard Avenue
Tulsa, Oklahoma 74135

All Rights Reserved
This book is sold for personal or library
reference use, or for regularly scheduled
university classes only. Use as a reference
or manual for adult training programs is
specifically reserved for Oil & Gas
Consultants International. This book or any
part of it may not be reproduced in any form,
nor transmitted, nor translated into a
machine language without the written
permission of the publisher.

The ideas and techniques presented herein are correct
to the best of our knowledge and ability. However,
we do not assume responsibility for results
obtained through their application.

Library of Congress Catalog Card Number: 77-13239
International Standard Book Number: 0-930972-05-8
Production Operations Set (Volumes 1 and 2) ISBN: 0-930972-03-1
Printed in U.S.A.
First Edition—March, 1978
Second printing—April, 1979
Third printing—March, 1981
Fourth printing—January, 1982
Second Edition—May, 1982
Second printing—February, 1984

Preface

The second edition of Production Operations includes revisions and updating of many chapters in volumes 1 and 2. In volume 2, a new chapter entitled "Paraffin and Asphaltene Deposition, Removal, and Prevention" has been added. Major revisions have been made in the Corrosion chapter, with extensive technical coverage of coatings and plastic pipe. Appendices covering Rock Mechanics and Oil Field Polymers have been added to the Fracturing chapter. Because an ever increasing number of universities and colleges in the world employ OGCI's *Production Operations* books as basic texts in production operations, the second edition includes additional clarification in major technical areas as well as required technical updating.

In an overview of the oil industry, well completion, workover, and stimulation seemingly plays only a small part. The same is true even if we limit ourselves to the exploration and production phases of the Industry. From our vantage, however, the focal point of exploration and production is a successful well completion that obtains, and maintains, effective communication with the desired reservoir fluids. The technology required for effective well completion involves many disciplines and many different types of talents. A well completion is not merely a mechanical process of drilling a hole, setting casing, and perforating a hydrocarbon section.

The importance of total reservoir description; the role of effective communication between the reservoir and the wellbore; the hazards of flow restrictions around the wellbore; the importance of knowing where fluids are and where they are moving to; and the rigors of excluding undesirable fluids all become more and more evident as we move deeper into the areas of enhanced methods of maximizing recovery of increasingly valuable hydrocarbon fluids.

In preparing *Production Operations*, Volume 1 and Volume 2, we have tried to logically separate well completion and well operation technology into packages to permit detailing the more important facets. Effective well completion and recompletion operations require consideration of specific problems using all available technology.

Volumes 1 and 2 are the product of some sixteen years of conducting training programs throughout the world for Industry groups, including engineers, managers, geologists, technicians, foremen, service company personnel, and others.

The question is often asked, "What's new in well completion technology?" Our answer must be that new technology per se is not the real issue in considering improvement in producing operations. "The key to optimizing oil and gas recovery and profits is the *effective application of proved technology*." This has been the theme of our Production Operations courses since our first effort in 1966, and is the theme of these two books on Production Operations. A primary objective of our technical training has been to assist operating groups reduce the length of time required for "proved techniques" to become routine field practice.

The business of well completion is continually changing. The learning process continues, technology improves, and just as important, the rules of the game change with the times and with the area. In many areas, effective and economic recovery of hydrocarbons from more and more marginal reservoirs is the name of the game. In other areas where costs are tremendous due to the complications of deep wells, offshore activities, or geographic location, high production rates, which are needed to provide sufficient return on the incomprehensible investment required, provide the winning combination.

Response to the first four printings of *Production Operations*, volumes 1 and 2, reflects Industry acceptance of our efforts. We anticipate that the improved second edition will be even more valuable for production operating personnel than the first edition of *Production Operations*. The widespread awareness of the need to update petroleum personnel at all levels in the application of proved technology provides OGCI with the incentive to invest time and money in providing new and improved training courses and books. To meet this need, OGCI is offering additional courses each year and is in the process of developing a series of technical books for the Petroleum Industry.

T. O. Allen
Alan P. Roberts

Tulsa, Oklahoma
May 1982

Acknowledgments

Authors always receive many different kinds of assistance from many different people. We wish to acknowledge a number of these contributions by name and also give our thanks to many others, as well.

We owe a special debt of gratitude to the various oil-well service companies. Without their help, there would be no oil industry. Nor would there be any book on Production Operations. We particularly want to acknowledge extensive help and counsel from Baker Oil Tools, Dowell Division of Dow Chemical Co., Dresser Industries, Halliburton Company, and Schlumberger.

For assistance on the first edition, special recognition is due Dr. Scott P. Ewing (deceased), who prepared the orginal writeup of the chapter on Corrosion (Vol. 2), and to Wallace J. Frank, who contributed a great deal to the first edition.

C. Robert Fast, OGCI; and C. C. McCune, Chevron Research, reviewed the chapter on Acidizing (Vol. 2) and made valuable suggestions. Dr. D. A. Busch, Dr. P. A. Dickey, Dr. G. M. Friedman, and Dr. Glenn Visher, OGCI, rendered valuable assistance in the preparation of the chapter on Geology (Vol. 1).

C. P. Coppel, Chevron Research, reviewed the chapter on Surfactants (Vol. 2). Norman Clark (deceased) and Dr. Charles Smith made valuable technical contributions to the Reservoir Engineering chapter (Vol. 1). Ray Leibach, OGCI, assisted in the preparation of the chapter on Sand Control (Vol. 2). G. W. Tracy, OGCI, reviewed the section on production testing, Well Testing chapter, Vol. 1. John E. Eckel, OGCI, contributed to the chapter on Downhole Production Equipment.

For work and suggestions on the second edition, Wayne Hower made valuable contributions in the revision of several chapters. We also wish to acknowledge the assistance of C. Robert Fast, Raymond E. Leibach, and Carl E. Montgomery, who had many helpful suggestions for the revision of the Hydraulic Fracturing chapter. Also, Raymond E. Leibach made a significant contribution on the revision of "Tubing Strings, Packers, Subsurface Control Equipment." And we wish to thank Dwight Smith for his assistance in updating and revising the Primary Cementing chapter.

Our thanks and appreciation go, too, to the various operating companies that have participated in our Production Operations course sessions. They have helped hone and refine many generations of lecture notes into this two volume textbook.

We appreciate the assistance of Gerald L. Farrer, publications editor of the first edition, and Patricia Duyfhuizen, publications editor of the second edition of the books.

Finally, we acknowledge the valuable contribution of Jewell O. Hough, who over a twelve year period prepared dozens of revisions, leading up to the publication of these books.

—The Authors

Contents Vol. 2

Chapter 1   **Problem Well Analysis**   1

Problem Wells 1
- *Limited Producing Rate 1*
- *Low Reservoir Permeability 1*
- *Low Reservoir Pressure 2*
- *Formation Damage 2*
- *Plugging of Tubing, Wellbore, and Perforations 2*
- *High Viscosity Oil 2*
- *Excessive Back Pressure on Formation 3*
- *Problems with Artificial Lift Wells 3*
- *Analysis of Problems in Rod Pumped Wells 3*
- *Analysis of Problems in Gas Lift Wells 4*
- *Typical Gas Lift Problems 4*
- *Analysis of Problems in Hydraulic Bottom-Hole Pumps 4*
- *Analysis of Problems in Electrical Submersible Pumps 4*
- *Water Production Problems in Oil or Gas Wells 5*
- *Gas Problems in Oil Wells 7*
- *Mechanical Failures 8*

Problem Well Analysis Checklist 8
- *Apparent Well Problem 8*
- *Analysis of Probable Productive Zones in Well 8*
- *Analysis of Well History 8*
- *Comparison of Well Performance with Offset Wells 9*
- *Reservoir Considerations 9*
- *Use Special Surveys, Cross Sections and Maps 9*

References 10

Chapter 2   **Paraffins and Asphaltenes**   11

Introduction 11

Paraffin and Asphaltene Chemistry 11
- *Analysis of Paraffins and Asphaltenes in Crude Oil 12*

Deposition Mechanisms 12
- *Paraffin Deposition 12*
- *Asphaltene Deposition 13*

Removal of Wax Deposits 13
- *Mechanical Removal of Wax 13*
- *Solvent Removal of Wax 14*
- *The Use of Heat for Removal of Wax 14*
- *Removal of Wax with Dispersants 15*

Preventing or Decreasing Wax Deposition 15
- *Laboratory Tests and Results 15*
- *Crystal Modifiers 16*
- *Plastic Pipe and Plastic Coatings 17*

Surfactants as Deposition Inhibitors 17
Downhole Heaters 17
Inhibiting Asphaltene Deposition 17
Production Techniques to Reduce Wax Deposition 18

Design for Wax Control 18

References 18

Chapter 3 — Squeeze Cementing—Remedial Cementing 21

Introduction 21

Theoretical Considerations 22
Fracture Mechanics 22
Cement Dehydration Process 22

Practical Considerations 23
Normal Squeeze Cementing Situation 23
Effect of Filtration Rate on Cement Placement 24
Basics of Low-Fluid Loss Cement Squeezing 24
Application of High Fluid Loss Cement 25

Special Squeeze Cementing Situations 26

Planning a Squeeze Cement Job 26
Type of Fluid 26
Plugged Perforations 26
Squeeze Packers 26
Cement Slurry Design 27
Cement Slurry Volume 28
Squeeze Pressure 28

Normal Operational Procedure 29
Mixing Cement 29
Surface Tests of Cement Slurry 29
Cement Placement 29
Filter Cake Buildup Process 29
Reversing Out Excess Cement 29
WOC 29
Evaluation 29

Other Operational Procedures 30
Circulation Squeeze 30
Block Squeeze 31

Plug-Back Operations 31

Special Cement Systems 32

References 33

Chapter 4 — Sand Control 35

Introduction 35
Definition of Sand Control 35
Critical Flow Rate Effect 35
Formation Strength Versus Drag Forces 35
Sand Control Mechanisms 36

Reduction of Drag Forces 36
Increasing Flow Area 36
Restricting Production Rate 36

Mechanical Methods of Sand Control 37
    Development of Design Criteria 37
    Formation Sand-Size Analysis 37
    Slot-Sizing Experiments 38
    Gravel-Sand Size Ratio 38
    Pack Thickness 39
    Fluctuating Flow Rate 40
    Mixing of Gravel With Sand 41
Practical Considerations in Gravel Packing 41
    Gravel Selection 42
    Screen and Liner Considerations 43
    Gravel Packing Fluids 44
    Inside Casing Gravel Pack Techniques 45
    Open Hole Gravel Pack Techniques 47
    Putting Well on Production 51
    Life of the Gravel Pack 51
Use of Screen Without Gravel 51
Resin Consolidation Methods 52
    Theory of Resin Consolidation 52
    Resin Processes 53
    Placement Techniques 53
Comparison of Sand Control Methods 55
References 56
Appendix A—Comparison Table of Standard Sieve Series 57
Appendix B—U.S. Sieve Series and Tyler Equivalents 58
Appendix C—Application Details for Eposand Resin 112
System for Sand Consolidation 59

Chapter 5

Formation Damage 63

Introduction 63
    Occurrence of Formation Damage 63
    Significance of Formation Damage 63
Basic Causes of Damage 63
    Plugging Associated With Solids 63
    Plugging Associated With the Fluid Filtrate 64
Classification of Damage Mechanisms 64
Reduced Absolute Permeability 65
    Particle Plugging Within the Formation 65
    Damage Reduction 67
Formation Clays 68
    Occurrence of Clays 68
    Clay Migration 68
    Clay Structure 68
    Effect of Water 69
    Control of Clay Damage 70
Asphaltene Plugging 71
Reduced Relative Permeability 71
    Increased Water Saturation 71
    Oil Wetting 71
    Corrective and Preventive Measures 71

Increased Fluid Viscosity (Emulsion) 73
Diagnosis of Formation Damage 73
References 73
Appendix—Formation Damage During Specific Well Operations 75

Chapter 6 — Surfactants for Well Treatments — 77

Characteristics of Surfactants 77
- Wettability 78
- Mechanics of Emulsions 78

Use and Action of Surfactants 79
- Action of Anionic Surfactants 79
- Action of Cationic Surfactants 79
- Action of Nonionic Surfactants 79
- Action of Amphoteric Surfactants 79
- Summary of Wetting Action by Anionic and Cationic Surfactants 79

Formation Damage Susceptible to Surfactant Treatment 79
- Oil-Wetting 80
- Water Blocks 80
- Emulsion Blocks 80
- Interfacial Films or Membranes 81
- Particle Blocks 81
- Change of Particle Size Affects Formation Damage 81
- High Surface and Interfacial Tension 81
- Key to Well Treating 81
- Susceptibility to Surfactant-Related Damage 82

Preventing or Removing Damage 82
- Workover Fluid Compatibility Test 82
- Surfactant-Selection Procedure to Prevent Emulsion 83
- Selection of an Emulsion Breaking Surfactant 83
- Visual Wettability Tests Based on API RP 42 83
- Requirements for Well Treating Surfactants 84

Well Stimulation With Surfactants 86
- Treatment Fluids Used 86
- Fluid Placement 86

Increasing Effectiveness of Rod Pumping 86
Prevention of Well Damage 87
References 87

Chapter 7 — Acidizing — 89

Acids Used in Well Stimulation 90
- Hydrochloric Acid 90
- Acetic Acid 90
- Formic Acid 90
- Hydrofluoric Acid 90
- Sulfamic Acid 90

Acid Additives 91

Surfactants 91
Suspending Agents 91
Sequestering Agents 91
Anti-Sludge Agents 91
Corrosion Inhibitors 92
Alcohol 92
Fluid Loss Control Agents 92
Diverting or Bridging Agents 93
Temporary Bridging Agents 93

Carbonate Acidizing 93
    Factors Controlling Acid Reaction Rate 93
    Retardation of Acid 94

Acidizing Techniques for Carbonate Formations 97
    Matrix Acidizing Carbonate Formations 97
    Fracture Acidizing Carbonate Formations 97
    Acetic-HCl Mixtures for Matrix and Fracture Acidizing 99
    Foamed Acidizing 99
    Specialty Acids 100
    Turflo Acid 100
    Use of High Strength HCl Acid 100
    Summary of Use of High Strength Acid 101

Sandstone Acidizing 101
    Reaction of HF Acid on Sand and Clay 101
    Planning HF Acid Stimulation 102
    Additives for Sandstone Acidizing 103
    Clay Stabilization 104
    Well Preparation Prior to Acidizing Sandstone Formations 104
    The Role of Mutual Solvents in Acidizing Sandstone 104
    Stimulation of Oil Wells in Sandstone Formations 106
    Stimulation of Gas Wells, Gas Injection Wells, and Water Injection Wells 107
    Procedures Contributing to Successful Sandstone Acidizing 108
    Emulsion Upsets from HF-HCl Acidizing 108
    In-Situ HF Generating System (SGMA) 108
    Sequential HF Process 108
    Clay Acid 109

Potential Safety Hazards in Acidizing 109

References 110

Chapter 8   **Hydraulic Fracturing**   113

Introduction 113
    Fracturing for Well Stimulation 113
    Fracturing Other Than for Stimulation 114

Mechanics of Fracturing 115
    Regional Rock Stresses 116
    Stress Distortion Caused by the Borehole 117
    Fracture Initiation 118
    Fracture Propagation 119
    Measuring Fracture Closure Pressure and Minimum Rock Matrix Stress 120
    Proppant Stress 121
    Vertical Containment of Fracture Growth 121
    Azimuth Prediction 123

Production Increase from Fracturing 124
   New Zones Exposed 124
   Bypassed Damage 124
   Radial Flow Changed to Linear 124
   Vertical Fracture Must be a "Super Highway" 124
   Near Wellbore Conductivity is Critical 125

Propping the Fracture 127
   Desirable Properties of Propping Agents 127
   Fracture Conductivity vs. Proppant Concentration and Load 128
   Placement of Proppant 129

Frac Fluids 131
   Fluid Properties and Modifications 132
   Fluid Loss Control 132
   Viscosity Control 133
   Rheology 137
   Proppant Carrying Ability 138
   Mixing, Storage and Handling 138
   Cost 139
   Formation Damage 139
   Frac Fluid Trade Names 141

Frac Job Design 142
   Select the Right Well 142
   Design for the Specific Well 142
   Optimize Design Over Several Jobs 143
   Utilize Calculation Procedures as a Guide 143
   Long Zones Need Special Consideration 143
   Frac Job Performance 146
   Logistical Problems 147
   Operational Problems 148

Frac Job Evaluation Techniques 149
   Extended Production Results 149
   Pressure Analysis 149
   Determination of Fracture Height 149

References 150

Appendix A—Checklist for Planning and Executing Frac Treatment 151

Appendix B—Rock Mechanics 154

Appendix C—Water Soluble Polymers 162

Chapter 9   **Scale Deposition, Removal, and Prevention**   **171**

Introduction 171
   Loss of Profit 171

Causes of Scale Deposition 171
   Tendency to Scale—$CaCo_3$ 171
   Tendency to Scale—Gypsum ($CaSO_4 \cdot 2H_2O$) or Anhydrite ($CaSO_4$) 172
   Tendency to Scale—$BaSO_4$ and $SrSO_4$ 173
   Tendency to Scale—NaCl 174
   Tendency to Scale—Iron Scales 174

Prediction and Identification of Scale 175
    Prediction of Scaling Tendencies 175
    Identification of Scale 175

Scale Removal 177
    Mechanical Methods 177
    Chemical Removal 177

Scale Prevention 178
    Inhibition of Scale Precipitation by Inorganic Polyphosphates 178
    Inhibition of Scale With Polyorganic Acid 178
    Inhibition of Scale With Organic Phosphates 179
    Inhibition of Scale With Polymers 180
    $CaCO_3$ Scale Prevention by Pressure Maintenance 180

Summary 180

References 180

Chapter 10   Corrosion Control   183

The Corrosion Process 183
    Types and Causes of Corrosion 185
    Carbon Dioxide (CO_2) or Sweet Corrosion 186
    Hydrogen Sulfide (H_2S) or Sour Corrosion 186
    Oxygen Corrosion 187
    Differential Aeration Cell 187
    Erosion Corrosion 187
    Corrosion Resulting from Bacteria 188
    Corrosion Fatigue 188

Detection and Measurement of Corrosion 188
    Finding Corrosive Environments 188
    Identify Potential Sources of Corrosion 188
    Tests for Corrosive Conditions 189
    Chemical Tests 189
    Tests for Bacteria 189
    Electrochemical Tests 189
    Tests for Pipelines 189
    Test of Current Flow in Well Casings 189
    Measurement of Corrosion Rate 189
    Visual Inspection 190
    Caliper Surveys 190
    Casing Inspection Log 191
    Ultrasonic Thickness Tests 192
    Metal Loss Rate Tests Using Coupons 192
    Hydrogen Probes 193
    Electrical Resistance Method (Corrosometer) 194
    Linear Polarization Resistance Method (Corrosion Rate Meter) 195
    Chemical Test for Corrosion Rate 195
    Corrosion Records 196

Corrosion Control 197
    Select Proper Materials to Reduce Corrosion Rate 197
    Corrosion Control Through Original Design 197
    Galvanic Cells and Corrosion-Resistant Metals 198
    Insulating Flanges or Nipples 199

CO_2 Enhanced Oil Recovery Projects 200
Coatings 200
Internal Coatings and Linings for Tubulars 201
Immersion (Tank) Coatings 203
External Pipeline Coatings 204
Inhibition With Chemicals 206
Gas Well Inhibition 206
Batch-down Tubing for Gas Wells 207
Nitrogen Squeezes for Gas Wells 207
Oil Well Inhibition 208
Removal of Corrosive Gases 209
Chemical Scavengers 209
Vacuum Deaeration 210
Gas Stripping 210
Combination Vacuum Deaeration and Gas Stripping 210
Cathodic Protection 210
Design of Cathodic Protection 211
External Protection of Well Casing 211
Cathodic-Protection Installation 213
NonMetallic Materials 215

References 216

Glossary of Corrosion Terms 217

Chapter 11 — Workover and Completion Rigs Workover Systems 219

Conventional Production Rigs 219
- Rig Utilization 219
- Production Rig Selection 220
- Depth or Load Capacity 220
- Braking Capacity 220
- Derrick Capacity 221
- Drawworks Horsepower 221
- Operational Efficiency 224
- Move In, Rig Up, Tear Down 224
- Pulling and Running Tubing 224
- Rotating and Circulating 225

Non-Conventional Workover Systems 226

Concentric Tubing Workovers 227
- Equipment for Concentric Tubing Workovers 227
- Hoisting and Rotating Unit 227
- Blowout Preventers 228
- Work String 228
- Pump 228
- Bits and Mills 228
- Coiled Tubing System 228
- Injector Hoist Unit 229
- Continuous Tubing 229
- Concentric Tubing Operating Practices 230
- Squeeze Cementing 230
- Perforating 230
- Sand Washing 230
- Sand Control 231
- Acidizing 231

Deepening 231
Through Flow Line Maintenance and Workover Techniques 232
Locomotive 232
Auxiliary Tools 232
Workover Operations 233
Surface Control Equipment 233

References 233

Chapter 12   **Workover Planning**   235

Reasons for Workover 235
   Problem Well Analysis 235

Workover to Alleviate Formation Damage 235

Workover for Low Permeability Well 236

Workover of Wells in Partially Pressure-Depleted Reservoirs 236

Workover to Reduce Water Production in Oil and Gas Wells 237

Workover to Reduce Gas Production in Oil Wells 237

Workover for Sand Control 238

Workover to Repair Mechanical Failure 238

Workover to Change Zones or Reservoirs 239

Workover to Multicomplete 239

Workover to Increase Production in Wells Producing High Viscosity (Low Gravity) Oil 239

Workover Economics 240
   Consider Workovers on a Program Basis 240
   Success of Most Workover Programs Directly Related to Preplanning and Study 240
   Economics 240

Summary 241

Symbols and Abbreviations 243

Contents Vol. 1

Chapter 1   **Geologic Considerations in Producing Operations**   1

Introduction 1
- *The Habitat of Oil and Gas 2*
- *Traps for Oil and Gas Accumulation 3*
- *Fractures and Joints in Reservoir Rocks 4*

Sandstone Reservoirs 5

Geologic Factors Affecting Reservoir Properties in Sandstone Reservoirs 10
- *Porosity and Permeability 10*
- *Shale Break Prediction from Outcrop Studies 11*
- *Effect of Silt and Clay Content on Sandstone Permeability 11*
- *Effect of Mineral Cement on Sandstone Permeability 12*
- *Permeability Variations with Texture Changes 12*
- *Relation of Permeability to Irreducible Water Saturation 12*
- *Identification of Channel or Fluviatile Sandstones 13*
- *A Comparison of Channel and Bar Deposits in a Deltaic Environment 14*
- *Geologic Control Summary for Sandstone Reservoirs 15*

Application of Geologic Concepts in Specific Sandstone Reservoirs 15
- *Reservoir Description in the Niger Delta 15*
- *Practical Use of Sedimentological Information in Oil and Gas Field Development 18*
- *Reservoir Description in the Elk City Field, Oklahoma 20*

Carbonate Reservoirs 27

Application of Geologic Concepts in Carbonate Reservoirs 30
- *Variations in the San Andres Reservoirs Significant in Well Completions and Well Stimulation 30*
- *Application of Carbonate Environmental Concepts to Well Completions in Enhanced Recovery Projects 32*
- *Geological-Engineering Team Changes Peripheral Flood to a Pattern Flood 34*
- *Detailed Reservoir Description—Key to Planning Gas-Driven Miscible Flood 37*

References 39

Glossary of Geologic Terms 40

Chapter 2   **Reservoir Considerations in Well Completions**   43

Introduction 43

Hydrocarbon Properties of Oil and Gas 43
- *Components 43*

Phases 44
Molecular Behavior 44
Pure Hydrocarbons 45
Hydrocarbon Mixtures 46
Retrograde Condensate Gas 47
Gas 48
Practical Uses of Hydrocarbon Data 48
Correlation of Properties of Oils 51

Characteristics of Reservoir Rocks 51
Porosity 51
Permeability 52
Relative Permeability 53
Wettability 55
Fluid Distribution 55

Fluid Flow in the Reservoir 57
Pressure Distribution Around the Well Bore 57
Radial Flow Around the Well Bore 58
Linear Flow Through Perforations 59
Causes of Low Flowing Bottom-Hole Pressure 60

Effects of Reservoir Characteristics on Well Completions 61
Reservoir Drive Mechanisms 61
Reservoir Homogeneity 65

References 66

Chapter 3   **Well Testing**   69

Well Production Testing 69

Introduction 69

Periodic Production Tests 69
Oil Wells 69
Gas Wells 70

Productivity or Deliverability Tests 70
Oil Wells 70
Gas Wells 72

Transient Pressure Tests 74
Basis for Transient Pressure Analysis 74
Limitations and Application 74
Description of Transient Pressure Tests 75

Drill Stem Testing 79

Background 79
Objective 79
Reservoir Characteristics Estimated 79
Basics of DST Operations 79
Pressure vs Time Plot 79
Field-Recorded DST Pressure Chart 81

Theory of Pressure Buildup Analysis 81
Reservoir Parameters Obtained by Buildup Analysis 83
Reservoir and Fluid Anomaly Indications 84
Radius of Investigation 85
Depletion 87

Reservoir Parameters (Gaseous Systems) 87
Analysis of DST Data Using Type-Curve Methods 88

Recommendations for Obtaining Good Test Data 88
Closed Chamber DST Technique 88
DST's in High Capacity Wells 90
DST's from Floating Vessels 90
Wireline Formation Tester 91

"Eyeball" Interpretation of Pressure Charts 92

References 97

Chapter 4   **Primary Cementing**   99

Introduction 99

Cementing Materials 99
Function of Cement in Oil Wells 99
Manufacture, Composition, and Characteristics of Cement 99
Selection of Cement for Specific Well Application 101
Required Properties and Characteristics of Oil Well Cements 103

Cement Additives 105
Functions of Cement Additives 105
Cement Accelerators 105
Cement Retarders 106
Fluid Loss Control for Primary Cementing 107
Slurry Viscosity Control 108
Light-Weight Additives 109
Heavy-Weight Additives 112
Lost-Circulation Additives 112

Cement Bonding 113
Bonding Measurements in the Laboratory 113
Pipe-Cement Bond 113
Formation-Cement Bond 113

Flow Properties of Primary Cements 114
Formulae for Making Flow Calculations 117

Displacement Mechanics 118
Factors Affecting Annular Flow and Mud Displacement 118
Laboratory Studies Point the Way 119
Contact Time 120
Computer Solutions 121
Summary of Important Factors 121
Practicalities 121

Cost of Primary Cementing 122

Special Problem Situations—New Developments 122
Cementing High Temperature Wells 122
Cementing in Cold Environments 124
Cementing Through Gas Zones 125
Salt Cement 126
Delayed Setting Cement 127
Expansive Cements 128
Densified Cement 128
Nylon-Fiber-Reinforced Cement 128
Thixotropic Cement 128

Primary Cementing Practices 128
Practical Considerations Before Cementing 128
Considerations During Cementing 131
Considerations After Cementing 132

References 133

Appendix—Cement Additives Comparison Chart 134

Chapter 5   **Well Completion Design**   135

Factors Influencing Well Completion Design 135
Reservoir Considerations 135
Mechanical Considerations 136
Method of Completion 136

Conventional Tubular Configurations 137
Single-Zone Completion 137
Effect of Tubing and Packer 138
Multiple-Zone Completion 140

Unconventional Tubular Configurations 142
Multiple "Tubingless" Completions 142

Sizing Production Tubulars 142
Inflow Performance Relation 142
Pressure Drop in Well Tubing 144
Pressure Drop Through Tubing Restrictions and Wellhead 147
Pressure Drop in the Flow Line 147

Completion Interval 147

References 150

Appendix—Tubingless Completion Techniques 150

Permanent Well Completion (PWC) 150

Tubingless Completions 150
Casing and Cementing Practices 151
Completing Tubingless Wells 152
Artificial Lift 152
Workover of Tubingless Wells 154

References 154

Chapter 6   **Tubing Strings, Packers, Subsurface Control Equipment**   155

Tubing Strings 155

Steel Grades 155

Tubing Connections 155
Standard API Coupling Connections 155
Extra Clearance Couplings 155
Integral Joint Connections 156
Connection Seals 156

Makeup of API Threaded Connection 157

Design of Tubing Strings 158

High Strength Tubing 158

Physical Properties of Steel 158

Contents Vol. 1

Sensitivities of High Strength Tubing 158
Tubing Inspection 160
    Methods 160
    Inspection Criteria 160
Tubing Handling Practices 160
    Mill Coupling—Buck-On Practices 160
    Loading, Transportation, and Unloading 161
Equipment for Use With High-Strength Tubing 161
    Tongs 161
    Elevators and Spiders 161
Tubing Running Practices 161

API Publications List 162

Packers and Subsurface Control Equipment 167

Production Packers 167
    General Considerations in Packer Selection 167
    Effect of Pressure and Temperature Changes 169
    Permanent Buckling or "Corkscrewing" 171
    Unseating of Weight Set Packer 172
    Retrievable Packers 172
    Permanent Packers 176
    Packer Bore Receptacle 178
    Cement Packers 179

Subsurface Control Equipment 180
    Safety Systems 180
    Surface-Controlled Safety Valves 183
    Operation Considerations 184
    Bottom-Hole Chokes and Regulators 185
    Subsurface Injection Safety Valves 186

References 162, 187

Chapter 7   **Perforating Oil and Gas Wells**   **189**

Introduction 189
Types of Perforators 189
    Bullet Perforators 189
    Jet Perforators 189
    Other Perforating Methods 191
Evaluation of Perforator Performance 191
    Development of Flow Index System 192
    Effect of Perforating in Various Fluids 192
    Effect of Formation Strength on Perforator Performance 193
    Downhole Evaluation of Perforations 196
Factors Affecting Gun Perforating Results 197
    Perforation Plugging 197
    Cleanout of Plugged Perforations 197
    Effect of Pressure Differential 197
    Effect of Clean Fluids 198
    Effect of Compressive Strength 198
    Perforation Density 198
    Cost 198
    Pressure and Temperature Limitations 198

Well Control 199
Casing and Cement Damage 199
Need for Control of Gun Clearance 200
Depth Measurements 201
Oriented Perforating 201
Penetration Versus Hole Size 201
Limitations in The Use of Exposed-Charge Jet Perforators 202

Perforating in a Clean Fluid with a Differential Pressure Into Wellbore 202

A Comparison Between the GEO-VANN System and Other Perforating Methods in Wells Equipped with Packers 204

Summary of Optimum Perforating Practices 205

References 205

Appendix 206
API Tests for Evaluation of Perforators 206
Recommended Use for API RP 43 Test Data 207

Chapter 8 — Completion and Workover Fluids — 211

Functions-Requirements-Selection Criteria 211
Fluid Density 211
Solids Content 211
Filtrate Characteristics 211
Fluid Loss 212
Viscosity-Related Characteristics 212
Corrosion Products 212
Mechanical Considerations 212
Economics 212

Formation Damage Related to Solids 212
Complete Solids Removal 212
Complete Fluid Loss Control 213

Oil Fluids—Practical Application 213
Crude Oil 213
Diesel Oil 213

Clear Water Fluids—Practical Application 213
Source of Water 213
Salt Type and Concentration for Prepared Salt Water 214
Emulsion—Wettability Problems 214
Viscosity Control 214
"Viscosity" Builders 214
Fluid Loss Control 216
Field Applications 217
Salt Solutions Where Increased Density is Required 217
Care and Maintenance of Clean Salt Water Fluids 220

Conventional Water-Base Mud 221

Oil Base or Invert-Emulsion Muds 221

Foam 221

Perforating Fluids 222
Salt Water or Oil 222
Acetic Acid 222
Nitrogen 222
Gas Wells 222

Packer Fluids 222
   Criteria 222
   Corrosion Protection 222
Well Killing 223
References 224
Appendix A—Commercial Completion Fluid Products 224
Appendix B—Fluid Density Conversion Table 226

Chapter 9 — Through-Tubing Production Logging — 229

Introduction 229
Logging Devices 230
   High Resolution Thermometer 230
   Inflatable Packer Flowmeter 231
   Continuous Flowmeter 231
   Full-Bore Flowmeter 233
   Gradiomanometer 234
   Radioactive Tracers and Gamma Ray Detectors 235
   Noise Log 236
   Pulsed Neutron or Thermal Decay Time 237
   Carbon Oxygen Ratio Log 237
   Through-tubing Neutron Log 237
   Through-tubing Caliper Log 238
   Bore-hole Televiewer 238
   Combination Tools 238
Application of Through-Tubing Production Logging 238
   Physical Condition of the Well 239
   Fluid Movement—Single Phase 240
   Fluid Movement—Multiphase 240
   Reservoir Fluid Saturations 241
   Reservoir Parameters—Formation Damage 241
Field Examples of Production Logging Techniques 241
   Location of Tubing Leaks 241
   Casing Leak in Pumping Well 243
   Flow Behind Casing 243
   Producing Zone Evaluation—Reservoir Management 244
   Water Injection Well Profiling 251
Primary Cement Evaluation 257
   Principle of the Bond Log Device 257
   Mechanics of Sonic Wave Travel 258
   Attenuation of the Casing Signal 259
   Interpretation of Cement Evaluation Logs 260
   Interpretation Problems 263
   Application of Cement Evaluation Logs 264
References 264
Appendix—Interpretation of the Temperature Log 265

Symbols and Abbreviations 271

Chapter 1 Problem Well Analysis

Limited producing rate
Artificial lift
Water production problems
Gas production problems
Mechanical failures
Problem well analysis checklist

What is a problem well? Depending on the economics of a particular situation, a problem well may be related within specific limits to low oil or gas production, high GOR, high water cut, or mechanical problems. Problems in injection or disposal wells may be related to high injection pressure and low volume or mechanical problems. Prior to considering individual wells, the analyst should make certain that a well problem exists and that it is not a reservoir problem. Problem well analysis may be handled on a reservoir basis, an area basis, or by study of an individual well.

The conclusion of such a study will usually result in one of the following recommendations: (1) workover, (2) continue to produce well until oil or gas production declines to a predetermined volume or to the economic limit, (3) pressure maintenance, (4) enhanced recovery operations, or (5) abandon.

Probably the greatest pitfall is to start analyzing a well problem after a workover begins. Careful analytical work should be completed before a workover rig is moved to the location. Because a comprehensive problem well analysis is usually the least costly part of a workover operation, profits from the well, lease, and reservoir can often be improved by increasing technical work and time allocated to problem well analysis. Preferably, a current area or reservoir study should be available before a problem well analysis is initiated.

PROBLEM WELLS

Problems may usually be categorized as limited producing rate, excessive water production, excessive gas production for oil wells, and mechanical failures. Gas well problems are similar; however, high water production is more difficult to handle in producing gas wells.

Limited Producing Rate

Limited producing rate may be a result of (1) low reservoir permeability, (2) low reservoir pressure for depth, (3) formation damage, (4) wellbore or tubing plugging, (5) high viscosity oil, (6) excessive back pressure on formation, (7) inadequate artificial lift, or (8) mechanical problems.

Low Reservoir Permeability

Low reservoir permeability may be an overall reservoir characteristic, or it may be limited to a specific area. If low permeability has been proved as a cause of limited production, this problem should be considered along with other possible causes of low productivity.

Characteristically, in a low permeability reservoir, well productivity declines rapidly as fluids near the wellbore are produced. If available geologic and reservoir data do not readily prove low reservoir permeability, production tests and pressure buildup tests may aid in differentiating between low permeability and formation damage. For a pressure buildup and drawdown test to be valid as a diagnostic tool, it is usually necessary to determine whether all layered, porous zones selected for production are actually in communication with this wellbore. Through-tubing flowmeters, radioactive tracer surveys, or straddle packers may be used to determine formation-wellbore communication for each interval.

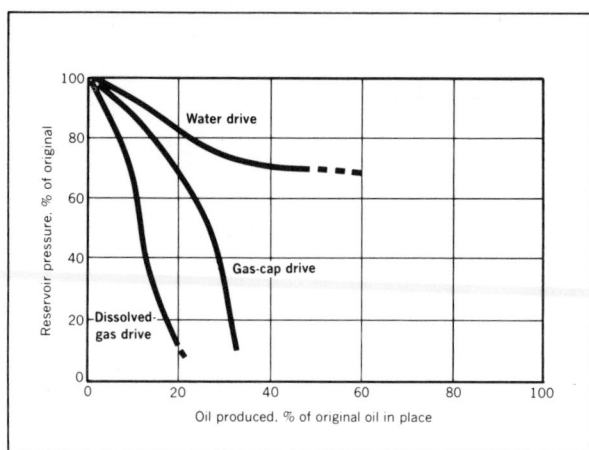

FIG. 1-1—*Typical pressure-production history for various reservoir-drive mechanisms.*[1] *Permission to publish by The Petroleum Engineer.*

Low Reservoir Pressure

If reservoir pressure measurements have been carried out on a routine basis, reservoir pressure history should be well documented. The next step is to consider the dominant reservoir drives in a particular reservoir and how these drive mechanisms are associated with the real or apparent well problem being investigated. Figure 1-1 illustrates the typical pressure-production history for various drive mechanisms.

Formation Damage

Formation damage may be defined as any impairment of well productivity or injectivity due to plugging within the wellbore, in perforations, in formation pores adjacent to the wellbore, or in fractures communicating with the wellbore. All wells are damaged. The problem is to determine the degree of well damage, probable causes of well damage, and finally, approaches to alleviate any serious damage.

Formation damage may be indicated by production tests, pressure buildup and drawdown tests, comparison with offset wells, and careful analysis of production history, including prior completions, workovers, and well servicing operations. If multiple zones are open in a single completion, production logging surveys run in flowing or gas lift wells will often show some permeable zones to be contributing little or nothing to production. Major zones of permeability, especially natural or induced fracture permeability, may be plugged.

A reservoir study may be required to differentiate between (1) production decline due to gradual formation plugging and (2) decline due to loss of reservoir pressure. Comparison of offset wells may not be sufficient to detect gradual plugging because all wells may be subject to similar damaging conditions.

Figure 1-2 illustrates the difference in pressure drawdown in a normal well as compared with a well with serious "skin" damage. In a relatively high permeability well with severe skin damage, reservoir pressure measured in the well may stabilize within a few hours. If reservoir permeability is low, days or weeks may be required for reservoir pressure to stabilize; under these conditions, it may be difficult to determine "skin" damage. "Skin" damage calculations using pressure buildup and drawdown analyses are carried out in many areas prior to planning well stimulation or reperforation.

Plugging of Tubing, Wellbore, and Perforations

When low productivity is indicated in an artificially-lifted oil well with a history of high productivity, a first consideration should be to check for proper operation of artificial lift equipment.

For all types of wells, the probability of tubing, wellbore, or perforation plugging should be evaluated. Plugging may be caused by gravel pack or frac sand, fines, mud, formation rock, paraffin, asphalt, scale, pipe dope, gun debris or other junk, and collapsed tubing or casing. For flowing oil or gas wells, gas-lift wells, or injection wells, tubing should always be left open-ended so that wireline tools and production logging tools can be run to check for plugging in tubing, wellbore, or perforations. If tubing is not open-ended, it may be necessary to lower tubing to check for fill in the bottom of the hole. Analysis of bottomhole samples is helpful in determining the cause of plugging.

High Viscosity Oil

High viscosity oil may be normal for a particular reservoir. If the reservoir is being produced by dissolved gas drive, oil viscosity will increase somewhat as gas is released from the oil.

If well producing problems are due to high viscosity water-in-oil emulsions in or near the wellbore, it may be economical to either break or invert the emulsion with surfactants to lower viscosity of produced fluid.

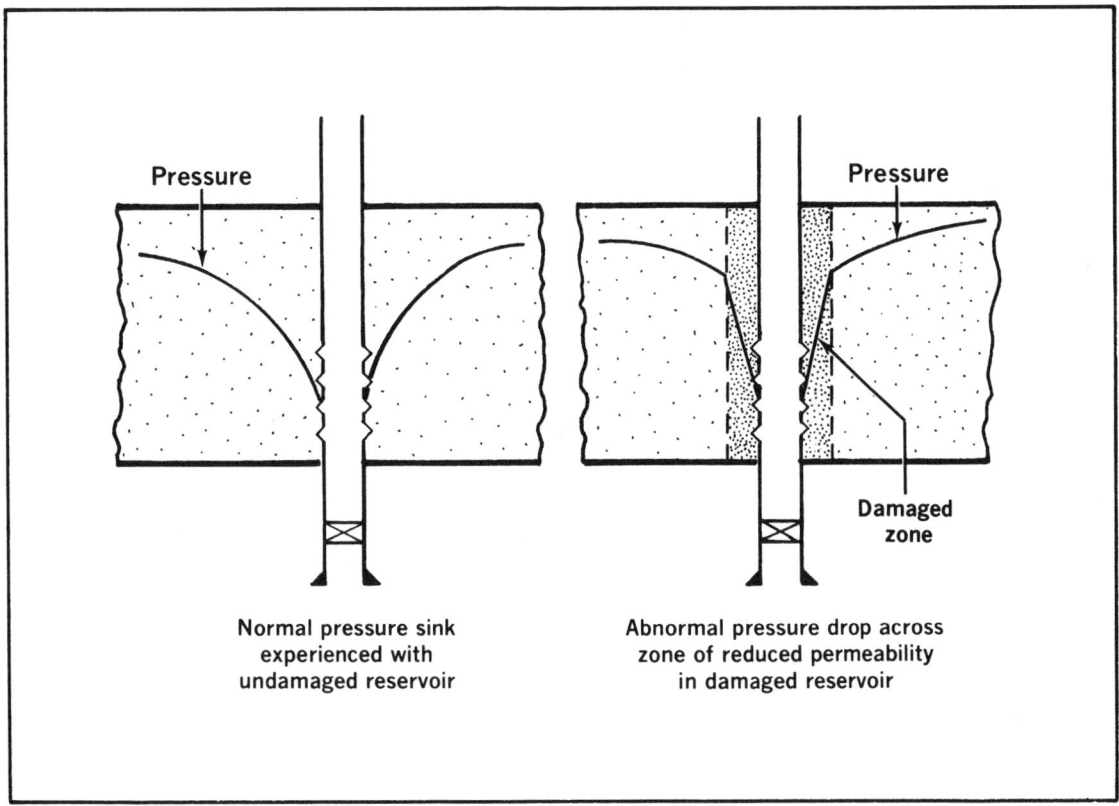

FIG. 1-2—*Difference between pressure drawdown in undamaged and damaged reservoirs.*[1] *Permission to publish by The Petroleum Engineer.*

Excessive Back Pressure on Formation

Excessive back pressure can appreciably lower producing rates in wells producing from reservoirs nearing pressure depletion. Excessive formation back pressure may be due to limited or plugged perforations; partially plugged wellbore, tubing, or flowline; subsurface or surface chokes; undersized gas-oil separator, flowlines, tubing, or casing; or excessive back-pressure setting on casinghead gas gathering system, or on gas-oil separator. Remedies include:

—For high capacity wells, the usual approach is to increase the size of tubing, flowline or separator.

—In oil reservoirs having appreciable loss of reservoir pressure, artificial lift plus reduction of separator, tubing, or casing pressures will increase production.

—If tubing, wellbore, or perforations are partially plugged, removal of restrictions by cleaning out will increase production. Reperforating is frequently the best approach.

Problems With Artificial Lift Wells

If decline in production of oil wells is due to insufficient bottom hole pressure in relation to the weight of the flowing fluid column, artificial lift is usually the first approach. If lift has been installed, improper design, improper application, or malfunction of equipment is a frequent cause of reduced oil production.

If excess water is the problem, workover is a possible solution. In a flowing well with low surface pressure, fluid slippage or fall back in the tubing may be the problem. Therefore, it may be necessary to swab or lift the well for several days to determine the correct water-oil ratio. In layered formations, early "fingering" of water into the well may give an erroneous indication of total movement of the water front into the well. Under these conditions, waterfree zones may still exist within the perforated interval.

Analysis of Problems in Rod-Pumped Wells

1. Determine fluid level with sonic device or other means and calculate back-pressure on formation.

2. If well is pumped off, the problem may be plugging of wellbore, perforations, or formation; low reservoir pressure; low permeability; or pump set too high.

3. If the well is not pumping off, run a dynamometer survey. Analysis of the survey may show one or more of the following problems: defective pump, leaky tubing, inadequate balance of pumping unit, a partially plugged mud anchor, gas-locking of pump, or undersized pumping unit and/or pump.

4. Time cycle may require adjustment for the most efficient pumping operation. If pumping units are engine-powered, the profitability of changing to electric motors should be considered. This would allow easier time-cycle control.

5. If low fluid level is apparently due to wellbore or formation plugging, the first step is usually to check for fill in the wellbore. The next approach is to investigate the probability of scale, paraffin, or asphalt in perforations or around the wellbore.

6. If reservoir pressure depletion is the primary problem, pressure maintenance or improved recovery is the only long range solution.

Analysis of Problems in Gas Lift Wells

Gas lift is basically a system to reduce the weight of the producing fluid column and back-pressure on the formation. Gas lift usually operates best in high fluid level, high capacity wells. Gas lift can be a very economical trouble-free system for high capacity wells requiring additional energy to lift the desired volume of fluid. Also the API gravity of medium or low gravity oil may be increased by absorption of gas lift gas. Low capacity wells present more of a problem on gas lift. Rod pumps are usually more applicable in low capacity oil wells.

To investigate gas lift problems, the most direct approach is to run pressure and temperature surveys inside the tubing to evaluate gas lift valve operation. Temperature surveys are often run to check for valve leaks, tubing leaks, and casing leaks.

Typical gas-lift problems—Some typical gas-lift problems are:

—Leaky valves, tubing, or casing

—Improper time cycle

—Improper valve setting or location for current condition of well

—Low gas pressure

—High back pressure on formation

—Scale, paraffin, or asphalt in flowline, gas lift valves, tubing, wellbore, or perforations

Analysis of Problems in Hydraulic Bottom-Hole Pumps

Hydraulic bottomhole pumps are usually operated with a power oil; however, some are operated with water. Hydraulic pumping operates best on wells producing a relatively high volume of fluid. It may also be the best choice in wells with an operating fluid level at relatively great depth, and in small diameter casing. If wells produce appreciable sand, the relatively close tolerance pumps may cause trouble. Economics are usually improved if more than one well can be operated on the power-oil or power-water system for closely spaced wells.

Operating costs can be very high if deep well pumps must be pulled frequently. An excellent pump maintenance service located near the field will aid appreciably in maintaining low cost operation.

If design and installation of the system, along with pump maintenance, is satisfactory, the most prevalent problem is dirty power oil and plugged power oil system. Scale, paraffin, asphalt, sand, and other plugging agents in the pump, tubing, and wellbore can cause considerable difficulties.

Analysis of Problems in Electrical Submersible Pumps

The use of electric submersible pumps to lift large volumes of oil and water from relatively high fluid level wells is increasing. Electric pumps are usually preferable to rod pumps for use in relatively high-volume directional wells. Assuming correct application and design, care must be exercised in running the pump to setting depth to prevent cable damage. Operation of electric pumps is usually more trouble-free if polyethylene-covered cable protected by a galvanized steel sheath is used. Reliable and convenient repair and maintenance service is imperative for operation of very high-volume tandem pumps.

The recording ammeter is key to detecting early trouble in electric pump. Compare each day's ammeter record with charts for the previous week. A badly worn pump may be detected by drastic reduction in electric current. If this is the problem, the pump should then be stopped to prevent damage to electric motor. If the pump fails, leave ammeter record on well at time of failure to aid serviceman in problem diagnosis. Gas-oil ratio problems can sometimes be reduced by lowering the pump to increase intake pressure and thereby reduce break out of solution gas.

Water Production Problems in Oil or Gas Wells

Water problems may result from:

1. Natural water drive or waterflood aggravated by fingering or coning.
2. Extraneous sources including casing leaks or primary cement failure.
3. Fracturing or acidizing into adjacent water zones.

Water encroachment is normal in a water drive reservoir. Three water-oil contacts, illustrated in Figure 1-3, may usually be defined in oil wells:

Original Water-oil Contact—Depth below which no oil is found.

Producing Water-oil Contact—Depth below which no producible oil is found. This depth rises with oil production in a water drive reservoir.

Completion Water-oil Contact—Depth below which first water production appears. This depth rises with production in a water drive reservoir or in a waterflood.

Oil wells completed in the transition zone between the completion water-oil contact and the producing water-oil contact cannot be expected to produce clean oil.

Gas wells completed in the transition zone between the completion gas-water contact and the producing gas-water contact cannot be expected to produce water-free gas.

Water encroachment is complicated by stratified or layered permeability. Fingering, illustrated in Figure 1-4, is differential water encroachment through the more permeable zones. Because fluids will move faster through zones of high permeability, these zones will usually be watered out first.

In stratified zones, early water breakthrough may not cause abandonment, but a large volume of water is often produced before oil or gas is depleted from the remaining zones. An economic question arises as to whether to produce each major zone separately to reduce lifting costs and increase zone recovery. The alternative is to complete in all zones to reduce workover costs resulting in increased lifting costs and possible reduced oil or gas recovery in some reservoirs which have a wide variation in permeability between zones.

In shallow marine sandstone oil reservoirs with continuous horizontal permeability over all or a large part of an entire structure, it may only be necessary to selectively produce each zone in the highest wells to insure high ultimate recovery, provided there is no gas cap.

If zones are very permeable, water production may be in excess of capacity of available artificial lift, thus causing premature abandonment unless high permeability water producing zones can be sealed off.

"Fingering" of water, defined as upstructure water movement in the more permeable zones of a multizone completion, is rate-sensitive. Premature fingering may be reduced by reducing total fluid production rates or increasing the permeability feet open to production. Location of water producing zones may be difficult. Production logs are useful to determine volume and type of fluid produced from each zone.

Detailed geologic knowledge of stratified permeabilities and the location and extent of barriers to crossflow between stratified, permeable layers is essential to successful water shutoff.

Water coning in oil or gas wells is defined as vertical movement of water across bedding planes in a producing formation. Figure 1-5 illustrates coning in an oil reservoir. Coning for more than a few feet is relatively rare in most unfractured sandstone reservoirs.

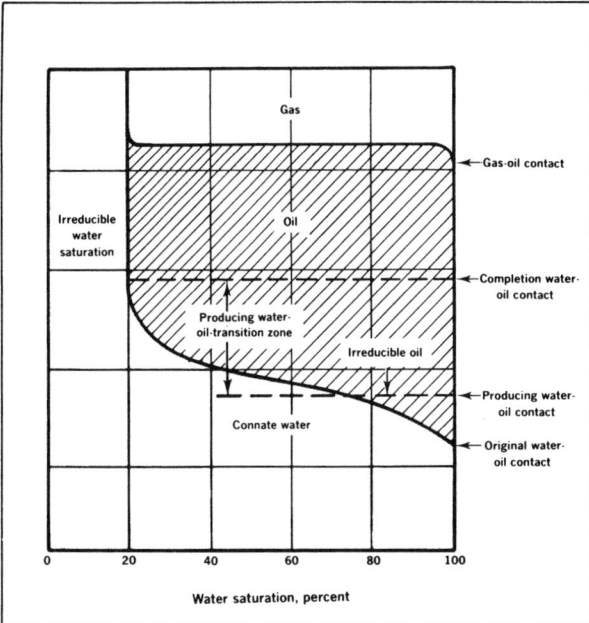

FIG. 1-3—*Fluid distribution in a uniform sand reservoir containing connate water, oil, and a gas cap.*[1] *Permission to publish by The Petroleum Engineer.*

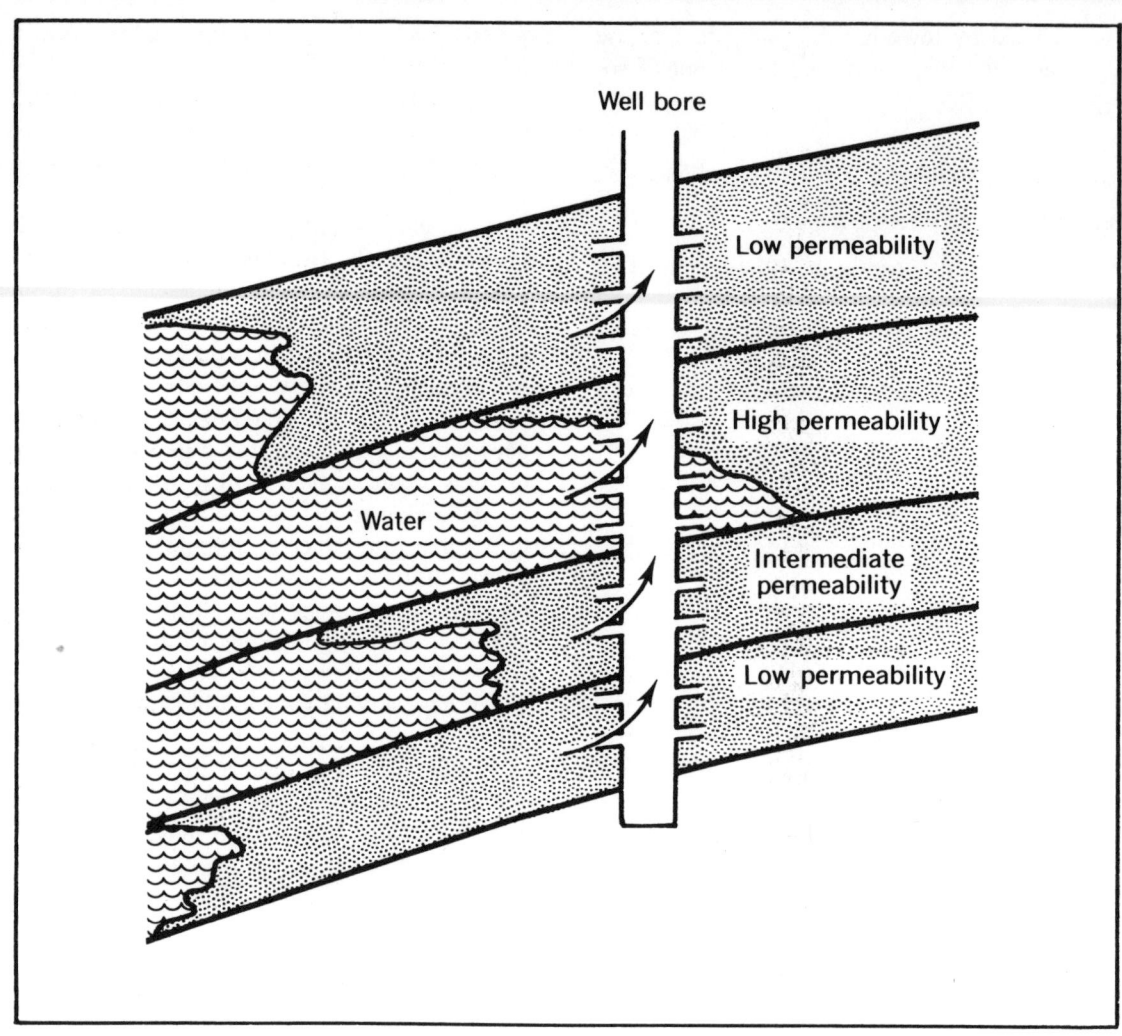

FIG. 1-4—*Irregular water encroachment in heterogeneous reservoir can result in early water breakthrough.*[1] *Permission to publish by The Petroleum Engineer.*

Water or gas coning will not cross barriers to vertical permeability unless these barriers are broken by natural or induced fractures. For vertical communication to exist through fractures, these fractures must be held open by tectonic forces, secondary deposition, or propped open with sand or glass beads; fractures in carbonates may be etched with acid to form flow channels. Major fracture systems tend to be nearer vertical than parallel to bedding planes.

Coning may be very severe in reefs, or other reservoirs having continuous vertical permeability.

Water coning may be aggravated by high fluid withdrawal rates and, once established, may become relatively stable due to increased relative permeability to water.

Water will channel or cone vertically through a faulty cement-to-casing or cement-to-formation bond.

To evaluate coning, a reservoir or area study may be necessary to determine the current location of the oil-water or gas-water contact.

Elimination of a cone usually requires reduced production rates. Recompletion or repair may alleviate the problem if water is coning through a cement channel or through vertical fractures.

Well diagnosis can be made to differentiate between coning, water block, and emulsion block.

Diagnosis of Coning—If coning is the problem, increasing production rate will usually increase the percentage of water produced; decreasing production rate or shutting in the well for one to three months

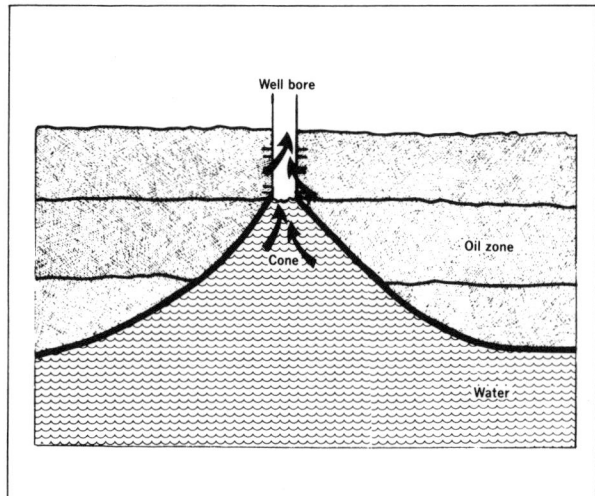

FIG. 1-5—*Coning of water.[2] Permission to publish by The Petroleum Engineer.*

will usually decrease water coning.

Diagnosis of Emulsion Blocking—If an emulsion block exists, the calculated average well permeability as determined by injectivity tests will be much higher than the average permeability determined from production tests. This provides a reliable way to predict emulsion blocks and is often called the "check valve" effect. Increasing or decreasing producing rates will not appreciably change the water percentages in an emulsion-blocked well.

Diagnosis of Water Blocking—A temporary shift of relative permeability in favor of water as the mobile fluid causes water blocking. Under these conditions, oil production will decrease, and the water percentage will increase. Water blocking is usually caused by circulating or killing the well with water. A water block may also develop if water is allowed to remain on the producing zones for several days or weeks.

Water percentages will usually decline with time as the well is produced. Increasing production rate may accelerate removal of a water block. Water blocking can usually be prevented by adding about one-tenth percent of a properly selected surfactant to all water used in well workovers, well killing, or well circulating operations (see Chapter 6, Volume 2, for details of surfactant selection).

Diagnosis of Water Problems Caused by Hydraulic Fracturing—Temperature surveys run in conjunction with hydraulic fracturing or acid fracturing will often give a clue as to whether subsequent water production is due to primary cement failures, fracturing or acidizing into water, casing leaks, or normal water encroachment.

Gas Problems in Oil Wells

The primary sources of gas in oil wells are: (1) gas dissolved in the oil, (2) primary or secondary gas caps, and (3) gas flow from zones or reservoirs above or below the oil zone. Normal gas-oil ratio behavior corresponding to the drive mechanism for any particular reservoir, illustrated in Figure 1-6, must be considered in problem well analysis.

In a dissolved gas drive reservoir, gas saturation increases as oil withdrawals continue and reservoir pressure declines. When this gas is released from the oil, gas flows to the wellbore, and if pressure decline continues, gas tends to become the dominant mobile fluid until the gas is dissipated.

If there are no barriers to vertical flow in a reservoir with a gas cap, a decline in reservoir pressure may allow gas to expand into the oil producing interval. With high pressure drawdown at the wellbore, gas coning may occur in wells with continuous vertical permeability without appreciable decline in reservoir pressure.

In stratified reservoirs, premature fingering of gas may occur with high pressure drawdown at the wellbore. Fingering is more prevalent in reservoirs where permeabilities vary appreciably between zones.

In layered reservoirs, gas flow from zones above or below the oil zone may be due to (1) casing leaks,

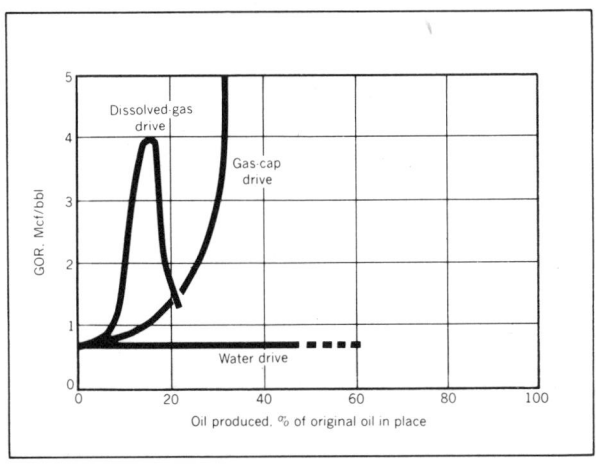

FIG. 1-6—*Characteristic gas-oil performance for various reservoir-drive mechanisms.[1] Permission to publish by The Petroleum Engineer.*

(2) cement bond failure, (3) natural or induced fractures communicating with gas zone, or (4) acidizing into gas zone.

Mechanical Failures

A large number of different types of mechanical failures can cause loss of production and/or increase costs in well operation. Some of the more common failures are: (1) primary cement failure, (2) casing, tubing, and packer leaks, (3) failure of artificial lift equipment, and (4) wellbore communication in multiple completions. Prior to moving a rig to a costly repair job, sufficient study should be carried out on the well to determine whether other remedial work is justified at the time of mechanical repairs.

Casing-cement-formation bond failures are frequently caused by applying fracture pressure on a well during a matrix acid job.

In locating casing leaks, water analyses are useful in differentiating between casing leaks and normal water encroachment. Temperature surveys and other production logs are beneficial in locating casing leaks. Exact location of leaks is usually made with packers or packer-bridge plug combinations.

Wellbore communications in conventional multiple completions can usually be detected by packer leakage tests, by abrupt changes in producing characteristics, or by observing equal shut-in pressure on two or more completions.

PROBLEM WELL ANALYSIS CHECKLIST
Apparent Well Problem

1. Study the indicated problem based on well performance.
2. Determine whether similar problems exist on offset wells, the same reservoir, field, or geologic trend.

Analysis of Probable Productive Zones in Well

1. Study productive zones currently open to wellbore.
2. Examine additional productive zones or reservoirs not currently open to the wellbore.

Analysis of Well History

1. Consider procedures used in drilling of productive zone, including fluids used and any indication of lost returns. Was well damage indicated from drilling or completion fluids?
2. Evaluate primary cement job, using the following procedure:
   a. What was the overall cementing plan, including chemicals used, location of float collar, centralizers, and scratchers?
   b. What mud conditioning and preflush plan was used in preparation for cementing the casing?
   c. Was initial primary cement job a failure? Were there any problems, including lost returns and squeeze jobs?
   d. Was this a two-plug cementing operation?
   e. What type of fluid was used to pump down top plug?
   f. Was casing rotated or reciprocated during cementing operation?
3. Study available data on initial completion, including:
   a. Date of completion.
   b. Well test data.
   c. Completion interval.
   d. Open hole completion—details.
   e. Perforated completion.

   —Type, size, and name of perforating gun and charge and charge phasing.

   —Projected hole size and penetration.

   —Perforating fluid.

   —Direction of differential pressure during perforating.

   —If thru-tubing perforation used, was perforator centralized, decentralized or uncontrolled?

   —Indication of perforating problems.

   f. Details of well stimulation and results.
   g. Final completion details, including fluid used.
   h. Size and arrangement of tubular goods and downhole equipment. Was tubing set open-ended above potential producing zones?
4. Workover, recompletion, well stimulation, well servicing, and chemical treatments. Analyze details on all jobs, including workover or well service fluids and job results.
5. Production history
   a. Current production test.
   b. Record of tests of oil, water, and GOR on a periodic basis, and at any time there was a significant change in production.

TABLE 1-1
Comparison of Basic Data on Problem Well and Offsets
(Alternate to Stratigraphic Cross Section)

Operator, lease & well no.	Well locations with respect to problem well		Production interval ft—subsea		Approximate contacts in ft—subsea		Reservoir limits ft—subsea		Current well tests			
	Distance ft	Approx. direction	From	To	G-O	W-O	Top	Bottom	Choke size in.	Bbl/day oil	Bbl/day water	Gas-oil ratio cu. ft/bbl
(Problem Well)	0	—	6200	6231	6150	6265	6140	6270	$7/64$	25	75	10,000
Co. A, Well 1	1320	N	6150	6170	6150	—	6050	6180	$6/64$	100	0	5,000
Co. A, Well 2	1320	E	6210	6220	6150	6265	6145	6285	$9/64$	70	0	1,000
Co. B, Well 21	1320	S	6240	6260	—	6265	6240	6370	$10/64$	60	0	1,000
Co. C, Well 67	1986	SW	6250	6275	—	6265	6200	6330	$11/64$	50	10	1,000

c. Estimated additional oil and gas recovery from this well. Is well required for recovery of reserves or is it needed only to accelerate income?

6. Reservoir pressure

a. Study annual subsurface pressure surveys. Note any changes in production concurrent with reservoir pressure changes during any period.

b. Compare subsurface pressure history of well with that of offset wells, the area, and reservoir.

Comparison of Well Performance with Offset Wells

1. Consider structural positions and completion interval including permeability feet open in each well.

2. Prepare cross sections when needed. As an alternate, a cross section type of data table may be prepared as shown in Table 1-1.

Reservoir Considerations

1. Permeability, porosity, water saturation, and relative permeability.

2. Log data and core data to show lithology changes, including continuity of barriers to vertical permeability.

3. Character of reservoir rock.

4. Maps and other data showing depositional and post depositional history of reservoir rock.

5. Effectiveness of various reservoir drive mechanisms.

6. Forecasted future reservoir behavior.

7. Potential changes in reservoir behavior through pressure maintenance or improved recovery.

8. All previous reservoir or area studies.

Use Special Surveys, Cross Sections and Maps

1. Examine structure, isopachous, isobaric, water percentage, and gas-oil ratio maps as available.

2. Study stratigraphic cross sections. Table 1-1 may be effectively used in place of cross sections to compare basic data on problem wells and offsets. This system can be quite useful for comparing well data on reservoirs with continuous horizontal permeability over some distance.

3. Obtain a current static subsurface pressure.

4. Consider running wireline tools to check for wellbore plugging.

5. Run production logging surveys when indicated.

6. Conduct diagnostic well tests such as pressure build-up, pressure fall-off, and injectivity tests. Review production decline curves.

7. Check for wellbore communication, mechanical failures.

8. For pumping wells, check fluid level with sonic devices and make dynamometer survey.

9. For gas lift wells, consider checking valve operation with producing subsurface pressure surveys, temperature surveys, and two-pen surface pressure charts.

10. For wells that have been shut-in for some time, conduct a new production test. Justification for a new test is based on the fact that a gas or water finger or cone may have dissipated, reservoir pressure may have increased or decreased, or the wellbore may be plugged with scale or debris.

11. Compare recent and initial water sample analysis.

12. Check bottomhole sludge sample analysis for scale, paraffin or asphaltenes. Analyze scale from

pump, mud anchor or tubing.

13. Check corrosion and casing leak history, including tubing or casing caliper records. A casing thickness survey may be needed.

14. Consult with pumper, field foreman, or field superintendent for sudden changes in well behavior characteristics that may not have been reported over the life of a well.

REFERENCES

1. Allen, T. O.: "Thin Oil Column Completion and Production Practices," *World Oil* (Apr. 1954) 218–225.

2. Clark, Norman, Schultz, W.: "Analysis of Problem Wells," *Petroleum Engineer*, Sept., 1956.

3. Allen, T. O.: "Guidelines to Problem Well Diagnosis," *Petroleum Engineer*, Dec., 1968.

4. Meehen, D. N. and Schell, E. J.: "An Analysis of Rate-Sensitive Skin in Gas Wells," SPE 12176, Oct., 1983.

5. Reed, R. N., and Wheatley, M. J.: "Oil and Water Production in a Reservoir with Significant Capillary Transition Zone," SPE 12066, Oct., 1983.

Chapter 2 Paraffins and Asphaltenes

Paraffin and asphaltene chemistry
Methods of wax removal
Preventing deposition
Design for wax control

INTRODUCTION

Paraffin and asphaltenes are constituents of most crude oils. Deposits of paraffin or asphaltenes in surface equipment and downhole are a major problem in production operations. Severity of deposition varies widely, depending on crude oil composition, well depth, formation temperature, pressure drop, and producing procedures. Any organic deposit associated with crude production is often called paraffin or wax. While paraffin compounds are usually the major component in these deposits, they are frequently a mixture of paraffin and asphaltenes. Many low API gravity crudes, such as those found at shallow depths in California, eastern Venezuela, and Trinidad have asphaltenes as their primary constituent.

C. E. Reistle's "Paraffin and Congealing Oil Problems,"[1] published in *Bulletin USBM* in 1932, is a classic study of paraffin deposition and the resulting producing problems. Reistle listed the following as the most significant reasons for separation of paraffin from crude oil:

—the cooling produced by the gas in expanding through an orifice or restriction
—cooling produced as a result of the gas expanding, forcing the oil through the formation to the well and lifting it to the surface
—cooling produced by radiation of heat from the oil and gas to the surrounding formations as it flows from the bottom of the well to the surface
—cooling produced by dissolved gas being liberated from solution
—change in temperature produced by intrusion of water
—loss in volume and change in temperature due to the evaporation or vaporization of the lighter constituents.

These conditions all result in cooling of the oil, which is the primary cause of paraffin deposition.

Since paraffin characteristics and asphaltene content vary significantly from reservoir to reservoir, production problems and their solutions also vary. Wax removal and prevention methods that are effective in one producing system are not always applicable in other reservoirs or in various wells within the same reservoir.

PARAFFIN AND ASPHALTENE CHEMISTRY

Paraffins and asphaltenes differ significantly in chemical structure. Paraffins are normal (straight-chain) or branched alkanes of relatively high molecular weight and are represented by the general formula C_nH_{2n+2}. This class of hydrocarbons is essentially inert to chemical reactions and is, therefore, resistant to attack by bases and acids. Paraffin deposits chiefly contain alkanes having a carbon chain length from 20 (melting point of 98°F) to 60 (melting point of about 215°F). Crude paraffin deposits are mixtures of these alkanes and consist of very small crystals that usually agglomerate to form granular particles. The wax varies in consistency from a soft mush to a hard, brittle material. Paraffin deposits will be harder if long chain alkanes are prevalent. Paraffin deposits can also contain asphaltenes, resins, gums, crude oil, and inorganic matter such as fine sand, silt, clays, salt, scales, and water.

Asphaltenes are the black components present in crude oil. Their molecular weight is relatively high, and they are normally polar chemicals because of the presence of oxygen, sulfur, nitrogen, and various metals in their molecular structure. Chemically, asphaltenes consist of polycyclic, condensed, aromatic ring compounds. They are soluble in aromatic solvents such as carbon tetrachloride and carbon disulfide but are insoluble in distillates such as kerosene and diesel oil. Asphaltenes are also insoluble in other low molecular weight hydrocarbons such as propane and butane.

Witherspoon and Munir[7] investigated the size and shape of asphaltic particles found in several crude oils. Generally, these particles are spherical in shape, 30 to 65 angstroms in diameter, with molecular weights of 10,000 to 100,000. Densities range to 1.22 g/cc. Witherspoon and Munir theorized that asphaltenes are located in the nucleus of an immense collection of molecules called a micelle. Gathered around the nucleus are lighter and less aromatic constituents that resemble the alkanes found in paraffins. This general description indicates that asphaltene particles form colloids.

Nelson[13] reported that the concentration of asphaltic material in crude oils from California, Venezuela, West Texas, and the Middle East was greater in the lower API gravity oils. The asphaltene content of 9° API crude averaged 82% compared to 3.4% for 41° API crudes.

Analysis of Paraffins and Asphaltenes in Crude Oil

Analysis to determine paraffin and asphaltene constituents in crude oil should be conducted during the early development of a reservoir in order to predict deposition problems and to develop methods of minimizing deposition.

ASTM D 2500-66 details the method for determining cloud point of oil, the temperature at which paraffin begins to come out of solution. The paraffin cloud is visible in light-colored crudes. However, darker crudes may mask the cloud; in this case, the cloud point is estimated from measurements or observation of the inflection point on a cooling curve. After the cloud point occurs, additional cooling causes more wax to come out of solution. These crystals may nucleate to form agglomerates that gradually develop an interlocking network. Eventually, the crude becomes so thick that it will not pour; the temperature at which this occurs is the pour point. ASTM Standard D 97-66 should be used to determine pour point.

Paraffin and asphaltic deposits may resemble each other closely after a few days. However, paraffin melts over a narrow temperature range when heated. The hot liquid is very fluid and has a low viscosity. Asphaltene melts slowly, gradually softening to give a thick, viscous liquid. Asphaltenes burn with a smokey flame and frequently leave a thin ash or carbonaceous ball. Paraffin deposits burn rapidly with less smoke and leave little residue.

DEPOSITION MECHANISMS

A paraffin or asphaltene problem may be defined as a condition in which a predominately organic deposit hampers the production of crude oil. The loss of crude production from the well depends on the amount and location of the deposition.

Paraffin Deposition

Paraffin can precipitate from crudes when equilibrium conditions change slightly, causing a loss of solubility of the wax in the crude. The point of deposition in a well's producing system is normally determined by how close the crude is to its solubility saturation point and the amount of wax in the crude. Loss of wax solubility, however, does not necessarily cause deposition. Wax crystals normally have a needle-like shape, and if they remain as single crystals, they tend to disperse in the crude instead of depositing on a surface. A nucleating material is usually present that gathers wax crystals into a bushy particle that is much larger than single crystals; these agglomerates may then separate from the crude and form deposits in the well's producing system. Asphaltenes are frequently the nucleating material that causes paraffin crystals to agglomerate. Other nucleating materials are formation fines and corrosion products.

Temperature reduction is probably the most common cause of wax deposition because wax solubility in crudes decreases as the temperature is lowered. Cooling of crudes occurs in numerous places in the well system. Expansion of oil and associated gas at the formation face, through casing perforations, or through a bottomhole screen causes cooling. Additional cooling occurs as the crude moves upward through the tubing, through the surface choke, through the flowline, and through the gas-oil separator. At some point in the system, the temperature drops below the wax melting point, and wax may start to accumulate on production equipment. Loss of gas and light hydrocarbons from crude also decreases wax solubility. This effect contributes to wax deposition in surface lines and tanks. High gas-oil ratio magnifies paraffin deposition problems.

Certain signs can indicate the start of paraffin deposition. Congealed paraffin can be separated from oil by centrifuging. A change in crude appearance, such as cloudiness, indicates that paraffin is coming out of solution. Accumulation of paraffin in stock tanks usually indicates that paraffin deposition may be expected soon in the flowline, tubing, and possibly later in the wellbore. Paraffin buildup in the tubing can lead to overload of rod-pumps and cause rod breaks. Finally, production decreases in wells producing paraffinic oil may be caused by paraffin deposition.

Paraffin deposition may be effected by changes in water-oil ratio. However, the primary effect is related to changes in the rate of fluid production with changes in water-cut. Increased fluid production rates will mean higher wellhead temperature and less wax deposition, whereas decreased fluid production would have the opposite effect.

Water may increase the water wettability of metal surfaces, thus reducing the probability of wax and crude contact with the metal surfaces. Water-wet metal surfaces signal the onset of corrosion, so the benefit of preventing wax deposition may be more than offset by corrosion costs. If the increase in water percentage does not cause the metal surface to become water-wet, wax deposition may continue. Many wells producing appreciable water are treated with oil-wetting corrosion inhibitors, thus aiding the accumulation of wax on the tubing. Also, emulsions may be formed and stabilized by the wax as it cools during movement through the producing system.

Injecting cold fracturing or acidizing fluids into an oil reservoir can cause a significant cooling of the crude and formation. If the crude is cooled below its cloud point, paraffin can precipitate in flow channels. If all of the wax is not redissolved after formation temperature is restored, oil production may be limited or even blocked. If the formation temperature is significantly greater than the wax melting point, no lasting effect will result. Heating the fracturing or acidizing fluid above formation temperature will prevent paraffin deposition during hydraulic fracturing.

Paraffin on tubing or casing can be scoured from these metal surfaces and forced into perforations or into the formation during fracturing, acidizing, well workovers, or paraffin removal operations. Many wells have been severely damaged or totally plugged in this manner. Once damage has occurred, restoring a well to full rates is frequently difficult to achieve.

Clean injection tubing or casing is essential where a well stimulation or fluid injection procedure is being conducted.

Asphaltene Deposition

Asphaltenes can form micelles that have polar characteristics. Their deposition in well systems is not as prevalent as that of paraffin, but problems can be very severe in wells producing high asphaltene-content oils. Asphaltenes often separate from asphaltic oil and deposit in the formation around the wellbore. Tests conducted by Katz and Beu[2] show that asphaltene deposition may be an electrical phenomenon. In one test, platinum electrodes were placed in freshly-filtered crude from a well in which asphaltenes were being deposited in the gas-oil separator. An electrical potential of 220 volts was applied to the electrodes for several days. At the conclusion of the test, the negative electrode was clean, but the positive electrode was covered with a black material similar in all respects to the asphaltic sludge found in the separator. The coating of the positive electrode by asphaltenes indicates that the asphaltene micelle is negatively charged.

Fluids flowing through capillaries or porous media can develop an electrical charge through the streaming potential phenomenon. In a flow test in which a black crude was forced through a sandstone core, a streaming potential of approximately 39 millivolts was developed during the flow process. Examination of the crude after the test showed the presence of asphaltene particles, although none could be detected in the crude prior to the test.

Deposition of asphaltenes on the formation sand grains near the wellbore will oil-wet the sand, reducing the relative permeability to oil and reducing oil production. Physical plugging with asphaltene further reduces production.

REMOVAL OF WAX DEPOSITS

Paraffin deposits vary greatly from one reservoir to another, and differences have even been noted in wells in the same reservoir. The most common methods of removing paraffin from wells are (1) mechanical, (2) solvents, (3) heat, and (4) dispersants.

Mechanical Removal of Wax

Scrapers and cutters are used extensively to remove paraffin from tubing. These techniques are relatively economical and usually result in minimal formation damage. However, scraping can cause perforation plugging if it is necessary to circulate scraped paraffin down the tubing and out of the casing. If frequent cleanout is required, mechanical cleaning becomes more costly, especially when the value of lost production is added to cleanout costs.

One widely-used paraffin removal method in flowing or gas lift wells employs a scraper attached to a wireline. Although most wireline units are operated manually, some scraper units are controlled automatically by a timing device. Another system requires shutting in the well long enough for a scraper to fall to the bottom of the tubing; when production is resumed, the scraper opens up or expands and scrapes the paraffin from the tubing as the scraper moves to the surface. To operate this tool, wells may be shut-in and opened manually or controlled with a timing device.

Paraffin may be removed from the tubing of gas lift wells with free pistons, installed to improve the efficiency of gas lift. Also scrapers may be attached to sucker rods to remove paraffin as the well is pumped.

Flowline deposits may be mechanically scraped by forcing soluble or insoluble plugs through the lines. Soluble plugs are made of microcrystalline wax or naphthalene, which dissolve over a period of time. Insoluble scraper plugs are usually hard rubber or sharp-edged plastic spheres. Automatic dispensers are sometimes used to periodically inject these spheres into the flowline; scrapers are trapped at the end of the line and reused.

Solvent Removal of Wax

The use of solvents is relatively common, but care must be observed in solvent selection. Chlorinated hydrocarbons, such as carbon tetrachloride are excellent paraffin solvents. However, they are not generally used in the United States because they can have an adverse effect on refinery catalysts. Carbon disulfide has been called the universal paraffin solvent. Unfortunately, it is expensive, extremely flammable, and toxic. It should be handled only by persons thoroughly trained in its use and hazards.

Snavely[17] has described the use of certain water-soluble organic compounds which decompose to form carbon disulfide. The compounds are mixed with water and introduced into the tubing or annulus and fall to the bottom of the well. These aqueous solutions are somewhat safer to handle than carbon disulfide.

Condensate, kerosene, and diesel oil are commonly used to dissolve paraffin in wells in which the asphaltene content of the deposit is very low. Asphaltenes are not soluble in straight-chain hydrocarbons such as kerosene, diesel oil, and most condensates. However, some condensates contain aromatic components that enable them to dissolve asphaltic deposits. Aromatic chemicals such as toluene and xylene are excellent solvents for asphaltenes as well as paraffin deposits. The solvent power of these chemicals can be enhanced as much as ten times by the addition of a small amount of a specific primary or secondary amine, such as Halliburton's Targon, to the solvent. These solvents also help dissolve paraffin that may be deposited with the asphaltenes. Moderate heating of the solvents will hasten deposit removal. Care should be taken during warming because of the relatively low flash points of toluene and xylene.

Selection of a solvent for any application should be based on its cost effectiveness in dissolving a specific organic deposit. Solvent application must be adapted to fit well conditions. One procedure is to circulate solvent down the annulus and back through the tubing. Soaking or surging of the solvent over a period of time will usually dissolve the maximum amount of paraffin per gallon of solvent. If the formation is partially plugged with wax, squeezing solvent and surfactant into the formation and soaking for 24 to 72 hours is very effective. Severe paraffin buildup in the tubing of rod-pumping wells often makes rod removal very difficult. In these instances, pumping a solvent down the tubing softens paraffin and facilitates rod pulling. Solvent selection can be accomplished by simple field tests. A small amount of paraffin is immersed in the solvent in a clear glass container. Side-by-side comparison of available solvents will usually enable selection of the best chemical within minutes.

The Use of Heat for Removal of Wax

Hot oiling is one of the most popular methods of paraffin removal. Paraffin is both dissolved and melted by the hot oil, allowing it to be circulated from the well and the surface producing system. Lease crude or other oil is heated to a temperature significantly greater than that of the formation. Hot oil is normally pumped down the casing and up the tubing. Where lift equipment permits, or in flowing wells, hot oil may be circulated own the tubing and up the casing or to the formation face. Billingsley[12] reported that paraffin could be removed from shallow pumping wells by periodically circulating hot oil down hollow rods to a depth below the lowest paraffin deposition in the tubing.

There is evidence that hot oiling can cause permeability damage if melted wax enters the formation, particularly in wells having a reservoir temperature of less than 160°F. There is much greater risk that perforations or formation pores will be plugged when hot oil is circulated down the tubing rather than down the casing. Formation damage may occur if melted or oil-saturated paraffin enters a formation in which the reservoir temperature is less than the cloud point of the hot oil or below the melting point of paraffin. The formation will cool the hot oil, causing paraffin to be deposited in pores of the rock.

Hot water is sometimes used to clean wells completed in low temperature reservoirs. Hot oil or hot water can be effective in removing paraffin, but care must be exercised to prevent melted paraffin from entering the formation and reprecipitating.

Paraffin deposits frequently contain scales and formation fines that are released when the paraffin is dissolved or melted. Well productivity can be reduced if these solids are forced into perforations or formation pores and fractures.

Steam has been used to melt paraffin or asphaltenes in the flowline, tubing, casing, wellbore, or formation. This method should be used carefully in downhole applications because melted paraffin forced into the formation may congeal before it can be produced with formation oil. If steam is injected for a number of days, as in the case of "huff and puff" steam stimulation, the huge quantities of heat normally will prevent wax reprecipitation in the formation.

Any application of heat to remove paraffin should be carried out before large deposits have accumulated. Effective removal of paraffin from tubing that is almost plugged is very difficult since loosened or partially-melted paraffin may bridge in the tubing. If large deposits have accumulated, mechanical removal of some of the paraffin may be advisable prior to heat application. The use of hot oil at regular intervals has proved to be effective in wells in which paraffin buildup rates are known.

Removal of Wax with Dispersants

Water-soluble dispersants can be used to remove paraffin deposits. Halliburton's Parasperse, a water-soluble dispersant, is used in 2% to 10% chemical concentrations, depending on the amount of paraffin to be removed. It does not dissolve paraffin but disperses paraffin particles to be circulated from the well. The Parasperse solution is more effective if it is heated to about 120°F before treating a well. Since the system is 90% to 98% water, it is relatively inexpensive and constitutes no fire hazard. Laboratory tests have indicated that, on a gallon to gallon basis, Parasperse is capable of removing fifty times as much paraffin as the best solvent.

Many wells have been treated with 30 to 50 barrels of Parasperse solution. In low pressure wells, the solution may be poured or pumped down the annulus and then pumped out with oil production. Where paraffin is very hard and dense, a two- to four-hour soaking period is suggested prior to returning the well to production. Surface lines can also be cleaned of paraffin by circulating this dispersant through the system.

PREVENTING OR DECREASING WAX DEPOSITION

Laboratory analysis of wax should be carried out as a basis for selecting the most economic system for preventing wax deposition from wells completed in each reservoir. Wax should be removed from the well producing system prior to applying any method for preventing deposition.

Laboratory Tests and Results

Test procedures have been developed to determine paraffin deposition properties of crude oils. Similar tests are used for screening of paraffin inhibitors. One method involves circulating a crude or a kerosene solution of paraffin through a system in which a cool jacket or cool finger is a part of the flow circuit. The oil is maintained at a temperature above its cloud point while the cool jacket or finger is kept at a temperature several degrees below the cloud point of the oil. Paraffin buildup on the jacket or finger is measured to provide data on deposition rates. Paraffin solvents and inhibitors also have been tested with this equipment.

Sutton[16] conducted tests in which filtered crude oil was circulated through cores at temperatures below the cloud point. In one test, crude at 13°F below the cloud point was pumped through a limestone core having an initial permeability of 4.2 md. A loss in permeability indicated that paraffin was being filtered from the oil. In this case, permeability was reduced

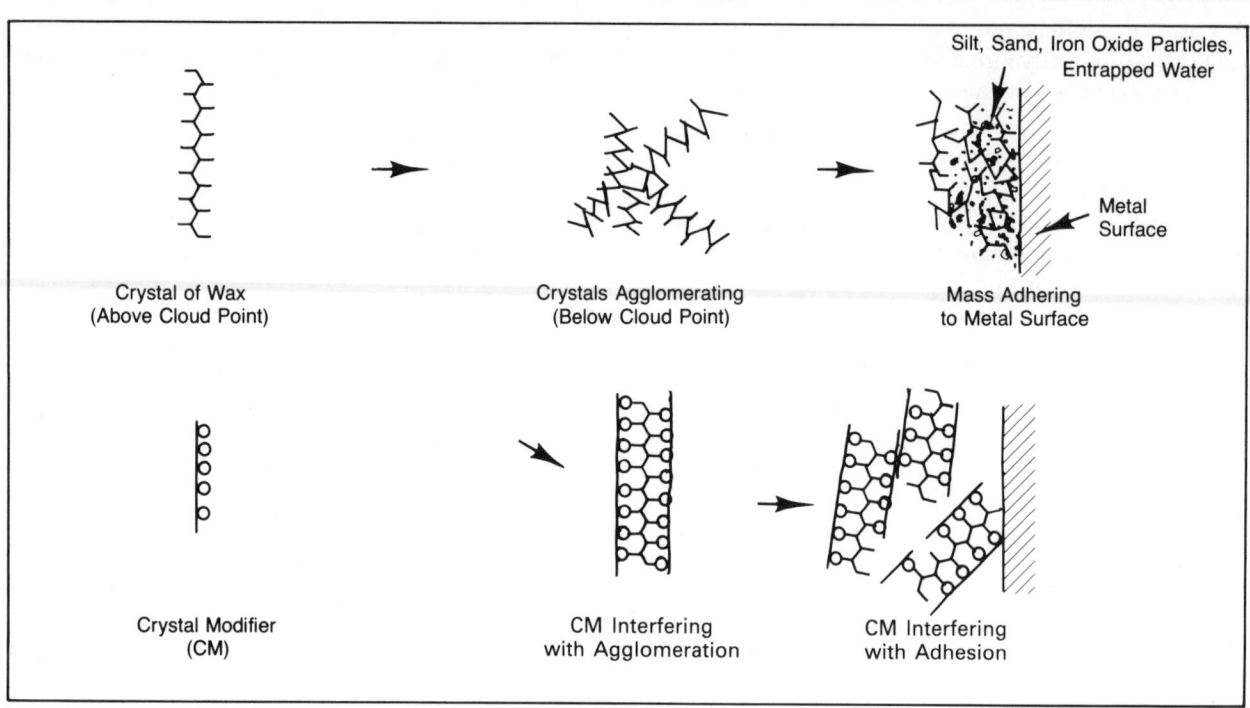

FIG. 2-1—*Effect of Crystal Modifier on Paraffin Deposition.*

88% after only 0.5 pore volume had passed through the core.

In another test, crude at 11°F below the cloud point was pumped through a 435 md sandstone core. In this case, the 88% reduction in permeability was reached after 3.5 pore volumes had passed through the core.

Other tests confirmed this trend, indicating that permeability damage occurs more quickly in the lower permeability cores. Recovery of permeability by circulating the same oil through the damaged cores at a temperature 22°F above the cloud point was fairly successful. Hot oil may be less successful in restoring permeability in a wellbore partially plugged with paraffin because hot fluid may not enter all of the plugged zones. Fluid contact with all of the paraffin-plugged pores is almost impossible to achieve, thereby preventing complete wax removal from all producing zones.

Crystal Modifiers

Paraffin comes out of solution as single crystals, which tend to agglomerate around a nucleus to form relatively large particles. Removal of the nucleating agent will prevent agglomeration of paraffin crystals and prevent deposition on metal surfaces as illustrated by Figure 2-1. After conducting numerous tests, Knox[11] described a class of chemicals that prevented agglomeration by keeping asphaltenes, the nucleating agent, in solution. Photomicrographs of crude produced from one well showed agglomerates of paraffin under normal conditions. When the crystal modifier was present, only single paraffin crystals were detected.

This crystal modifier, Halliburton's Parachek, is based on fused-ring chemicals such as naphthalene. Parachek has been effective *only* where asphaltenes are the nucleating agent for forming crystal agglomerates. Parachek can be placed as small balls in baskets near the bottom of the tubing, and by dropping sticks down the tubing. Liquid Parachek may be injected on a continuous basis into the well annulus or flowline or into the power oil in a hydraulic pumping system. It can also be placed in the formation during a frac job.

Polymers have been used successfully as crystal modifiers in some areas; their use should expand as more effective polymers are developed.

Laboratory tests should be conducted with each paraffin to determine the character and type of nucleating agent and the manner in which the modifier changes crystal growth and agglomeration. Then crystal modifiers can be chosen for field trial. Field tests can be utilized to choose the best chemical from those selected by laboratory screening and to determine the required treatment volume.

Field tests should start by using high concentrations of crystal modifiers, which are then reduced until the level is established at which paraffin once again begins to form. One company begins field tests at 1000 ppm and reduces the rate by 200 ppm increments at one to two week intervals.

Wellhead and power oil pump pressures, pressure drop in the flowline, producing rate, and removable nipples or spool pieces are all used to monitor depositon during the test.

Inhibitors seldom eliminate the requirement for hot oiling or scraping. Costs must be carefully monitored. Since typical treating concentrations are in the range of 250–1000 ppm, the interval between hot oil jobs must be increased considerably to justify the inhibitor cost. Total cost of an inhibitor treatment plus paraffin removal must be compared to costs for removal only. The value of lost or deferred production must be considered for both methods.

Plastic Pipe and Plastic Coatings

Smooth plastic pipe and plastic-coated pipe are used in some areas to reduce the rate of paraffin deposition. However, plastic pipe or plastic-coated pipe is usually selected to prevent corrosion.

Although deposition of wax on plastic surfaces is much slower than on steel, accumulation will continue at the same rate as on steel pipe after the plastic pipe has been covered with a layer of wax. Because wax eventually must be cleaned from plastic pipe or plastic-coated steel pipe, cleaning problems must be considered. For example, solvents or hot oiling will damage PVC pipe. Hot oiling fiberglass-reinforced plastic pipe may blow the plastic pipe apart at the joints. Therefore, with fiberglass-reinforced plastic pipe, considerable care must be exercised to maintain pump pressure well below the working pressure rating of the pipe during hot oiling. Also, hot oil temperature must be considered in determining the allowable working pressure of the plastic pipe. In general, working pressure ratings must be decreased at temperatures above 150°F. Thin film phenolic coating (5 to 9 mils thick) on steel pipe may pay out the coating costs as a result of reduced corrosion and less wax accumulation. However, so-called paraffin coating with a phenolic coating thinner than 5 mils is not recommended because of the greater probability of holidays in the coating. Phenolic coatings can be hot oiled up to 300°F without damage. Hot oil temperature of epoxy-coated lines should not exceed 150°F.

Surfactants as Deposition Inhibitors

Surfactant use to reduce wax deposition has been limited to a few areas. In one use, a surfactant is employed to water-wet the pipe surface. The water film, which must be continually maintained by the addition of surfactants, acts as a barrier to prevent paraffin contact with the pipe. Wells producing water are the best candidates for this type of treatment. However, if a well produces large volumes of fluid, surfactants may not be needed, because the temperature of the fluid may be above the cloud point of the wax. If the water-oil ratio is high, the steel may be water-wet without surfactants.

Some surfactants can act as solubilizing agents for the nucleating agent and thus prevent paraffin agglomeration. In this use, the surfactant must be added continuously to the production stream. Tests can be conducted to select the best surfactant for wells producing from a specific reservoir to inhibit wax deposition and also to prevent emulsions.

Downhole Heaters

Electric heaters can be employed to raise crude temperature as it enters the wellbore. Ideally, the oil temperature should be sufficiently high that it will *not* cool to its cloud point before reaching the surface. This technique is limited by economics, maintenance costs of the heating system, and electrical power availability. Further, if oil temperature at the heater is too high, the crude may coke and plug the wellbore.

Inhibiting Asphaltene Deposition

Asphaltene deposition is difficult to inhibit, particularly in the formation. One approach is to squeeze a solution of the naphthalene chemical into the formation at regular intervals to increase asphaltene solubility in the crude. If deposition occurs in tubing, the same chemical may be added to the production system by slowly dissolving balls or sticks of chemical in the flow stream. Balls may be placed in a screen basket on the tubing bottom or by dropping them down the annulus. Sticks of chemicals may be dropped down the tubing at regular intervals.

Another approach is to use a flowline bypass feeder to inject a small amount of oil production through a bed of inhibitor balls and into the casing-tubing annulus. These placement procedures are also effective in reducing paraffin deposition from crudes where asphaltenes are the nucleating agent for paraffin crystals.

Production Techniques to Reduce Wax Deposition

If a well's producing rate is high enough to maintain flowing temperature above the cloud point of the crude, deposition of paraffin will *not* occur. In addition, higher flow rates help prevent precipitated wax from adhering to a metal surface because of the shearing action on the wax particles at the deposition surface. High flow rates can also selectively remove the lower-melting, softer fractions from the growing deposit. The remaining deposit may accumulate more slowly, but may be more difficult to remove. Certain crudes have shown increasing deposition rates when the flow was in the laminar regime. Turbulent flow reduced the deposition rate in most cases.[6]

DESIGN FOR WAX CONTROL

Tests to determine cloud point, pour point, paraffin content, and asphaltene content of crude should be conducted as part of the planning phase of equipment design for a new field. Such information will indicate whether wax problems can be expected. It will also reveal the point in the system that wax will start to come out of solution.

Plastic coating of tubing may be advisable if downhole paraffin deposition is predicted. In some cases, the selection of artificial lift may be influenced by paraffin problems. Hydraulic pumping systems are particularly amenable to continuous inhibition, as well as solvent or thermal treatment for paraffin removal. A number of patented downhole treating devices are available that allow chemical or hot oil to be pumped into the tubing via the annulus in wells equipped with a tubing packer.[14,15,18]

Paraffin scrapers attached to rods will reduce the frequency of hot oiling in rod-pumped wells. The presence of paraffin may require higher strength or larger rods. Larger tubing may be beneficial because of additional clearance between the rods and the tubing.

Where paraffin or asphaltene content is so high that the crude gels completely as it cools, special provisions must be made in the oil gathering system. Insulation, with and without heat tracing, has been used to maintain flowline temperatures high enough to prevent gelling of high pour-point crude. Myers[19] describes the use of "Skin Effect Current Tracing" in a North Dakota field. Some Indonesian crude, which becomes a solid below 80°F, is moved through a double-walled subsea pipeline having urethane foam insulation in the annulus. One large field in Argentina produces a highly asphaltic crude which freezes at ambient temperature. In this case, all gathering and power oil lines are insulated and steam traced using steam generated in boilers.

Paraffin deposits in stock tanks frequently appear as bottom sludge that may contain several percent of water and lesser amounts of clay, silt, sand, rust, and scale. A circulating pump is used to pump this sludge from the stock tank and through a heater treater. After the paraffin is separated from the BS and water, it can be sold with the crude. This usually requires heating the stock tank oil to a temperature above the cloud point. The most common approach is to continue circulating the oil through the heater treater until the entire contents of the tank can be sold. Some tanks are fitted with heating coils to facilitate sludge removal from tanks. Insulating the stock tanks will save considerable energy where it is necessary to heat the sludge in the tanks.

Manually cleaning the tanks may be the best solution when a large volume of wax is present. In instances in which hot oiling is anticipated, a valve should be installed on the stock tank as a connection for the hot oil unit. These valves should be three or four feet from the bottom of the tank to reduce the amount of paraffin-rich sludge or "bottoms" transferred to the hot oil units or to tanks for use in completion, workover, or well stimulation operations.

REFERENCES

1. Reistle, C. E.: "Paraffin and Congealing Oil Problems," Bulletin USBM, p. 348, (1932).

2. Katz, D. L. and Beu, K. E.: "Nature of Asphaltic Substances," Ind. and Eng. Chem., vol 37, no. 2, (1945), p. 195.

3. Nathan, C. C.: "Solubility Studies on High Molecular Weight Hydrocarbons Obtained from Rod Waxes," Trans. AIME (1955), p. 151.

4. Nathan, C. C.: "How to Evaluate Paraffin Inhibitors," Petr. Engr., vol. 27, no. 3 (1955), p. 66.

5. Shock, D. A., Sudbury, J. D., and Crockett, J. J.: "Studies of the Mechanism of Paraffin Deposition and its Control," Journal Petr. Tech., vol. 7 (1955), p. 23.

6. Jesson, H. W. and Howell, James N.: "Effect of Flow Rate on Paraffin Accumulation in Plastic, Steel, and Coated Pipe," Petr. Trans., vol. 213, TP 8010, (1958).

7. Witherspoon, P. A., and Minir, Z. A.: "Size and Shape of Aspahltic Particles in Petroleum," AIME Fall Meeting in Los Angeles, California, October 16–17, 1958.

8. Cole, R. J. and Jessen, F. W.: "Paraffin Deposition," The Oil & Gas Journal, vol. 64, (1960), p. 87.

9. Dwiggins, C. W., Jr. and Dunning, H. N.: "Separation of Waxes from Petroleum by Ultracentrifugation," J. Phys. Chem., vol. 64, (1960), p. 377.

10. Hunt, E. B., Jr.: "Laboratory Study of Paraffin Deposition," SPE 279, Production Research Symposium, SPE of AIME, Tulsa, Oklahoma (April 12–13, 1962).

11. Knox, J. A., Waters, A. B., and Arnold, B. B.: "Checking Paraffin Deposition by Crystal Growth Inhibition," SPE 443, 37th Annual Fall Meeting, Los Angeles, California, 1962.

12. Billingsley, D. L.: "How to Control Paraffin in Shallow Pumping Wells," World Oil, July 1963.

13. Nelson, W. L.: "Asphalt in Various World Crude Oils," The Oil & Gas Journal, December 7, 1964, pp. 170–171.

14. Roach, E. E., U. S. Pat. No. 3,329,211, (July 4, 1967).

15. Tomlin, D. R., U. S. Pat. No. 3,376,936, (April 9, 1968).

16. Sutton, G. D. and Roberts, L. D.: "Paraffin Precipitation During Fracture Stimulation," SPE 4411, Casper, Wyoming meeting, May 15–16, 1973.

17. Snavely, Earl S, U. S. Pat. No. 3,724,552, (April 3, 1973).

18. Alexander, Harvey C. and Hudson, Ray E., U. S. Pat. No. 4,011,906, March 15, 1977.

19. Myers, Ralph: "An Electrically Heated Buried Gathering System Transfers High Pour Point Crude," SPE 6797 (1977).

20. Coppel, C. P., Newberg, P. L., "Field Results of Solvent Stimulation in a Low-Gravity-Oil Reservoir," *J. Pet. Tech.* (Oct., 1972) 1213–1218.

21. Hirshberg, Avraham, de Jong, Lon. N. J., Schipper, Bas. A., Myers, J. G.: "Influence of Temperature and Pressure on Asphaltenes Flocculation," SPE 11202 (Sept., 1982).

22. McClaflin, G. G. and Whitfill, D. L.: "Control of Paraffin Deposition in Production Operations," SPE 12204 (Oct., 1983).

Chapter 3 Squeeze Cementing— Remedial Cementing

Objectives, applications, techniques
Theoretical considerations
Practical considerations
Planning a squeeze job
Operational procedures
Plug back operations
Special cement systems

INTRODUCTION

The technique of Squeeze Cementing was developed before there was much knowledge of fracture mechanics of earth materials, or much ability to adjust the properties of neat cement slurries. As a result misconceptions developed concerning where the cement goes on a squeeze job, what well conditions and cement slurry properties are required for effective squeezing, and what can be expected of a squeeze cementing operation. This being the case, it is well to review the basics of squeeze or remedial cementing.

Squeeze cementing is an operation where a cement slurry is forced under pressure to a specific point in a well for remedial purposes. The objective is to fill all perforations, or channels behind casing with cement to obtain a seal between the casing and formation.

To accomplish this objective requires only a relatively small volume of cement, but it must be placed at the right point in the well. Sometimes a major difficulty is confining the cement to the wellbore. Cement placed in a vertical fracture, away from the wellbore, is of no value.

Applications include: seal off undesired perforations; plug primary cement channels to exclude water or gas from the oil zone; repair damaged casing; and supplement the original primary cement job. An important point is that squeeze cementing can only be expected to solve communication problems in the annulus between the casing and the borehole. It cannot solve a problem of vertical communication within the formation.

Two general techniques are recognized for accomplishing the remedial cementing objective:

1. The *high pressure* technique involves fracturing the formation and pumping cement slurry into the fracture until a particular surface pressure is reached and "holds" without bleed off. Normally, neat cement (very high fluid loss) is used. This technique has inherent disadvantages, most of which are overcome by the low-pressure technique.

2. *Low pressure* technique, or more descriptive, the "low-fluid-loss-cement" technique involves spotting cement over the perforated interval and applying a pressure sufficient to form a filter cake of dehydrated cement in the perforations and in channels or fractures that may be open to the perforations. Low fluid-loss cement (50 to 100 cc API) and clean workover fluids must be used. It is not necessary or desirable to hydraulically fracture the formation.

Improved knowledge in two areas has benefited our understanding of the squeeze cementing process, and has directed us toward the low pressure technique:

—The mechanics of fracturing rock and the resulting fracture orientation.
—The filter cake buildup, or dehydration characteristics of a cement slurry pressured against a permeable media.

THEORETICAL CONSIDERATIONS
Fracture Mechanics

Sedimentary rock is inherently a low tensile strength material primarily held together by the weight or compressive forces of the overlying formations. The compressive forces act in all directions to "hold" the rock together but do not usually have the same magnitude in all directions, Figure 3-1.

When sufficient hydraulic pressure is applied through the wellbore against a particular formation, the formation rock fractures along a plane perpendicular to the direction of the least compressive stress. Usually this plane is vertical, and a vertical fracture results, often having a preferred azimuth or direction. The geometry of this fracture depends on the rate of fluid injection, the fluid loss and viscosity characteristics of the fluid, and the volume of fluid injected.

Based on this view of fracture mechanics:

1. Obviously a large vertical fracture extending through gas or water contacts, and the natural barriers to vertical flow, is undesirable when squeeze cementing to exclude gas or water from an oil zone.
2. The horizontal "pancake" of cement extending outward from the wellbore is usually a misconception.
3. A vertical fracture does not necessarily "find" a channel in the original primary cement job.

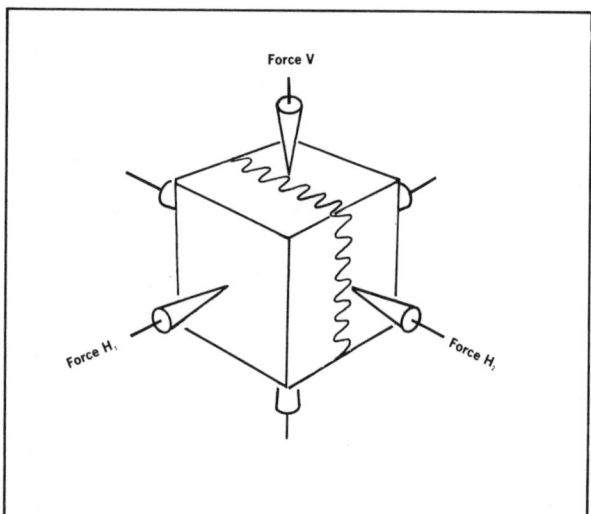

FIG. 3-1—*Triaxial loading of rocks.*

Injecting 100 sacks of cement into a formation might, depending on fracture width and barrier zones, create a fracture along a vertical plane fanning out 40 to 50 ft in the vertical direction and a similar distance in the horizontal direction as shown in Figure 3-2.

Cement Dehydration Process

Cement particles having a size range from about 20–75 microns are too large to enter a normal formation pore structure. Thus, cement particles "plate out" on the face of a formation having granular permeability.

The mechanism of dehydration, when cement is placed against a permeable media and exposed to differential pressure, is exactly the same as occurs with a drilling fluid. Water is forced from the cement slurry and a filter cake of cement is formed on the permeable media. Similar to mud cake, the cement cake is relatively soft and can be removed by jetting but is not pumpable—considerable pressure would be required to force it through a restricted area.

Thickness of the filter cake depends primarily upon the permeability of the cake or formation (whichever is lower), the fluid loss characteristics of the slurry, the magnitude of the differential pressure, and the length of time the differential is maintained.

The API filter loss of neat cement ranges from 600 to 2500 cc in 30 minutes; in fact, dehydration occurs so rapidly it is difficult to measure. Filter loss can be reduced to relatively low values, 25 to 100 cc in 30 minutes, by addition of bentonite and dispersing agents, or by polymers. Figure 3-3 shows a typical relation between fluid loss and increasing amounts of a polymer-type fluid loss additive.

The relation of fluid loss characteristics, filter cake permeability and rate of filter cake buildup—is shown in Table 3-1. These data are based on surface measurements and probably should not be applied quantitatively to a downhole situation. The data reflect the correct trend of downhole events, however.

The differential pressure applied to the slurry directly affects the rate of filter cake buildup but does not affect the permeability of the filter cake. Formation permeability also affects the rate of filter cake buildup, but not on a directly proportionate

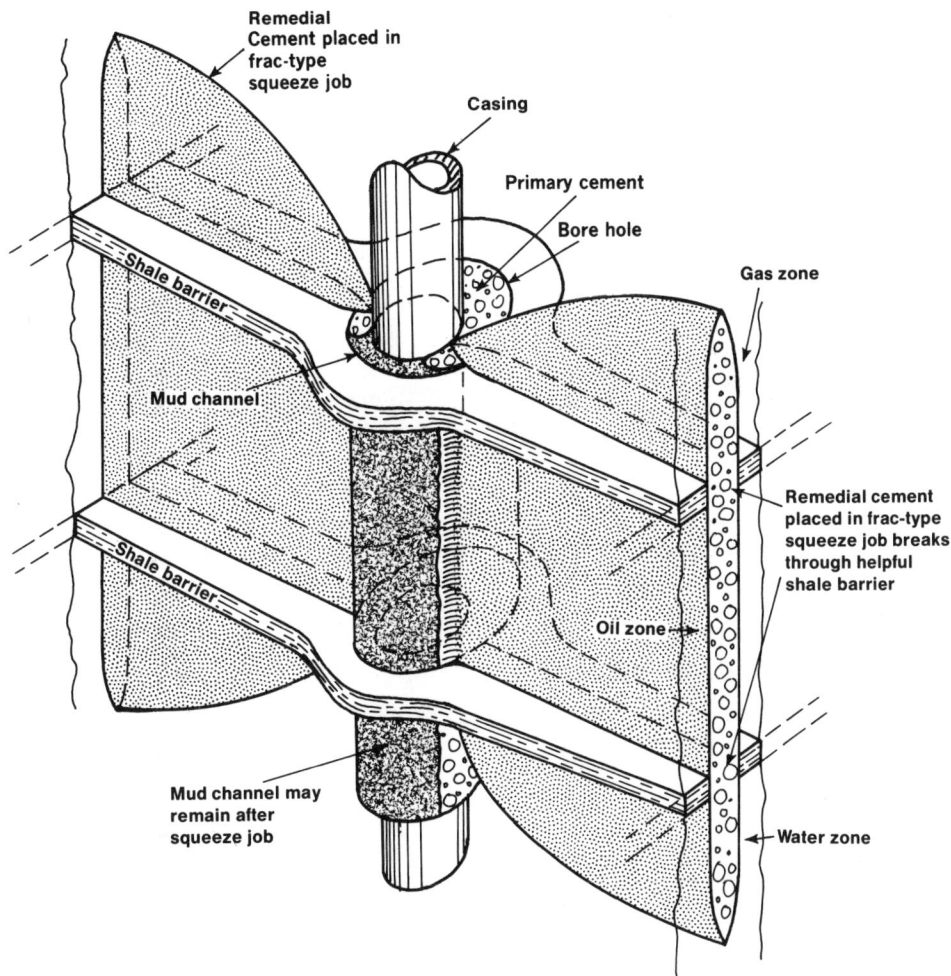

FIG. 3-2—*Probable result of fracture-type squeeze-cement job.*

basis. Surface tests show that it takes about twice as long to form the same thickness of filter cake against a 30 md formation as it does against a 300 md formation.

PRACTICAL CONSIDERATIONS
Normal Squeeze Cementing Situation

Based on current knowledge of fracture mechanics, and the fluid loss properties of cement slurries, we are able to visualize what must occur downhole during a "normal" squeeze cement job, and use this visualization to develop a reasonable technique to accomplish our squeeze cementing objective. A normal squeeze cement job is defined as one, wherein, it is desired to repair a primary cement channel, or merely close off the perforations opposite a producing zone having reasonable matrix permeability.

TABLE 3-1
Fluid Loss Vs. Filter Cake Permeability and Cake Buildup Rate

API fluid loss at 1,000 psi (cc/30 min.)	Permeability of filter cake at 1,000 psi (md)	Time to form 2-in. cake (min.)
1200	5.00	0.2
600	1.60	0.8
300	0.54	3.4
150	0.19	14.0
100	0.09	30.0
50	0.009	100.0
25	0.006	300.0

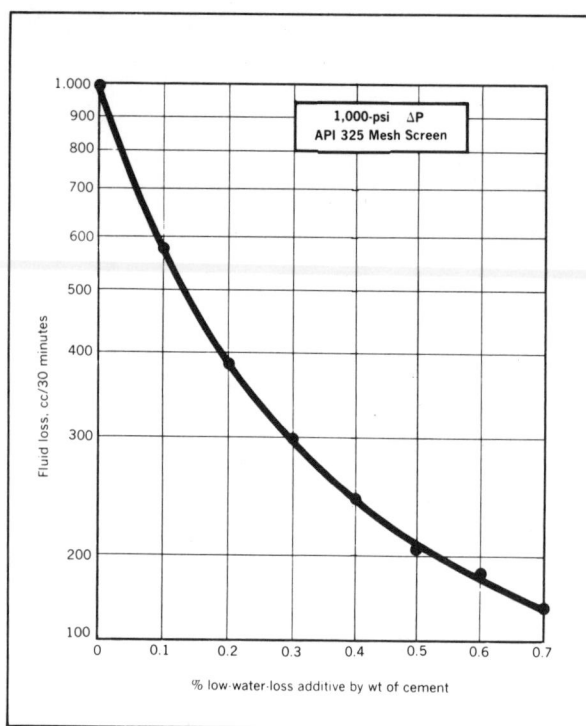

FIG. 3-3—*Fluid loss reduction by polymer additive.*[4] *Permission to publish by The Society of Petroleum Engineers.*

A reasonable time period is required for this process. Thus, the key to effectively filling the channel is cement *fluid loss control*, both to reduce the rate of filter cake buildup and to prevent an increase of slurry viscosity through loss of water, and *pumpability time* so that the operation can be accomplished before the cement takes an initial set. When the cement reaches the lower end of the channel opposite the water zone, the cement particles plate out, and the filter cake build process begins, ideally filling the channel with cement filter cake.

Basics of Low-fluid-loss Cement Squeezing—Clean workover fluids are an essential requirement of low-fluid-loss-cement squeezing. If a perforation is completely plugged with mud so that no permeability remains, it is impossible to form a filter

Effect of Filtration Rate on Cement Placement—High-fluid-loss cement pumped under a packer to "squeeze" a long perforated zone may (depending on the pumping rate) dehydrate so rapidly opposite the upper perforations that filter cake of dehydrated cement completely fills the casing and prevents application of cement or pressure to the lower perforations as shown in Figure 3-4. The same effect may occur in a channel, i.e., cement filter cake may quickly close the channel at one end and prevent filling the channel with cement. If the same zone is reperforated, the shattering effect of perforating probably re-opens the original channel to the perforations.

Cement with low filtration rate (50 cc in 30 minutes) forms a thin filter cake which thickens slowly. Most of the slurry remains "fluid" and pumpable, thus chances of filling a channel with cement or contacting each perforation with cement are very much improved, as shown in Figure 3-5. Water filling the channel from the lower water zone, is forced back into the water zone by the low-fluid-loss cement.

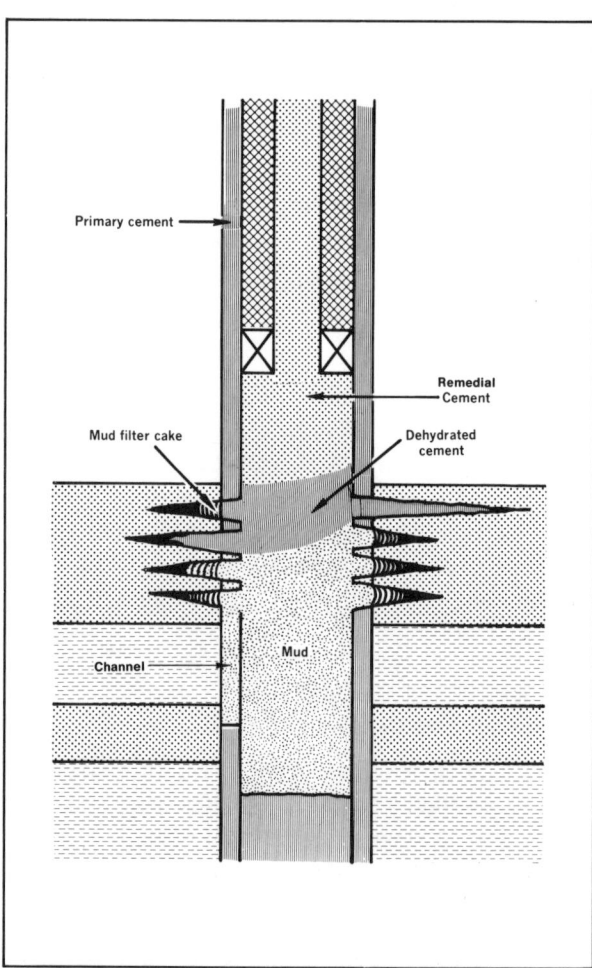

FIG. 3-4—*Probable result of high-fluid-loss cement squeeze.*

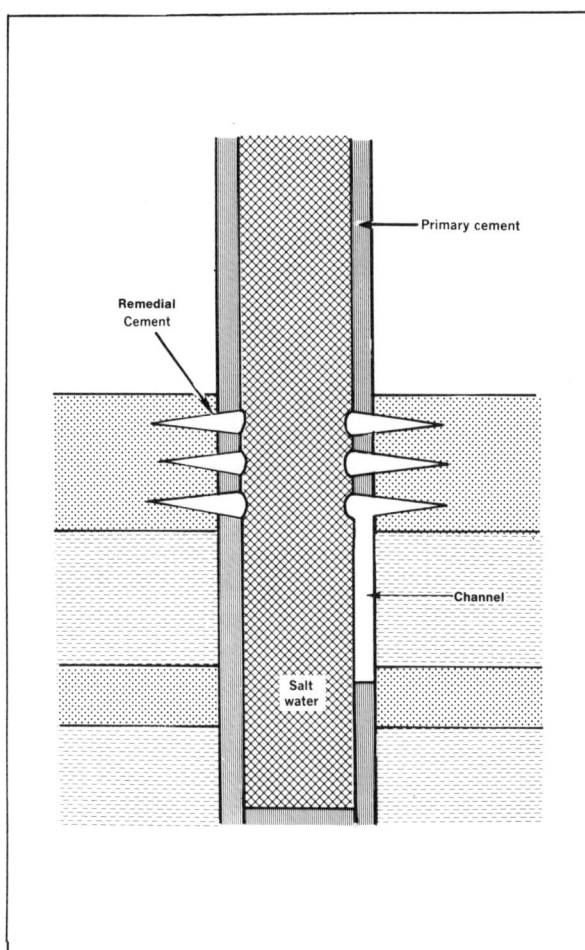

FIG. 3-5—*Probable result of low-fluid-loss cement squeeze.*

Bottom-hole pressure should be kept below formation fracture pressure since fracing is not necessary or desirable. Small fractures are probably not detrimental, but large fractures may extend vertically into gas.

Application of High Fluid-loss Cement—There is almost no advantage to using high fluid-loss cement for remedial cementing. A squeeze packer is required to protect upper casing sections from pressure. Maximum surface pressure is usually based on some "magic" formula, i.e., (0.4 × Depth + 500 psi). Large volume of cement must be used (100 to 200 sacks is typical). Cement must be drilled out following job. Large vertical fractures may be detrimental to control of formation fluids.

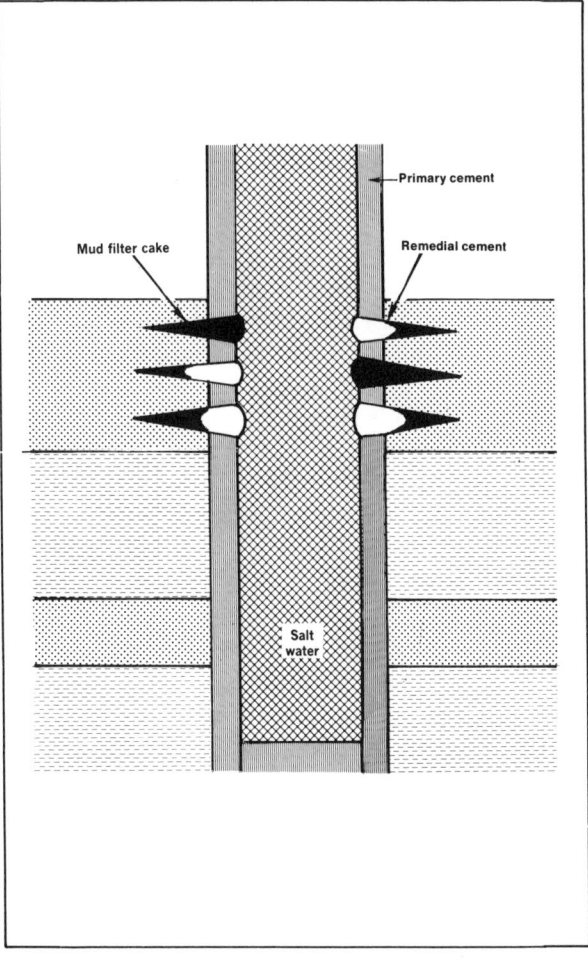

FIG. 3-6—*Mud-plugged perforations prevent cement filter-cake formation.*

cake of cement in that perforation, as shown in Figure 3-6. If perforations are plugged more than one squeeze job may be required.

A squeeze packer is not necessary unless upper perforations are open or casing is damaged. Cement slurry can be spotted over perforations prior to job and, if desired, can be washed out after job. With proper filter loss control there is very little danger of sticking tools in the hole due to filter cake buildup inside the casing. Ideally only a small cement node remains inside the casing opposite each perforation after washout. Volume of cement required for squeezing perforations is low (0.1 to 0.2 cu ft per perforation, plus the casing fill-up). Small low-volume, low-pressure pumping equipment can be used.

Special Squeeze Cementing Situations

Special squeeze cementing situations are defined as those cases where formation permeability is very high or very low.

Carbonate Formations—Fractured carbonate formations present a somewhat different situation than sandstone, since the matrix permeability may be very low, and actual, or effective permeability, consists of inter-connected voids or fracture systems. In this case, cement particles may move into the large voids or open fracture system. The problem then becomes one of confining the remedial cement slurry to the area of the wellbore.

It should be recognized that cement cannot solve a vertical communication problem in the formation back away from the wellbore. It should also be recognized that allowing cement to move into open fractures, vugs or voids in the carbonate productive zone may put the zones out of reach of the perforating gun. All we can hope to do with squeeze cementing is to repair communication along the original Primary Cement job in the casing-borehole annulus.

Thus, if the "permeability" of the flow system is quite high it may be necessary to use one of several possible techniques to confine the cement to the wellbore. Possibilities include a two-stage job with the first stage containing granular loss circulation material for bridging; higher fluid loss cement to form filter cake faster, a quicker setting slurry, a slurry with thixotropic properties such that viscosity increases with lower flow velocity, or a material which forms a stiff gel in contact with formation brine. After the first stage seals the formation face, a second stage of lower fluid loss cement can be used, if needed, to repair communication along the original primary cement job.

Shale Formations—Shale formations have essentially no matrix permeability; thus, a cement filter cake does not build up on a shale zone. Squeeze cementing becomes a process of attempting to place cement in a channel or other mud-filled annular space and allowing it to set up. In perforating opposite a shale section for a "block squeeze" this lack of a filter cake effect may make it difficult to close off the perforation.

PLANNING A SQUEEZE CEMENT JOB

Planning a squeeze job involves consideration of the following factors: fluid in the wellbore, condition of perforations, squeeze tools, type and quantity of cement, and the selection of the squeeze pressure.

Type Workover Fluid

Clean salt water is the preferred workover fluid for low (or high) pressure squeeze jobs. Considerations of formation damage often dictate the specific composition of the fluid.

Since formation fracturing is necessary to force mud filter cake away from each perforation, mud should not be used as the workover fluid unless absolutely necessary to control well pressures.

Plugged Perforations

If well was perforated in mud, or if perforations were ever contacted by mud, some mud plugs may remain, and more than one squeeze job may be needed even if a clean fluid is used. Even if squeeze pressures sufficient to fracture the formation are employed, there may still be perforations which do not accept cement, are not sealed off, and thus may "breakdown" when differential pressure into the wellbore is applied.

Attempts to clean perforations known to be plugged prior to a squeeze cement job, involve use of acid-surfactant formulations. To be effective some technique of contacting each perforation with the treating fluid must be used. One technique involves injecting a clean fluid above frac pressure, and using ball sealers to successively close off perforations taking fluid, until all perforations have been broken down.

This is somewhat time-consuming and costly since a squeeze packer may be required. It may well be less costly to do more than one squeeze cement job, particularly if concentric tubing workover techniques are used.

Squeeze Packers

Squeeze packers are not normally required for low pressure squeeze cementing. When there are other sets of perforations in a wellbore in addition to those which are to be squeezed or where anticipated squeeze pressures exceed the casing burst pressure allowance, then squeeze tools must be used.

For circulation type squeeze jobs, a retrievable

packer or a cement retainer set between upper and lower sets of perforations is required to circulate cement in the annulus behind the casing.

A drillable packer or cement retainer is used primarily to isolate lower perforations when upper perforations are to be cemented, or to cement lower perforations which are to be abandoned.

A retrievable squeeze packer may be advantageous, in that resetting permits flexibility (i.e., can locate casing leak and squeeze cement on same trip). Less rig time is required.

Cement Slurry Design

Design of a suitable squeeze cementing slurry is not complicated. Slurry density, volume, cost, and strength must be considered, as in primary cementing; but pumpability time and filtration rate are of primary importance.

Uniformity of the cementing material is important in squeeze cementing; thus API Class G or H basic cements, which are manufactured to more rigid specifications, are often good choices. Additives are then used to provide desired pumpability time and filtration rate.

Pumpability time required for low pressure squeeze cementing depends on several factors, including well depth, tubular size and pumping rate, differential pressure applied and the filtration rate of the cement slurry. In other words, pumpability time must exceed, by a suitable safety factor, the time to get the cement slurry to bottom, and to form the required cement filter cake. Two and one-half to four hours are typical pumpability times.

The Amoco Pressure-Temperature Thickening Time Tester should be used to determine pumpability time for critical conditions or where routine squeeze cementing formulations have not been established. To be meaningful, tests must be run using the same cement, water, and additives which will actually be present on the job.

Table 3-2 shows the standard API squeeze cementing well-simulation test schedules. For a critical job, if the standard conditions set up in the schedule do not fit anticipated job conditions, then the thickening time test procedure should be modified as necessary.

As a general rule the "hesitation process" of squeeze cementing reduces by a factor of two, the pumpability time of cement as measured using the regular API squeeze cementing schedule test procedure.

Filtration rate in the range of 50 to 125 cc on the API test for cement (1,000 psi, 325-mesh screen, ambient temperature) is satisfactory. Filtration rate is one factor that controls time required to form the cement filter cake; thus, if it is necessary to fill a long channel with cement, lower fluid-loss values should provide additional time.

If a naturally-fractured formation is being squeezed, higher fluid-loss values should reduce the penetration of cement into the natural formation fractures. As previously noted, granular materials may also be required in this case to provide a bridge

TABLE 3-2
API Squeeze and Plug-Back-Cementing Well-Simulation Test Schedules

1	2	3	4	5	6	7	8
Schedule no.	Depth, ft.	Mud density, lb. per gal.	Surface pressure, psi	Bottom-hole circulating temperature deg. F.	Time to reach circulating temperature min.	Bottom-hole pressure psi	Time to apply final squeeze pressure, min.
12	1,000	10	500	89	3	3,300	23
13	2,000	10	500	98	4	4,200	25
14	4,000	10	500	116	7	5,600	28
15	6,000	10	800	136	10	6,700	31
16	8,000	10	1,000	159	15	7,800	35
17	10,000	12	1,300	186	19	9,400	38
18	12,000	14	1,500	213	24	11,800	42
19	14,000	16	1,800	242	29	14,000	45
20	16,000	17	2,000	271	34	16,500	48
21	18,000	18	2,200	301	39	19,000	51

to form the cement filter cake against.

Desired filtration rate and pumpability characteristics can be obtained by any of several types of cement slurries. Twelve % bentonite "Modified" cement, (similar to that used in primary cementing) is a satisfactory squeeze cementing slurry. It has the advantage of lower density, 13.0 to 13.5 lb/gal, which may significantly reduce the back pressure applied against squeezed perforations if a long column of cement must be reversed out after the job.

It typically consists of: API Class A, B, G or H cement; 12 lb/sack Wyoming bentonite (must not be "peptized" or "beneficiated"); and 0.6 to 1.0% calcium lignosulfonate (HR-7, Lignox, Kembreak), depending on pumpability time requirements.

Fluid loss control and retarding effect can also be obtained by high molecular weight polymers such as Halad-9 or FLAC. Mixing is simplified. Tables 3-3 and 3-4 show properties of Halad-9 slurries.

A combination of bentonite and high molecular weight polymer is used in some areas. A typical slurry consists of Class H cement with 4% bentonite, 1.0% Halad-9 or FLAC, and 0.25 to 0.5% retarder.

In Canada a typical slurry consists of Class G Oil Well cement, with about 1.0% CFR-2 for fluid loss control.

Compressive strength of the set cement is not an overly important factor in slurry design. It should be noted that a perforation plugged with mud filter cake will often withstand tremendous differential pressures, either into the well, or into the formation. Cement with a 24-hour compressive strength of 500 to 1000 psi will satisfactorily plug a perforation.

Cement Slurry Volume

The optimum volume of slurry is the minimum amount required to seal perforations or channels. The slurry volume should completely fill the casing through the perforated interval, and should allow an excess of 0.1 to 0.2 cu ft per perforation. If channels are anticipated behind the casing, allow additional excess. Total volume on a low pressure job rarely exceeds 10 to 15 bbl.

If the formation is hydraulically fractured a tremendous volume of cement may be pumped outside the casing into the vertical fracture, and away from the well. However, there is no correlation between volume of cement pumped into a fracture and job success.

Squeeze Pressure

Squeeze pressures higher than necessary do not add to job success—only increase chance of fracturing the formation. With high fluid-loss cement, the pressure seen at the surface is not actually exerted on the cement in the perforations, but merely on dehydrated cement inside the casing many feet above the perforations.

Final squeeze pressure at the formation should be 200–300 psi less than the formation fracture pressure. With no other experience available the fracture gradient is often considered to be 0.75 psi/ft. It should be noted that a full column of 15.5 lb/gal cement often exerts sufficient pressure to fracture a formation even with no additional surface pressure.

If excess cement must be reversed out, final squeeze pressure should be 300–500 psi greater than reverse circulating pressure.

TABLE 3-3
Halliburton Halad-9—Fluid Loss Properties
API Class A Cement
Water Requirement—5.6 Gal/Sack

Percent Halad-9	Fluid loss: cc/30 Min.—325-mesh screen Pressure indicated, psi		
	100	500	1,000
0.60	96	178	250
0.80	24	70	100
1.00	14	32	60
1.20	10	24	43

TABLE 3-4
Halliburton Halad-9—Thickening Time Properties
API Class A Cement
Pressure-Temperature Thickening Time Test

Percent Halad-9	API squeeze tests			
	2,000 ft.	4,000 ft	6,000 ft	8,000 ft
0.00	3:15	2:09	1:12	0:52
0.60	4:00+	2:52	2:14	1:32
0.80	4:00+	4:03	2:30	2:17
1.00	4:00+	4:00+	3:05	2:20

NORMAL OPERATIONAL PROCEDURE

Operational procedure must be varied to suit the specific situation. Careful supervision is the key to job success.

Mixing Cement

Mixing of cement should be done by the batch method, rather than by the hopper method in order to obtain uniform slurry, exact proportioning of ingredients, and to permit measurement of slurry properties before it is put into well. Pre-mixed bulk materials are undesirable due to chance of mistake. Typical procedure is as follows:

1. Measure required water into tank.
2. Add fluid loss polymer and retarder to water and mix thoroughly before adding cement.
3. If high gel slurry is to be used, add one sack cement to the water to reduce bentonite yield; then add bentonite and remaining cement simultaneously.

Surface Tests of Cement Slurry

Surface tests of the cement slurry are imperative. After ingredients have been mixed and circulated for 5–10 minutes, sample should be weighed, and fluid loss should be checked. On the standard field model Filter Press (100 psi ambient temperature) fluid loss should be 25 to 50 cc in 30 minutes. A yellow color to the filtrate usually indicates the presence of a lignosulfonate retarder.

Cement Placement

Cement slurry should be spotted over perforations, particularly if long perforated sections are present. If salt water displacing fluid is used, the accelerating effect of the salt on the cement should be considered and, if necessary, the cement protected with a spacer fluid fore and aft. Tubing and packer (if used) should then be pulled above the top of the cement. However, with proper fluid loss control and sufficient clearance between tubing and casing, many jobs have been done with tubing through the perforated zone.

Filter Cake Buildup Process

To dehydrate cement in perforations, pressure should be increased slowly with occasional hesitations, as shown in Fig. 3-7. Rate of filter cake buildup is a function of pressure and time. If fracture occurs, pump should be shut down for several minutes, then started slowly. With the pump stopped, the rate of surface pressure bleed-off gives some indication of filter cake buildup—bleed-off rate decreases as filter cake builds up. When desired final pressure is reached, pump should be shut down, and pressure observed for 5–10 minutes. If bleed-off occurs, pressure should be built back to desired level. Job is complete when final pressure does not bleed back.

Reversing Out Excess Cement

Reverse circulation should be accomplished so as to maintain differential pressure into the perforations, but a somewhat lower differential pressure than the final squeeze pressure. All cement can be washed out of perforated interval if desired. However, sometimes success ratios are higher if cement is left through perforations, and drilled out later.

WOC

Waiting-on-cement time is governed by same factors as in primary cementing: well temperature, pressure, water-cement ratio, retarder, etc. Cement having a compressive strength of 200 to 250 psi should effectively plug a perforation. Four to six hours WOC time is often sufficient before proceeding to the next operation, if reasonable care is observed to avoid excessive differential pressure into wellbore.

Evaluation

Test of squeeze job should be determined by requirements of subsequent well operations. Don't overdo.

Apply pressure into formation at least as great as will be exerted by subsequent frac jobs. It should be noted that pressuring up inside casing alone is not a positive indication of a cement-sealed perforation, since a mud-plugged perforation may also withstand considerable differential pressure.

Apply differential pressure into wellbore at least as great as will be encountered in producing.

If a squeeze cement job fails with differential pressure into the wellbore, it is likely that mud

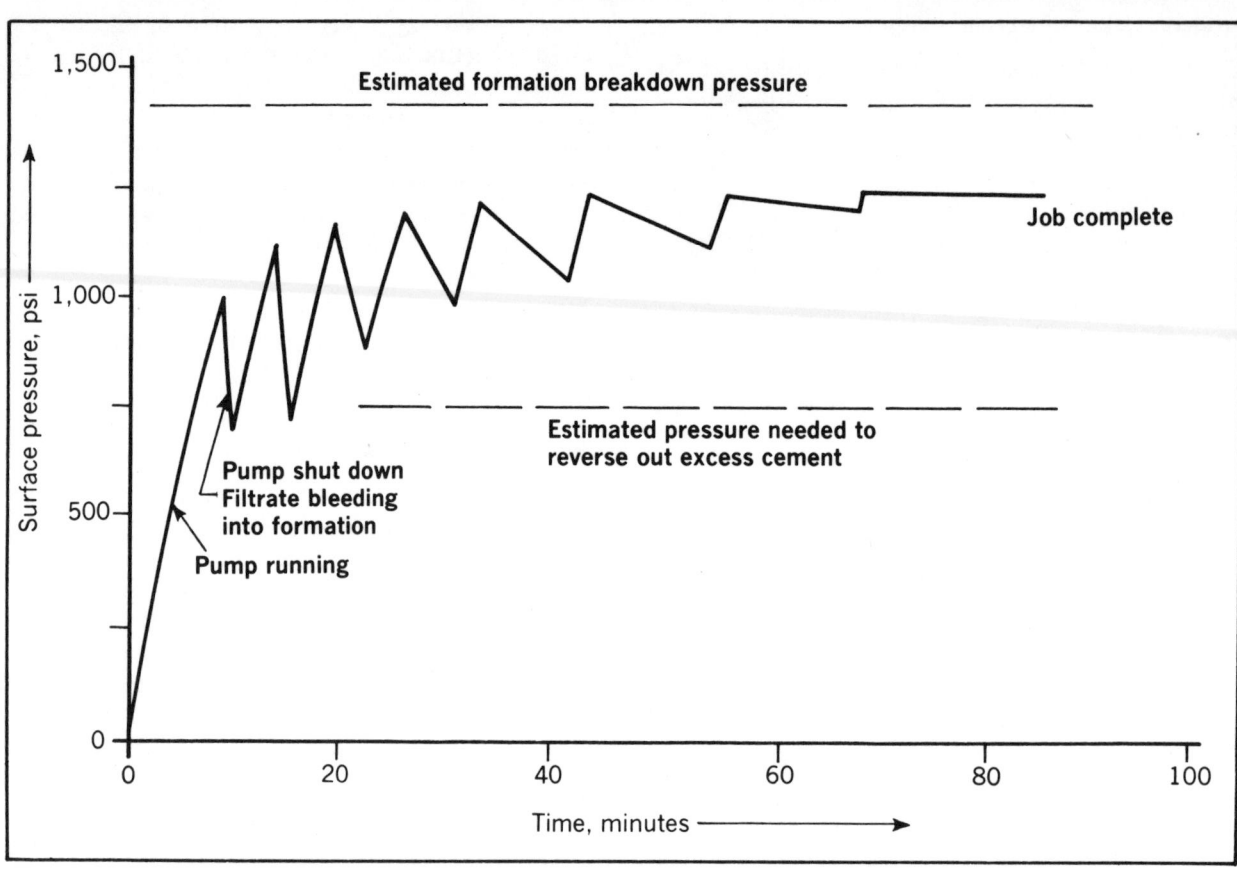

FIG. 3-7—Idealized recording-pressure chart during squeeze job using hesitation technique.

plugs popped out of one or more perforations. The solution is merely to do another squeeze job. This should not be taken as an indication that the cement slurry or job technique was at fault. With the low fluid-loss cementing technique, the cost of a second or third job is minimized.

OTHER OPERATIONAL PROCEDURES
Circulation Squeeze

Occasionally the Cement Evaluation Log may clearly indicate that there is essentially no cement in the annulus through a particular interval—or perhaps that the zone is above the top of the original primary cement. This situation is one where a circulation-type squeeze, Figure 3-8, may be a good solution. Basically, the technique requires perforating several holes at the top and bottom of the interval where a cement seal is desired.

The Cement Evaluation Log usually is a reliable indication of where perforations can be placed with a reasonable chance of circulating between zones.

A cement retainer is set between the perforations, clean fluid is circulated to remove the gelled mud, cement is circulated behind the casing with returns through the top perforation, and finally a low pressure squeeze is performed.

Several advantages and limitations are apparent. Advantages are:

—A chance is provided to clean out gelled mud through fluid circulation.
—Circulation usually permits placement of a longer column of cement than could be placed by squeezing "blind" from one set of perforations.

Limitations are:

—Volume of cement required can only be estimated. Cement returns from the upper perforations are desirable to obtain full fill-up of the annulus, and to close off these perforations.
—The upper perforations may have to be re-squeezed to effectively seal them.

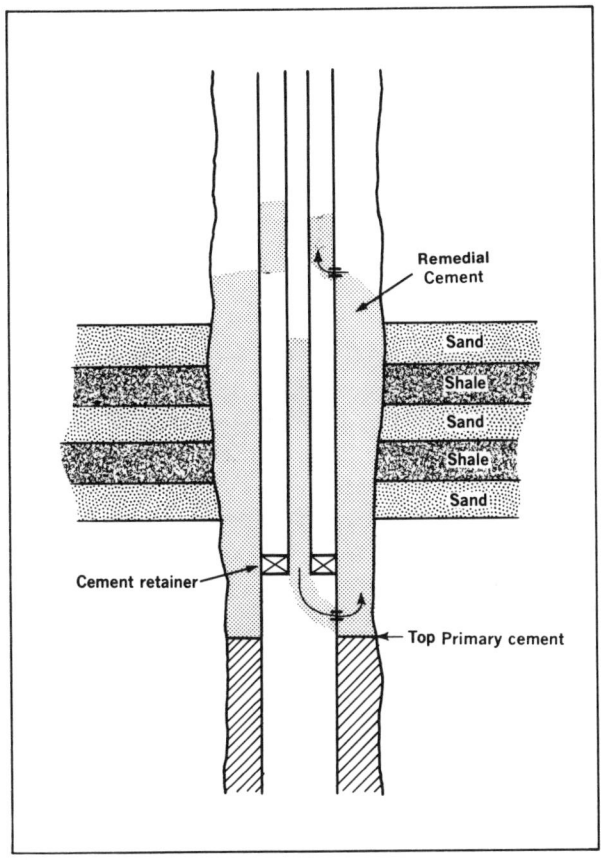

FIG. 3-8—*Circulation squeeze.*

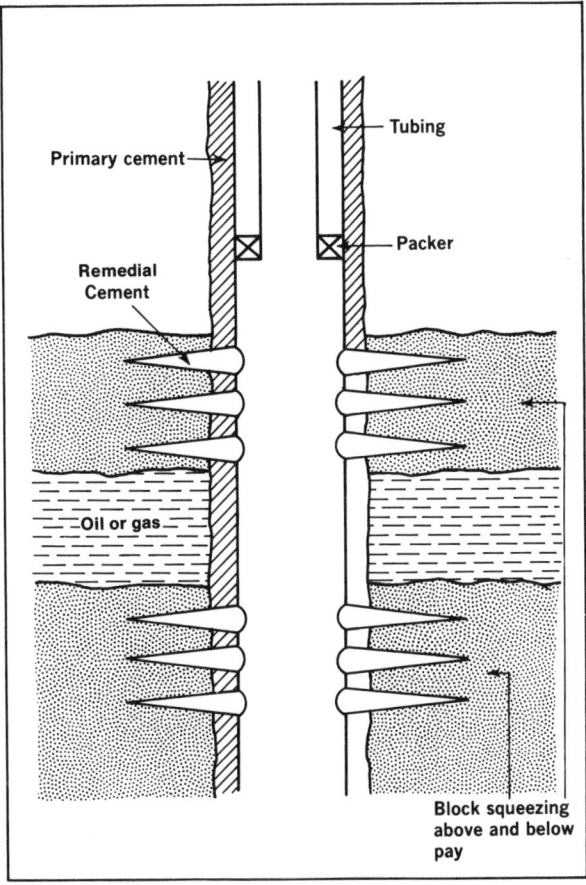

FIG. 3-9—*Block squeeze.*

Fluid loss control and pumpability time are again important factors to job success.

Block Squeeze

Block squeezing, Figure 3-9, is a technique wherein one or two perforations are placed above and below a prospective producing zone. Squeeze jobs are then performed through each set of perforations in turn as a precaution against possible channeling of water or gas into the zone.

This technique is a questionable practice. A primary disadvantage is that holes which may subsequently leak are placed in the casing closer to the undesired water or gas zones. It may also be an unnecessary operation. Better practice is to perforate the desired producing zone, test it to determine if the primary cement job is, in fact, defective. If it is defective, repair by squeezing those perforations.

PLUGBACK OPERATIONS

A plugback operation is designed to completely fill some selected portion of the hole, usually for abandonment. It is often associated with open hole, since bridge plugs can be used in casing to accomplish the same purpose. See Figure 3-10.

Cement slurry type is usually not critical for a normal plugback, but "Densified Cement" is often a good choice.

Densified cement, mixed with a reduced water-cement ratio, (3.4 gal per sack), and with dispersant (0.75 to 1.0% CFR-2), to control viscosity, will tolerate more mud contamination than normal cement.

Calcium chloride is sometimes used as accelerator in shallow wells. Cement slurry should be somewhat heavier than displacing fluid. Densified cement weighs about 17.5 lb/gal.

Cement placement can be best accomplished by circulation. The key is to place cement with as

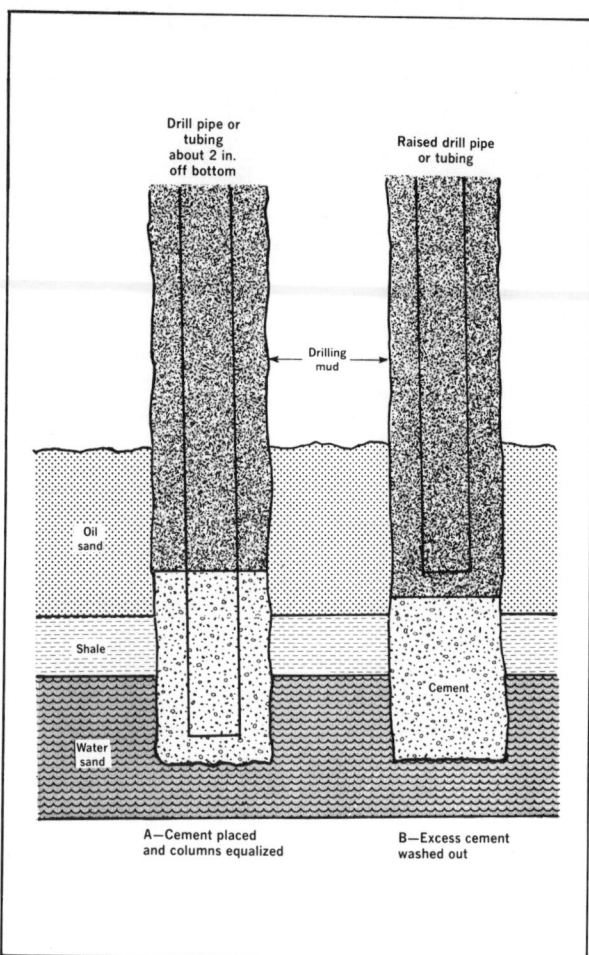

FIG. 3-10—*Plug-back cement job.*

little mixing as possible with wellbore fluids. Hydrostatic pressures must be balanced in the circulation method, i.e., length of cement column, spacer column, and mud column inside pipe must be matched by an equal length of each outside the pipe.

A novel plugback technique has been developed in Libya to shut off bottom water in long zone open hole fractured vugular carbonate completions without the use of a rig. The method involves filling the plugback zone with sand, topped with a short cement cap, as follows:

1. Locate hole bottom and tubing bottom with wireline.
2. Use cementing unit to load hole with water (or oil) and establish stabilized pump-in rate (4 to 5 bbl/min.).
3. Add sand to water (or oil) through cement tub or cone-type hopper at concentration of 0.5 to 1.0 lb/gal, and pump in the volume of sand required for desired fill-up less cement cap. Pump-in rate should be selected to screen out sand at the bottom of the hole (2 to 3 bbl/min.).
4. Displace sand mixture with calculated volume of fluid to place top of sand at desired fill-up depth. When sand clears tubing, reduce displacement rate to 0.5 to 1.0 bbl/min. to avoid washing sand into vugs or fractures.
5. Check top of sand fill with wireline. Place additional sand if necessary.
6. Batch mix slurry containing 80% neat cement and 20% frac sand. Slurry should weigh 18 or 19 lb/gal. Mix sufficient volume to provide a 10- to 20-ft cement cap on top of sand fill-up. Use 100% excess slurry volume.
7. Displace sand-cement slurry with water (or oil) to place top of plug at desired depth. Use low displacement rate to avoid washing sand plug into formation.
8. Check top of cement after WOC time of six hours. Add more sand-cement slurry if necessary.
9. Test well on small choke after 48 hours WOC. If no water is produced, stimulate upper section with acid to increase PI and reduce danger of coning around plug.

SPECIAL CEMENT SYSTEMS
Thixotropic Cement

Thixotropic cement is intended to control loss of whole cement into a fractured, vuggy or highly permeable zone, such that a seal around the wellbore can be obtained. It consists of a mixture of API Class G cement modified with about 6 to 8% sulfate, which provides a pumpable slurry as long as it is being moved—but which develops high gel strength rapidly when not agitated. In application, it is usually spotted over the zone to be sealed off, and a shortened "hesitation squeeze" technique used to obtain a pressure buildup. A more recent thixotropic type cementing formulation contains about 1 lb/sack of gum crosslinked with an organic material, dry-blended with API Class G Cement.

Two-Stage Sodium Silicate-Cement System

A recent technique designed to plug a fractured, vuggy or highly permeable zone involves a two-stage treatment. The first stage is a sodium silicate-like polymer solution containing fibrous material and/or sand for bridging. This mixture forms a stiff gel when it comes in contact with formation saltwater (or injected saltwater). This, in effect, temporarily seals the formation permeability, such that the second stage, consisting of a normal squeeze cement slurry, is contained in the vicinity of the wellbore to provide a permanent shut-off.

Resin Cement

Resin cement consists of a blend of Portland cement and water with liquid resins (plus a catalyst). Its primary use is for shutting off water in low pressure producing wells. Filtrate squeezed into the formation under pressure will actually set up as does the plug itself. It develops compressive strength very rapidly—Hydromite develops 1500 psi within 10 to 15 minutes after initial set. Temperature limits are 60°F to 200°F.

Application is limited and it is normally used only in small quantities.

Diesel Oil Cement

The purpose of Diesel Oil Cement is to control extraneous formation water. DOC consists of Portland cement mixed in diesel oil or kerosene (no water) with a surfactant to improve wetting of cement particles. The cement will not set up until it is contacted with water, hopefully from the extraneous water zone. Successful application of DOC requires that it be properly placed in contact with the water zone—and that it not contact water from the producing zone. As a practical matter, this requirement limits the usefulness of Diesel Oil Cement systems.

REFERENCES

1. Huber, T. A., Tausch, G. H., and Dublin, J. R., III: "A Simplified Cementing Technique for Recompletion Operations," J. Pet. Tech., Jan. 1954.

2. Howard, G. C. and Fast, C. R.: "Squeeze Cementing Operations," Transactions AIME (1950) *189* 53-64.

3. Goolsby, J. L.: "A Proven Squeeze-Cementing Technique in a Dolomite Reservoir," J. Pet. Tech., Oct. 1969.

4. Hook, F. E. and Ernst, E. A.: "The Effect of Low-Water-Loss Additives, Squeeze Pressure, and Formation Permeability on the Dehydration Rate of a Squeeze Cementing Slurry," SPE-2455, May 25, 1969.

Chapter 4 Sand Control

Sand control theory and methods
Effect of well completion and production practices
Mechanical sand control methods
Gravel pack design criteria
Practical gravel packing considerations
Gravel selection, screens, fluids
Inside casing gravel pack problems and techniques
Open hole gravel pack techniques
Screens for sand control
Plastic consolidation, processes, placement techniques

INTRODUCTION

Sand production is one of the oldest problems of the oilfield. It is usually associated with shallow formations of Tertiary Age, but in some areas sand problems may be encountered to depths of 12,000 feet or more.

It has been said that the best technique of sand control is no sand control. This means that well completion practices are a critical consideration in zones where there is a tendency for sand production. Often sand production problems are created by less-than-adequate completion practices.

Definition of Sand Control

In considering sand control, or *formation solids* control, it is necessary to differentiate between load bearing solids and fine solids associated with formation fluids which are not part of the mechanical structure of the formation. Some fines are probably always produced. This is, in fact, beneficial since if fines are free to move, and if they are not produced, they, along with other fines moving in behind, must eventually block the flow channel.

Thus in defining sand control we mean control of the *load bearing solids*.

As a practical matter, it has been suggested that:

—All produced solids smaller than the 90 percentile formation sand are probably interstitial fines.
—Produced solids between the 90 and 75 percentile range probably represent some of the smaller load bearing solids.
—Produced solids between the 75 and 50 percentile range certainly represent load bearing solids.

Critical Flow Rate Effect

The normal producing situation is that a well may make a rather uniform amount of sand or fines independent of production rate until some critical production rate is exceeded. Continued production above the critical rate results in increasing amounts of sand production.

Formation Strength Versus Drag Forces

Sand grains are stabilized by compressive forces due to the weight of the overburden, by capillary forces, and by cementation between sand grains. Causes of sand production are related to:

—*Drag forces* of flowing fluid which increase with higher flow rates and higher fluid viscosity.
—*Reduction in formation "strength"* often associated with water production due to dissolving or dispersion of cementing materials, or a reduction in capillary forces with increasing water saturation.
—*Reduced relative permeability* to oil, due to increased saturation, which increases pressure drawdown for a given oil production rate.
—*Declining reservoir pressure* which increases com-

paction forces and may disturb cementation between grains.

Sand Control Mechanisms

Basically sand production can be controlled by three mechanisms:

1. *Reducing Drag Forces*—This is often the cheapest and most effective. It should be considered along with any other method of control. It often is the natural outcome of proper well completion practices.

2. *Bridging Sand Mechanically*—This is the "old standby" and properly done, has wide application. It is more difficult to apply in multiple zones or small diameter casing.

3. *Increasing Formation Strength*—Sand consolidation has specialized application—it leaves a full-open wellbore and can be used in small diameter casing.

REDUCTION OF DRAG FORCES

Reducing drag or frictional forces is often the most effective and the simplest means of controlling sand. Fluid production rate causing sand movement must be considered as a rate-per-unit area of permeable formation open to the wellbore.

Increasing Flow Area

The first thing to consider is increasing flow area. For a fixed fluid production rate, flow rate per unit area may be reduced by:

1. Providing clean, large perforations through the existing producing section.
2. Increasing perforation density.
3. Opening increased length of section.
4. Creating a conductive path some distance into the reservoir by means of a packed fracture.

Often, good completion practices, use of clean fluids, and careful selection of perforating charges and perforating conditions can effectively reduce an otherwise serious sand control problem.

Restricting Production Rate

Where reservoir considerations and market demand will support higher production rates, determining the maximum rate or the critical producing rate above which sand production becomes "excessive" is an important economic question.

Excessive Sand Production—In the U.S. Gulf Coast the practical limit on sand production is often assumed to be 0.1% (600 lb/1,000 bbl) as measured in a shake-out graduate. In coarse formation sand areas such as Nigeria, much lower limits are considered to be necessary. Both areas tie their limits to mechanical and economic factors related to:

—Well sandup with attendant cost of clean out, loss of production, and possible formation damage.
—Cut-out of well head or surface equipment with possibility of physical danger or disaster.
—Casing collapse, due to slumping of higher formations into the zone weakened by sand production.

"Bean-up" Technique—In Nigeria, careful measurements of produced sand concentration versus production rate show relations idealized as Figure 4-1.

As fluid rate is increased step wise, sand concentration jumps at each increase, then tapers off to the original concentration. The surge effect apparently breaks unstable bridges which reform at the higher rate.

When the critical range is reached, bridges do not reform. The strength of the structure has been exceeded and sand production continues at higher rates.

Production rate must then be reduced significantly below the critical range to allow bridges to reform, after which rate can be increased to somewhat below the critical.

This procedure, carefully done over a period of several months, has been effective in establishing maximum production rates for wells with or without other sand control mechanisms.

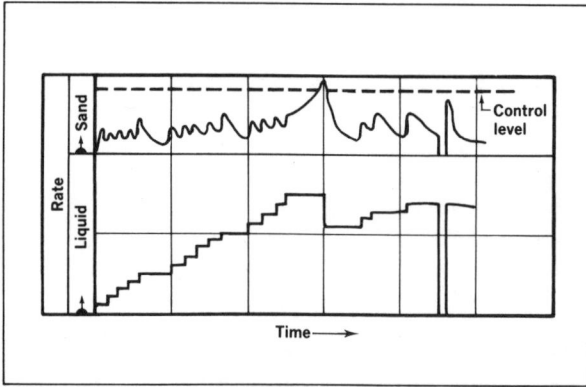

FIG. 4-1—*Sand production vs. flow rate relation. Nigeria.*

A side benefit of the "bean-up" technique is that the gradually increasing producing rates provide ideal clean up of induced and inherent fines in the pore channels around the wellbore. This often results in significantly higher P.I.'s.

Reducing "Bean-Up" Time—In some sand producing areas, it appears possible in a given zone to establish, for untreated wells, a maximum production rate (without excessive sand production) which is proportional to the length of the producing interval. The slope of this maximum rate line is a measure of the "strength" of the particular formation.

Maximum production rate without excessive sand appears to be related to three equally important factors: productivity index, length of the completion zone, and the formation strength or "drawdown factor."

Within an area, correlation of drawdown factor with sonic transit time (Δt) appears possible. Thus on a new well, measurement of sonic transit time and P.I. at a low rate should, together with the zone length, provide an estimate of maximum producing rate before sand problems.

In another approach based on sonic travel time measurements, equations have been proposed for determining the need for sand control. This technique appears to offer promise as more field experience is obtained.[9]

Which Well is the Sand Producer—A flow line probe or erosion tube has the twofold purpose of detecting which well is producing sand and also shutting the well in before surface fittings are destroyed.

A sonic device is available for temporary installation in a flow line to "hear" sand impinging on the detector.

MECHANICAL METHODS OF SAND CONTROL

Mechanical methods of sand control involve use of *gravel* to hold formation sand in place (with a screen to retain the gravel), or a *screen* to retain the formation sand (with no gravel).

The basic problem is how to control formation sand without an excessive reduction in well productivity.

Basic design parameters include:

1. Optimum gravel size in relation to formation sand size.
2. Optimum screen slot width to retain the gravel, or if no gravel, the formation sand.
3. An effective placement technique is, perhaps, most important.

Development of Design Criteria

Many investigators have studied mechanical sand control design parameters and many rules of thumb have been proposed. Although significant "art" remains, Industry opinion is more and more directed toward certain rather specific design criteria.

Formation Sand Size Analysis—The first step is getting representative samples. Sand-grain size distribution often varies through a particular sand body and certainly from one genetic zone to another. Thus, for representative measurements a number of samples are needed.

Full-diameter cores are best; rubber-sleeve core barrels may be needed for recovery. Side wall cores are good—some larger grains are crushed and drilling particles are included.

Samples from perforation washing or backsurging are acceptable. Produced or bailed samples are better than nothing.

Sieve analysis provides grain size distribution on percentile basis. But sieve analysis techniques have not been standardized for the oil industry. Reference 12, "Testing Screens and Their Uses," suggests lab techniques for performing reproducible sieve analyses. Normally for oil field work, the U.S. Standard Sieve Series (ASTM Spec. E1170) is used, and sieve analysis is reported in inches or millimeters. However, in some areas, Tyler Mesh designations are used. Appendixes 4A and 4B give comparisons of various sieve series.

Figure 4-2 shows a typical sand analysis distribution. Ten percentile sand size is defined as the point on the distribution scale where 10% by weight of the sand is of larger size and 90% of smaller size.

Grain size distributions vary considerably around the world, both in average size and in uniformity. Figure 4-3 presents typical distributions from various areas.

A specific size is needed to describe the distribution curve. It is convenient to describe the grain size distribution curve by the size at a specific percentile point. Several points on the percentile scale are used:

1. 10 percentile—typical of Gulf Coast
2. 50 percentile—Karpoff-Saucier-Halliburton
3. 10, 40 or 70 depending on slope—Schwartz

The Schwartz method[7] seems to have general application. A uniformity coefficient is determined by comparing the 40 percentile size (D_{40}) with the 90 percentile size (D_{90}):

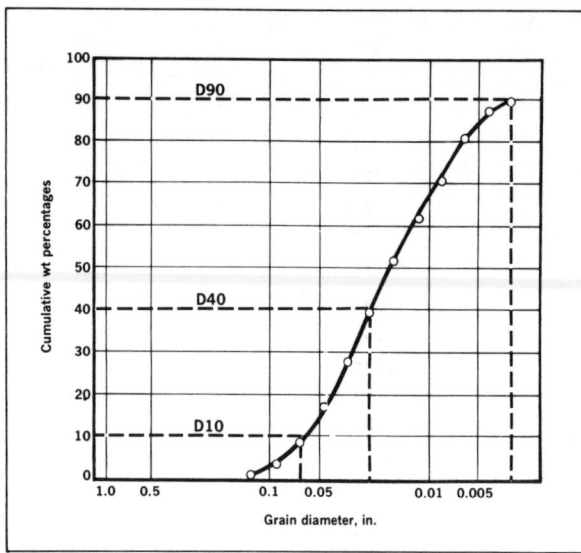

FIG. 4-2—*Typical sieve analysis.*

Uniformity Coefficient: $C = \dfrac{(D_{40})}{(D_{90})}$

If $C < 3$ —sand is uniform and is described by the (D_{10}) size

If $C > 5$ —sand is non-uniform and is described by the (D_{40}) size

If $C > 10$—sand is very non-uniform and is described by the (D_{70}) size

Screen Slot Size Experiments—Ideally, slot width should be as large as possible to retain sand grains but not restrict flow of fluids and interstitial fines.

Early experimental work by Coberly established the upper limit on slot width of not more than twice the 10 percentile sand size in order to bridge effectively.

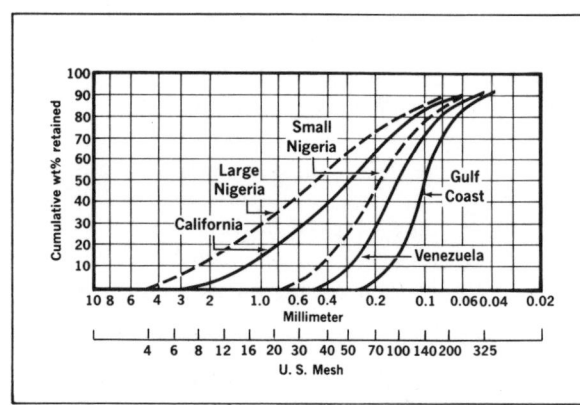

FIG. 4-3—*Variations in sand size distributions.*

This rule allows some movement of grains through the slot before bridges form. Also, changes in flow rate may cause bridges to fail, after which they must reform allowing more sand passage.

For retaining uniform formation sands where grains are more difficult to bridge and/or where frequent changes in flow velocity occur, experience dictates use of slot width equal to (not twice) the 10 percentile sand diameter.

Slots to retain gravel should not use the 2:1 bridging rule. Since it is imperative that all gravel be tightly placed and retained, screen slot width for a gravel pack must be smaller than the smallest gravel grain size (usually 2/3 to 1/2 the smallest gravel).

Gravel Size to Control Sand—Early work by Coberly in defining gravel sand size ratios considered only the problem of preventing movement of sand into the wellbore, and not the permeability of the gravel pack. This led to rather large gravel-sand ratios.

Later, it became clear that for maximum productivity, the formation sand must be stopped at the outer face of the gravel pack. If sand bridges occur within the gravel pack itself, permeability is significantly reduced. This thinking started the current Industry trend toward the lower G-S ratios.

The term "gravel-sand size ratio" has not been standardized. Early investigators Coberly, Hill, Wagner, Gumpertz meant:

$$\text{G-S ratio} = \frac{\text{Largest gravel size}}{10 \text{ percentile sand size}}$$

Saucier means:

$$\text{G-S ratio} = \frac{50 \text{ percentile gravel}}{50 \text{ percentile sand}}$$

Schwartz means:

$$\text{G-S ratio} = \frac{10 \text{ percentile gravel}}{10 \text{ percentile sand}}$$

$$\text{or } \frac{40 \text{ percentile gravel}}{40 \text{ percentile sand}}$$

Maly means:

$$\text{G-S ratio} = \frac{\text{Smallest gravel size}}{10 \text{ percentile sand size}}$$

Sand Control

The effect of G-S ratio on gravel pack permeability is best shown by lab work by Saucier,[10] Figure 4-4, indicating an ideal ratio in the range of 5 to 6.

Schwartz recognizes the effect of flow velocity, but makes essentially the same recommendation as Saucier:

1. Uniform sand (C less than 5) and with flow velocity less than 0.05 ft/sec:

G-S ratio: D_{10} gravel = 6 times D_{10} sand

2. Non-uniform sand (C greater than 5) and/or with flow velocity greater than 0.05 ft/sec:

G-S Ratio: D_{40} gravel = 6 times D_{40} sand

3. Flow velocity should be calculated:

$$\text{Flow velocity} = \frac{\text{Production Rate ft}^3/\text{sec}}{50\% \text{ of open area of slots ft}^2}$$

Concensus—G-S Ratio—Most laboratory work shows ideally that gravel-sand size ratios should be in the range of 5 to 6 (comparing similar percentile points on the gravel and the sand sieve analysis curves). This is a basic rule-of-thumb.

It should be noted that with a "tight pack," Figure 4-5, a G-S ratio of 6 implies that the reference sand grain is too large to move through the pores of the gravel pack. With a "loose pack," however, the reference sand grain can move through the pores between gravel grains, and we are forced to depend on Coberly's bridging rule to stop the sand from moving into the pack. Bridging is a statistical occurrence with many grains passing through before the bridge forms, and with any change in flow dynamics, a new bridge must form. Thus, with a "loose pack" using the G-S ratio of 6 may mean lower permeability and shorter gravel pack life.

Thickness of the Gravel Pack—In lab experiments a gravel pack thickness of four or five gravel diameters will control sand. In practice, much thicker packs are necessary.

With the practical problems of placing gravel, and with fluctuating flow velocities, a 3-in. thickness of gravel is considered a minimum. This means that

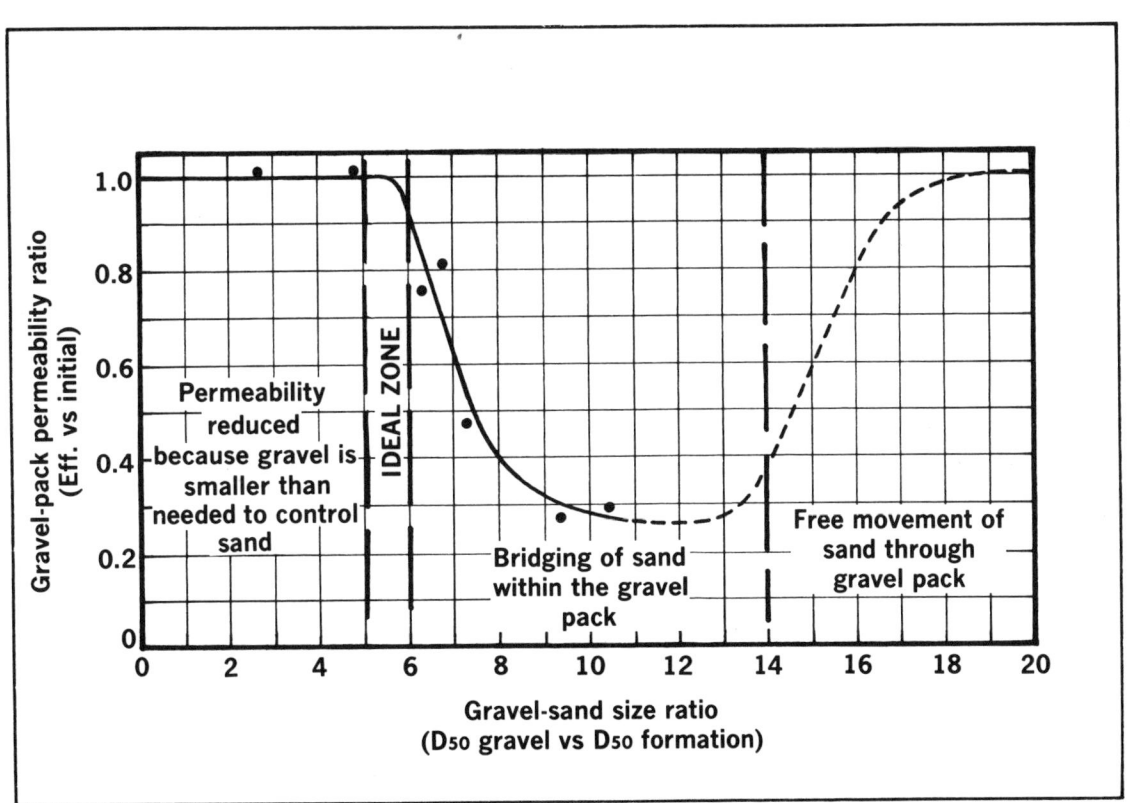

FIG. 4-4—Effect of gravel-sand ratio on gravel-pack permeability.[10] Permission to publish by The Society of Petroleum Engineers.

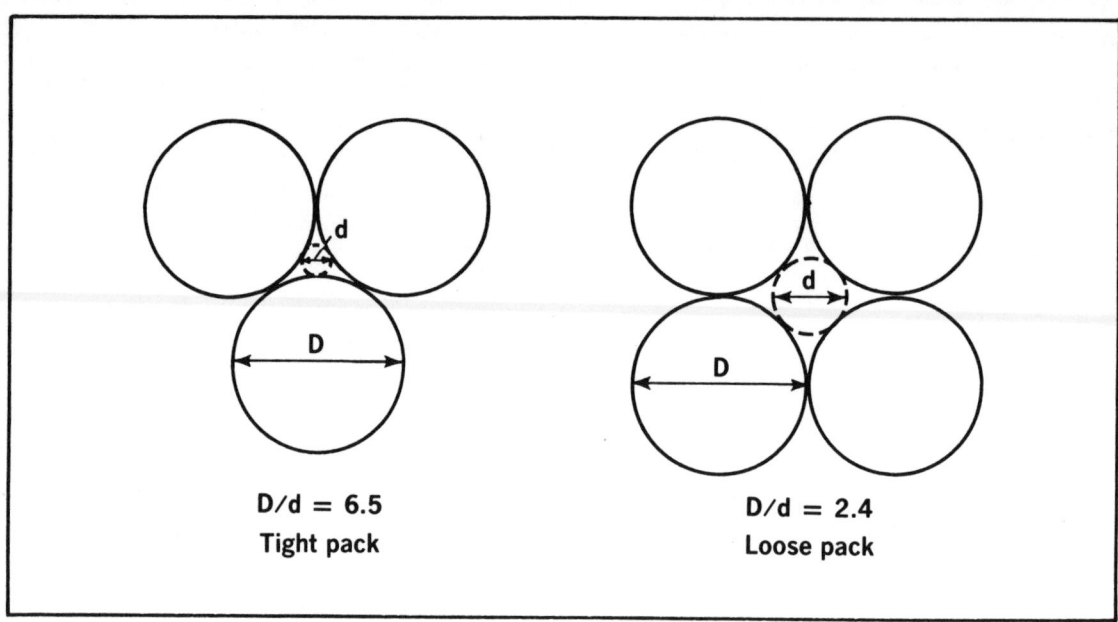

FIG. 4-5—*Ideal comparison of tight vs. loose pack.*

open hole must be underreamed to provide 3 in. on the radius between the screen and the formation. In perforated casing, gravel must be placed through the perforation tunnel and outside the casing.

Thicker gravel packs permit higher production rates. Sage & Lacey[3] developed the relation of Figure 4-6 describing rate-of-sand-passage through a 3-in. thick gravel pack vs. production-rate-per-foot-of-section. Based on this work, operators gravel-packing high volume wells in Lake Maracaibo have established 100 bpd per foot of section as the maximum production rate practical for a gravel packed well. As Figure 4-6 points out, thicker gravel packs should provide higher production rates per foot within acceptable sand production limits.

Fluctuating Flow Rate—Recent lab work by Saucier[10] demonstrated the importance of inertial effects on sand production. Either a sharp increase or a sharp decrease in flow rate through a 3-in. gravel pack caused a temporary increase in sand production, Figure 4-7. If the flow rate was held uniform after the change, sand production decreased, apparently indicating re-establishment of bridging effects. Gas evolution had a significant effect in increasing sand production.

Results of these tests indicate that rate changes may be more significant than the magnitude of the flow rate. Apparently for a given flow condition, sand bridges form that are stable for existing geometrics and hydrodynamic forces. As fluid forces are altered, bridges break down, and more sand is produced until, if possible, new bridges form under the new conditions of stability.

These tests also showed that gravel-sand ratios

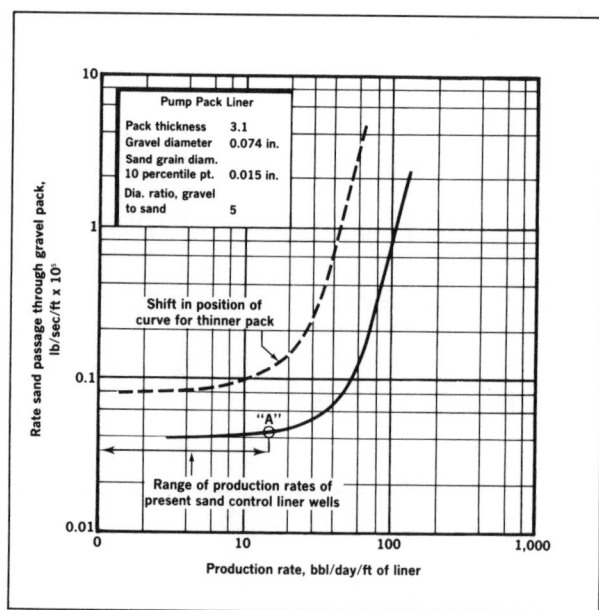

FIG. 4-6—*Production rate vs. sand passage rate.[3] Permission to publish by The Society of Petroleum Engineers.*

Sand Control

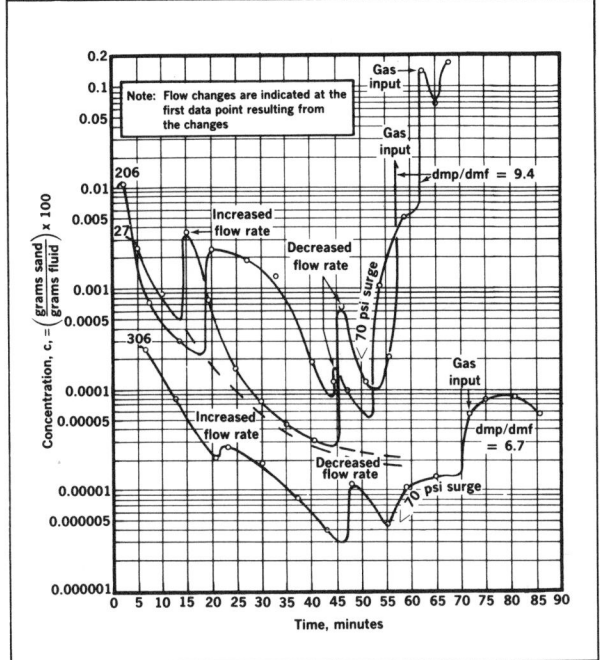

FIG. 4-7—*Sand movement into gravel pack vs. flow dynamics.*[10] *Permission to publish by The Society of Petroleum Engineers.*

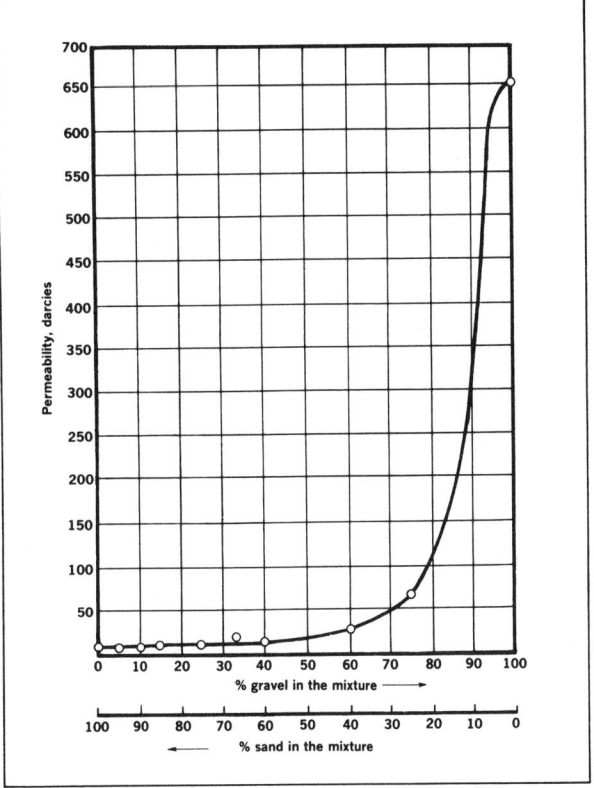

FIG. 4-8—*Permeabilities of 10–20 mesh gravel mixed with Oklahoma No. 1 sand.*[11] *Permission to publish by The Society of Petroleum Engineers.*

must be less than 6 to minimize sand production under disturbed flow conditions, or under high velocity conditions.

Mixing of Gravel With Sand—As shown by Sparlin[11] in Figures 4-8 and 4-9, mixing of high permeability gravel with formation sand, as may occur during placement of gravel, will significantly reduce the permeability of the resulting sand-gravel mixture. Fines in the gravel will also decrease gravel pack permeability.

Summary Rules-of-thumb—In summary, these rules are suggested:

1. Use as large a gravel as possible, but formation sand must be stopped at the outer edge of the gravel pack.

2. Gravel size (at 40 percentile point) should be 6 times the 40 percentile point on the sand analysis curve. For low velocities and uniform sands, the 10 percentile points can be used for sand and gravel reference.

3. Where sand grain analyses vary within the formation, pay more attention to the smaller sand sizes, particularly with higher flow velocity, more non-uniform sand, fluctuating flow rate, and high gas-oil ratios.

4. Pack the gravel tightly. G-S ratios are based on a tight pack.

5. Pack thickness should be at least 3 inches. Thicker packs permit higher flow velocity.

6. Don't mix gravel with formation sand in placement.

Practical Considerations in Gravel Packing

The keys to successful gravel packing are:

1. Selecting gravel of the proper size and quality.
2. Placing the gravel without contamination, at the proper location, as tightly as possible—then holding it in place for the life of the well.

The crux of the problem is to control the load-bearing solids without excessive loss of productivity. Where reservoir conditions are such that high production rates can be sustained, every trick must be employed to maximize productivity and reduce flow velocity per unit area.

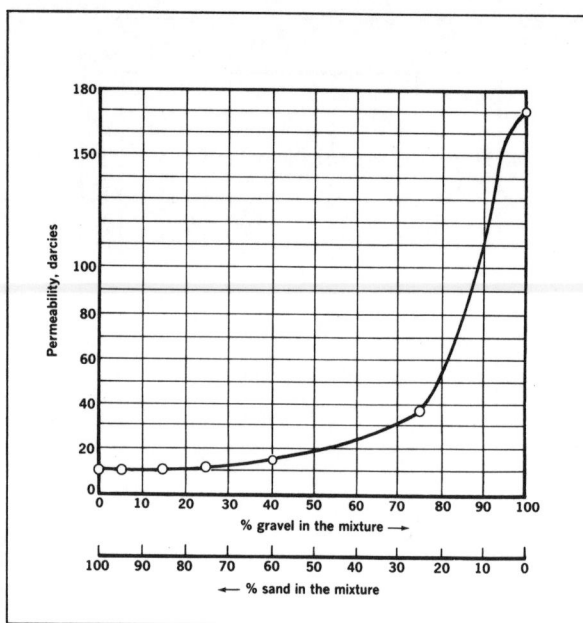

FIG. 4-9—*Permeabilities of 20–40 mesh gravel mixed with Oklahoma No. 1 sand.*[11] *Permission to publish by The Society of Petroleum Engineers.*

TABLE 4-1
Productivity Comparison of Sand Control Methods

	Productivity index, bpd/psi	
	LL-5 zone	LL-3 zone
Gun perforated casing (reference yardstick)	36.6	5.2
Open-hole gravel pack	48.4	6.4
Inside casing gravel pack, with sand oil squeeze	12.9	3.2
Inside casing gravel pack, without sand oil squeeze	4.0	1.7

An open-hole gravel pack provides maximum productivity. The much greater area open to flow of fluids into the well bore, and the corresponding reduction in flow velocity for an open hole gravel pack allows some margin for error—or "poor" practices compared with the inside casing pack.

Table 4-1 shows a comparison of Productivity Index (bpd/psi drawdown) for several types of long-zone sand control completion methods used in two Creole reservoirs in Lake Maracaibo's BCF area. PI's for gun perforated casing completions measured before a sand control method was applied serve as a reference yardstick. Production rates in the LL-5 and LL-3 zones average 1,200 and 400 bpd, respectively.

Based on the tremendous PI advantage, an open hole gravel pack should always be selected when the open hole is compatible with other completion considerations. Control of extraneous water or gas in the subsequent life of the well may, however, rule in favor of the perforated casing or inside casing gravel pack.

Gravel Selection—Gravel size is often specified in terms of "U.S. mesh" designation in the same manner as frac sand, Table 4-2. Probably, a better designation is in terms of thousandths of an inch to correspond with standard U.S. terminology for slot size.

Comparison of sizes in U.S. mesh, Tyler mesh, inches, and microns is given in Appendixes 4A and 4B.

Gravel size should be based as far as possible on the recommendation of Schwartz or Saucier. In many long zone wells, variations in formation sand size through the formation make compromise necessary, as shown in Figure 4-10.

If sonic travel time or other data show a particular zone to be weak, then obviously, gravel should be sized to control this zone. If no data are available, gravel should be sized to control the smaller formation sizes. As noted by Schwartz, lower flow velocity permits relatively larger gravel-sand ratios.

Quality Control Important—Suitability of a particular gravel depends on:

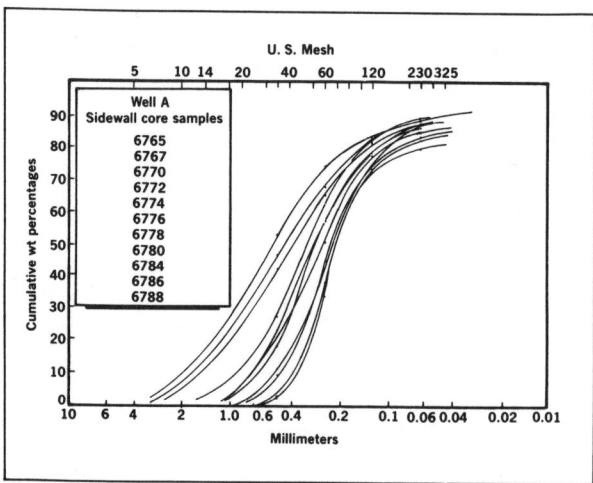

FIG. 4-10—*Comparison of grain-size analyses at various depths through sand body.*

Sand Control

TABLE 4-2
Recognized Gravel Packing Sand Sizes

Sand/gravel size in	US mesh size	Approx. median diameter in	microns
0.010 × 0.017	40/60*	0.014	350
0.012 × 0.023	30/50	0.018	450
0.017 × 0.033	20/40*	0.025	630
0.023 × 0.047	16/30	0.035	880
0.033 × 0.066	12/20*	0.050	1260
0.047 × 0.094	8/16	0.071	1770

*Primary gravel size

1. Roundness and sphericity: Krumbein scale of 0.6 or better. Flat or angular grains should be avoided.
2. Grain strength: depends on depth and formation stress level same as frac sand. In standard lab tests, fines generated by 2000 psi stress should be less than 4%.
3. Acid solubility: acid solubility should be checked. Gravel should be greater than 98% pure silica. Feldspar content should be nil since feldspar is completely soluble in HF acid. Glass beads are slowly soluble in HF acid.
4. Uniformity: the closer the limits on gravel grain size variation, the greater will be the permeability. Schwartz suggests a uniformity coefficient less than 1.50. Material finer than the lower size limit is particularly bad.
5. Clay size material: Presence of clay or silt can be determined by adding clear water to a bottle partially full of gravel. After vigorous shaking, turbidity indicates fines. Turbidity should be less than 1%.

A typical gravel sieve analysis is shown in Table 4-3.

To maximize relative permeability to oil, the gravel must be water-wet before it is added to the placement fluid. This is particularly true if the placement fluid is oil. Laboratory comparison of oil relative permeability for a water-wet pack versus an oil-wet pack is shown in Figure 4-11. Water wetting can be done by circulating water containing 1.0% water-wetting surfactant through the dry gravel.

Screen and Liner Considerations—Ribbed wire wrapped screens cost two or three times more than slotted pipe, but for the same slot width, have perhaps eight to ten times the open area. All-weld screen has twice the open area as the ribbed screen, but costs nearly twice as much, Table 4-4.

Wire-wrapped screens also have the advantage of more erosion and corrosion resistant materials. Vertically slotted pipe has more axial strength and bending strength than horizontally slotted pipe.

Flush-joint connections should be used inside casing to prevent bridging of gravel. Liner-casing size ratios should be:

TABLE 4-3
Typical Gravel Sieve Analysis

U.S. mesh	Sieve opening (microns)	Actual Sample Weight retained %	Cumulative retained %	Specification cumulative retained %
10	200	1.2	1.2	Less than 5.0
12	170	38.8	40.0	
14	140	45.4	85.4	
16	118	14.1	99.5	At least 98.0
18	100	0.5	100.0	100.0

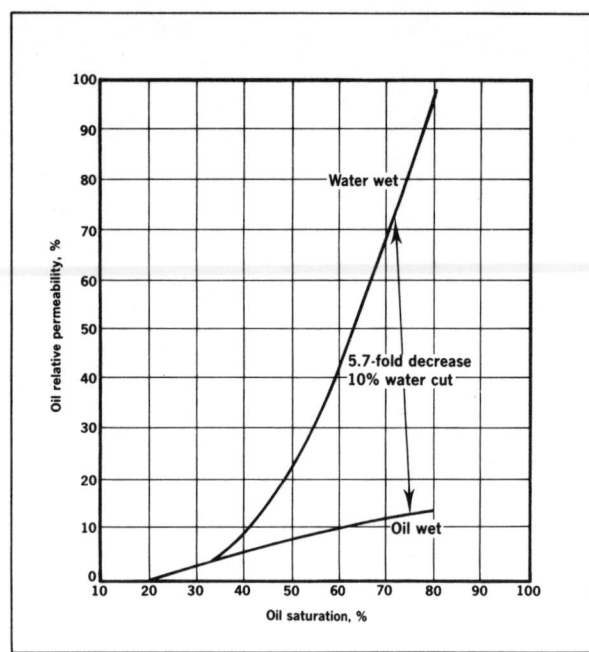

FIG. 4-11—*Effect of oil-wet gravel.*[8] Permission to publish by The Society of Petroleum Engineers.

Casing size in.	Inside csg. pack liner size	Open hole pack liner size
5½	2⅞	3½
6⅝	3½	4¼
7	3½	5
8⅝	4½	6⅝
9⅝	5	7

The liner must be centralized inside casing or in open hole. A liner packer is required to prevent flow of fluids and gravel upward around the outside of the liner.

Gravel-Packing Fluids—One of the most important considerations in gravel packing is proper attention to the gravel packing fluid.

Low-viscosity fluids may provide tighter gravel placement.

Salt water, crude oil, diesel, or HCl acid fluids are satisfactory for gravel placement provided:

1. They are tailored to minimize clay or wettability problems.

2. They are clean—or filtered through two micron filters—or non-gravel solids are acid or oil degradable.

Gravel concentrations of ½ to 1 lb/gal can be carried with pump rates up to 5 bbl/min. Squeezing low viscosity fluids into formation probably provides tighter gravel pack.

All the tricks must be used to insure clean fluids. It is absolutely necessary that a solid-particle filter cake not remain on the formation face in open hole gravel packs or in the perforations or cavity on inside casing gravel packs to be held in place by the gravel.

Minimum fluid system equipment should include curved-bottom compartmented fluid tanks with several clean-out doors in each compartment. Tanks should be scrubbed before use. Clean fluids should be stored in separate tanks and periodically used to displace "dirty" fluids from the well.

For wells capable of producing high flow rates per foot (above 20 bpd/ft in OHGP, or any IGP well) plugging becomes more critical, and filtered fluids should be used. For underreaming operations, a desilter should be used in connection with the filter.

TABLE 4-4
Screen and Liner Effective Inlet Areas (Sq. In.)

Pipe, od, in.	Slotted pipe slot size, in.			Wire-wrapped screen, rib type			All-weld screen		
	0.010	0.020	0.030	0.010	0.020	0.030	0.010	0.020	0.030
1¼	0.4	0.8	1.3	4.2	8.0	11.2	8.1	14.8	20.3
2⅜	0.5	1.1	1.6	5.7	11.4	17.0	10.8	19.7	27.1
2⅞	0.7	1.3	2.0	6.9	13.8	20.7	12.7	23.1	31.8
3½	0.9	1.8	2.7	8.4	16.8	25.3	15.1	27.4	37.7
4½	1.1	2.3	3.4	10.6	21.2	30.6	18.8	34.2	47.1
5	1.3	2.5	3.8	11.8	23.6	35.5	20.7	37.6	51.8
6⅝	1.6	3.2	4.8	15.5	31.0	46.5	26.8	48.8	67.1
7	1.7	3.4	5.2	16.4	32.9	49.3	28.3	51.3	70.7

Minimum safe hydrostatic overload above the formation pressure should be used.

Viscosity and fluid loss control are a problem. Viscosity builders and fluid loss control materials must cause plugging to function; thus, their use should be minimized.

In very uncemented sands, sloughing may occur with no-solids fluids. This must be controlled, and usually can be, by building a fluid loss particle bridge on the face of the formation, against which hydrostatic pressure can be applied.

A reasonable approach where fluid loss control must be applied is to use acid soluble or degradable material. Possibilities include (1) $CaCO_3$ for fluid loss control with an HEC polymer to provide some viscosity and hold the $CaCO_3$ in suspension or (2) Oil-soluble particle for fluid loss control with degradable polymer for viscosity and suspension.

Fluid density is also a problem where weights greater than 10 lb/gal are required to hold formation pressures. Fluid costs and complications increase significantly.

Clean fluids containing calcium chloride can be prepared weighing up to 11.5 lb/gal. Calcium bromide fluids are available up to 15.0 lb/gal or with zinc bromide up to 19.2 lbs/gal; but they are very expensive. Suspensions of finely ground calcium carbonate particles in brine with a polymer viscosity builder can provide fluid density of 12 to 14 lb/gal. If plugging results, the calcium carbonate can be dissolved with HCl. However, getting the HCl to the right spot may be a problem in a completely plugged section of the pack.

Viscous water fluids permit extreme gravel concentrations (15 lbs/gal of liquid or 9 lbs/gal of slurry). Thus, the gravel is placed more quickly and with less loss of fluid to the formation. Some laboratory tests show that carrier fluid viscosity minimizes gravity segregation of the gravel in placement and thereby minimizes plugging of the liner slots with gravel fines which develop in the placement process. Another advantage is that acid, HCl, or self-generating HF can be incorporated to subsequently remove acid soluble materials from the pack.

With viscous water fluids, two problems must be considered:

1. The type and concentration of viscosity breaker must be determined depending on formation temperature.
2. The "sump" volume below the screen must be minimized to reduce further settling of the gravel, thereby loosening the pack, after the viscosity has broken.

Inside Casing Gravel Pack Technique—Perforations are the basic problem. Everything must be aimed at obtaining large diameter clean perforations. Gravel cannot be carried into a perforation that does not accept fluid readily.

Even where clean perforating fluids are used, injectivity into a perforation is restricted by metallic particles from the jet charge and crushed formation particles. Restriction is particularly severe where perforating is done with differential pressure into the formation.

Big-hole jets or bullets are the best. Even with proper standoff conditions, the entrance hole size of many otherwise satisfactory jet charges is too small for effective gravel packing (i.e. $3/8$-in.). This is particularly true of through-tubing hollow carrier charges. Increased perforation density, six or even eight shots per foot is one approach; however, in unsupported J-55 pipe, casing damage may result unless steel carrier guns are used.

In low compressive strength formations, a bullet gun giving a $1/2$-in. hole is preferable to small hole jets.

Probably the best compromise between entrance hole size and perforation depth is the new big hole jet charges giving perforation tunnel diameters of $3/4$ to $7/8$ in.

The charge of Figure 4-12 shoots, at 0° phasing, 1 to 4 shots per foot. The effective core penetration is not large; however, the perforation hole of deep penetrating charges probably collapses, anyway, in soft sand-control problem formations. Six to eight shots per foot with this steel carrier gun should be considered in high rate producing areas.

Some perforation-cleaning technique must be used. In short zones (less than 25 ft), use of the back surge techniques to clean perforations before gravel packing has significantly increased productivity in South Louisiana field tests. Optimum air chamber volume is about 1 gal per perforation. The use of a painted tail pipe gives an indication of the effectiveness of the device in opening individual perforations. Where paint is removed, the perforation opposite that spot should be open.

The cup type of perforation wash tool, Figure 4-13, is a step in the right direction. Used with clean fluids it should do an excellent job of cleaning the

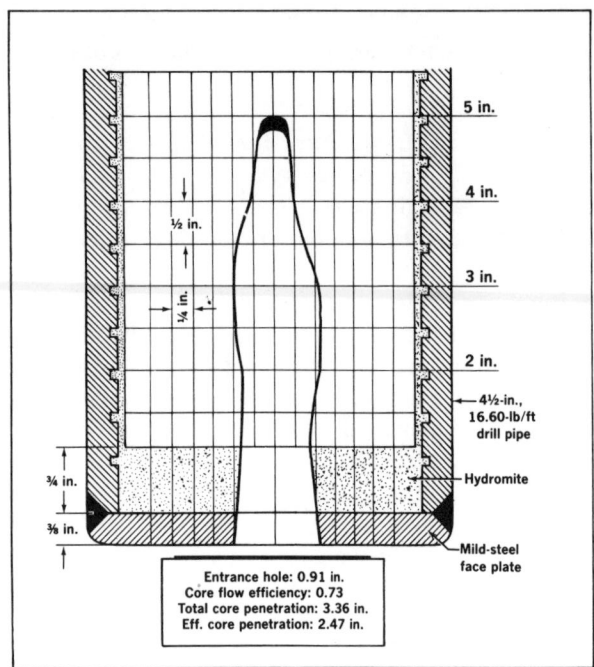

FIG. 4-12—*API test target, 4-in. big-hole charge.*

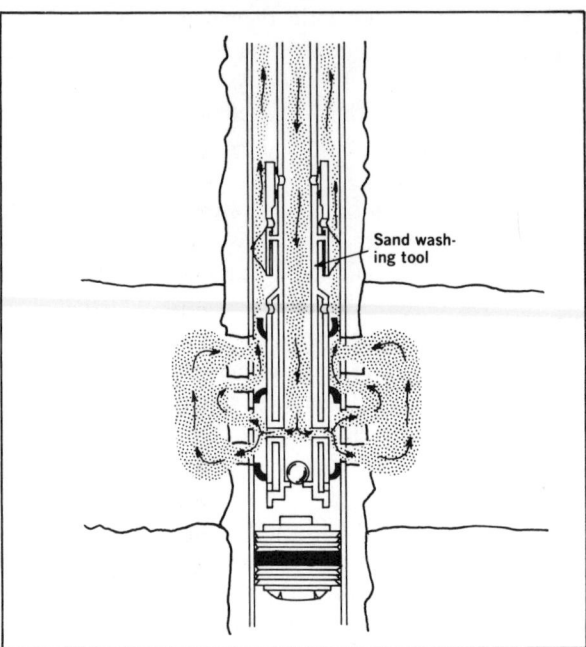

FIG. 4-13—*Cup-type perforation wash tool.*

perforation tunnel, perhaps "breaking away" some of the primary cement to reduce the "length" of the perforation tunnel. It may also create a cavity behind the casing to permit a greater thickness of gravel.

HF-HCl acid has also proven effective in removing metallic charge particles and crushed formation particles.

Producing the well for a period before gravel packing provides some improved results, even through it is not a very effective method of cleaning perforations.

The perforation tunnel must be filled with gravel or it will be plugged with sand. As shown by Saucier[10] in Figure 4-14, a perforation filled with even high permeability formation sand is effectively plugged. Gravel must be placed in the perforation "tunnel" through the casing and cement sheath and outside the casing.

Unfortunately, the linear form of Darcy's Law probably applies (assuming laminar flow) to fluid flow through the perforation tunnel:

$$q = \frac{1.127 \times 10^{-3}\, kA\Delta p}{\mu L} \quad \begin{array}{ll} q = bpd; & k = md \\ A = ft^2; & L = ft \\ \Delta p = psi; & \mu = cp \end{array}$$

Increasing the perforation hole area (A) or the permeability (k) of the material filling the perforation tunnel is a primary approach for reducing pressure drop for a given flow rate through a single perforation. In effect, perforation tunnel length (L) may be reduced by shattering the cement during perforating

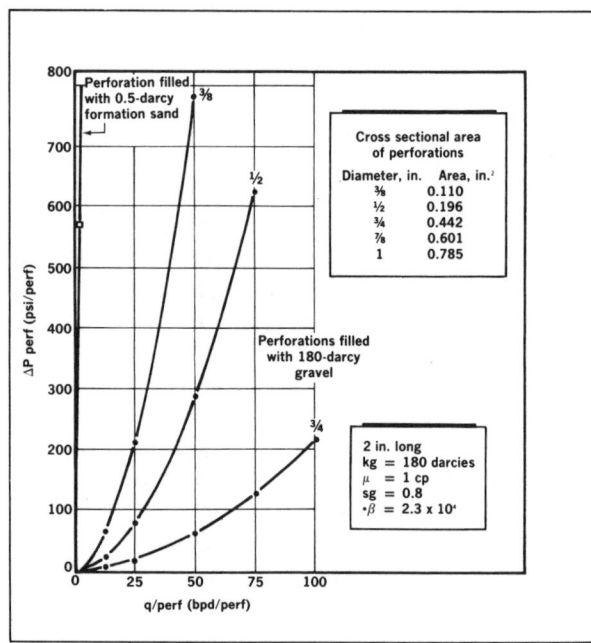

FIG. 4-14—*Importance of perforation-tunnel size and filling material.*[10] *Permission to publish by The Society of Petroleum Engineers.*

(particularly 6 to 8 shots/ft) and washing it out with a perforation washing tool.

Techniques for filling the perforation tunnel with gravel vary in effectiveness:

The two-stage sand-oil squeeze from Venezuela probably provides best results in long zones (200–300 ft). The first stage consists of a small frac job (15 bbl/min pump rate) using a clean oil carrying fluid (10–20 cp viscosity) with gravel sized to control the formation sand.

Ball sealers dropped randomly or *degradable* diverting agents may be used to insure that a high percentage of the perforations accept gravel.

During the second stage, as shown in Figure 4-15, a wire-wrapped screen or slotted liner assembly (with a cup type of crossover device at the top and a double check valve washdown shoe at the bottom) is run and washed down through gravel remaining inside the casing.

Additional gravel is placed around the liner through the crossover, until a surface pressure increase indicates that gravel has covered the telltale slots. The telltale slot section is positioned to give 30 to 50 ft of reserve gravel. Finally, the crossover and wash pipe assembly is pulled and the liner packer is set.

For shorter zones (50 ft or less) *the Wikker Dikker technique* temporarily plugs the top of the liner, and permits gravel to be placed at high rates (15 bpm or more) and at pressures above the fracture gradient, around the outside of the liner and through the perforation tunnels.

A clean oil fluid (10–20 cp viscosity) should be used to provide suitable gravel carrying capacity; however, a water fluid having similar characteristics with a degradable viscosity builder could be used.

This method, proved through extensive use in Venezuela, has two important advantages: (1) after gravel is placed, it is not necessary to run down through the perforated zone again to disturb or plug the gravel; and (2) mechanical procedures are simplified. Use of the Wikker Dikker method is shown in Figure 4-16.

The pressure pack method attempts to carry gravel through open perforations to fill a possible "cavity" behind the pipe and the perforation tunnel, using crossover equipment similar to that shown in Figure 4-17. During placement, fluid carries gravel down the annulus between the inside of the casing and outside the screen or slotted liner, with the gravel being deposited, and the fluid moving through the slots to the bottom of the liner assembly, and back up through the wash pipe.

Periodically, return flow into the wash pipe is restricted, forcing fluid through the perforations and into the formation at below fracture pressure, hopefully carrying gravel into the perforation tunnels.

Although this method is used on the U.S. Gulf Coast, it is not considered to approach the effectiveness of the two-stage sand-oil squeeze or Wikker Dikker methods in Venezuela.

The wash down technique has application in short low-productivity zones (10–15 ft). Gravel is placed through open ended tubing, then a screen is washed down through the gravel. Circulation is stopped, the gravel allowed to settle back around the liner, and a liner packer is set.

Dual Zone Gravel Pack Equipment—Equipment is available for gravel packing two zones in a dual completion, maintaining pressure separation between zones. Equipment and procedures are complicated, and every consideration should be given possible simplifications.

Through-Tubing Gravel Pack—Small diameter wire wrapped screens are available which can be run through $2\frac{3}{8}$ in. or $2\frac{7}{8}$ in. tubing. Thus, tubingless completions can be gravel packed to control sand. Also, small screens can be gravel packed inside casing below $2\frac{3}{8}$ in. or larger tubing using concentric through-tubing methods. Gravel placement is done by forcing all fluid and gravel into the perforations similar to the Wikker Dikker technique.

Open Hole Gravel Pack Technique—Due to the much higher area open to flow of produced fluids entering the well bore, the open hole gravel pack has an inherent margin for error (i.e., less clean fluids, slightly larger gravel, etc.).

The open hole section should be underreamed 4 to 6 in. on the diameter to provide needed thickness of gravel, and to remove drilled solids and mud cake from the face of the open hole. Underreaming fluids must be cleaned as the operation proceeds.

Equipment and placement technique are similar to inside casing, using the crossover reverse circulation method shown in Figure 4-15. The liner assembly must be centralized and a packer must be used.

A caliper survey may be needed to aid in estimating the amount of gravel required. If the open hole section "accepts" significantly less than the calculated amount, then gravel bridging in the annulus is probable. A sonic bond log is sometimes useful in detecting "open" sections within the pack.

Operators gravel packing long, highly deviated

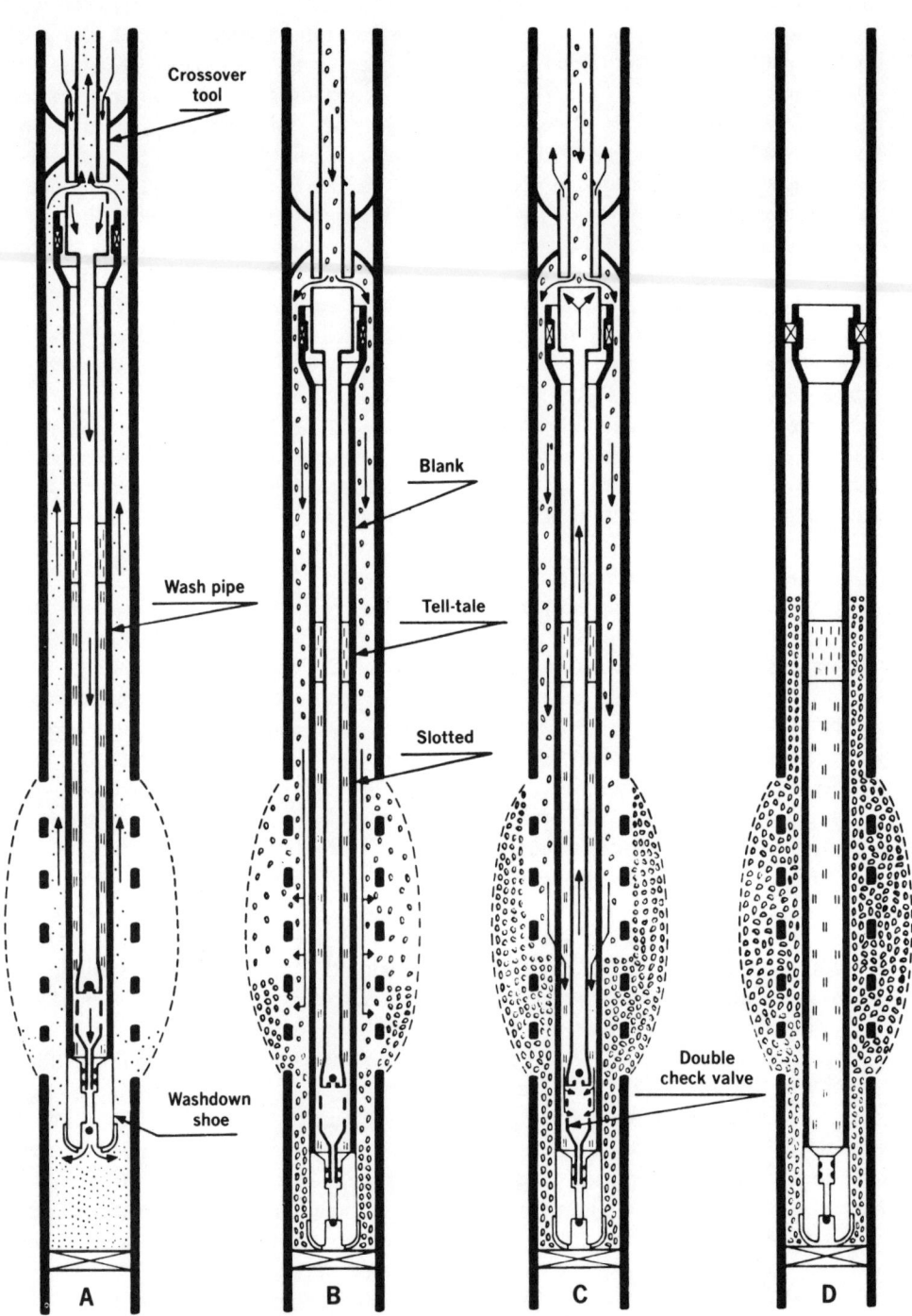

A—washing liner to bottom
B—Squeeze packing the formation
C—Gravel packing the liner
D—Completed gravel pack

FIG. 4-15—*Second stage of two-stage squeeze gravel pack. After Leibach.*

Sand Control 49

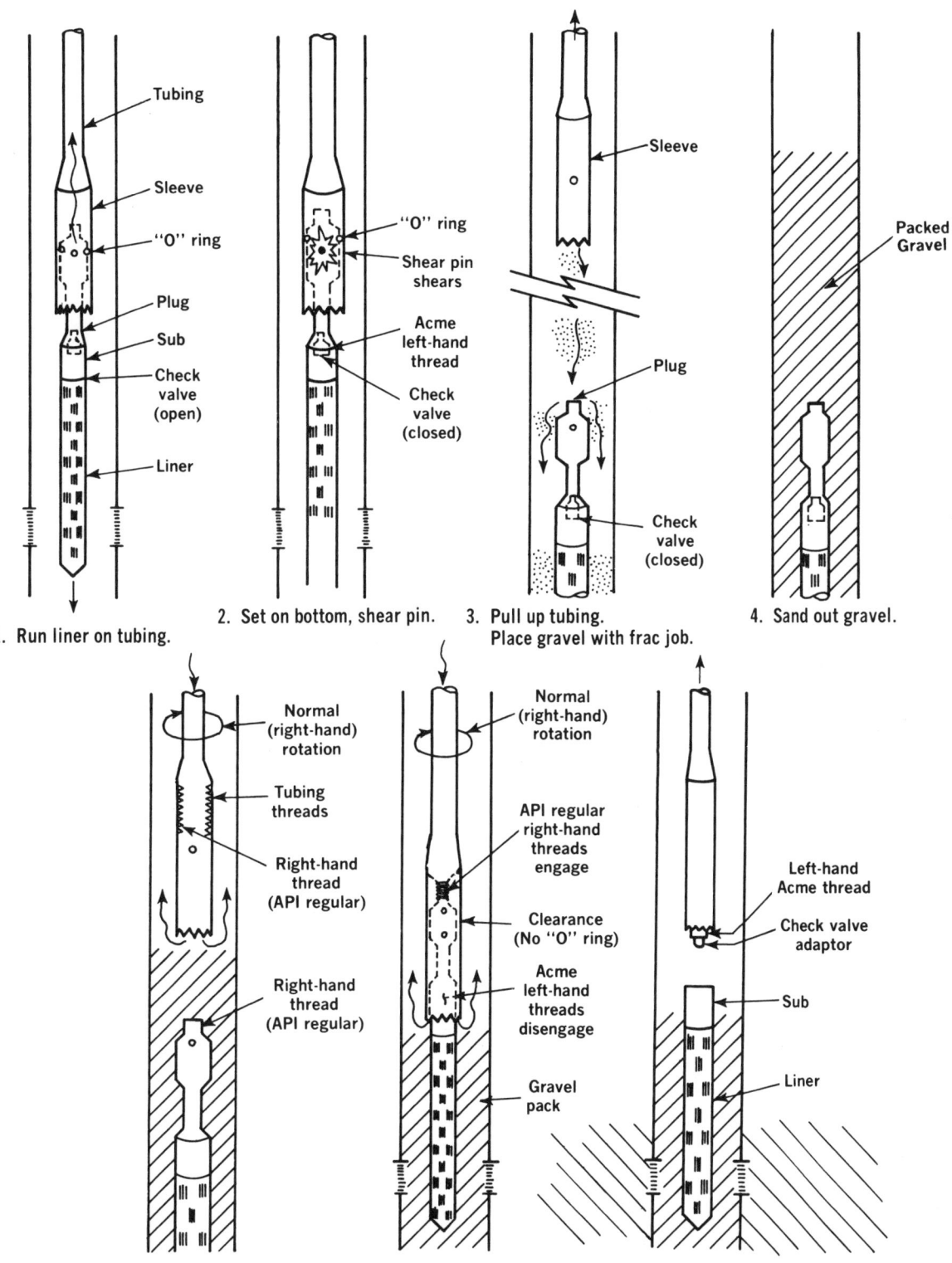

FIG. 4-16—Wikker-Dikker gravel-placement technique.

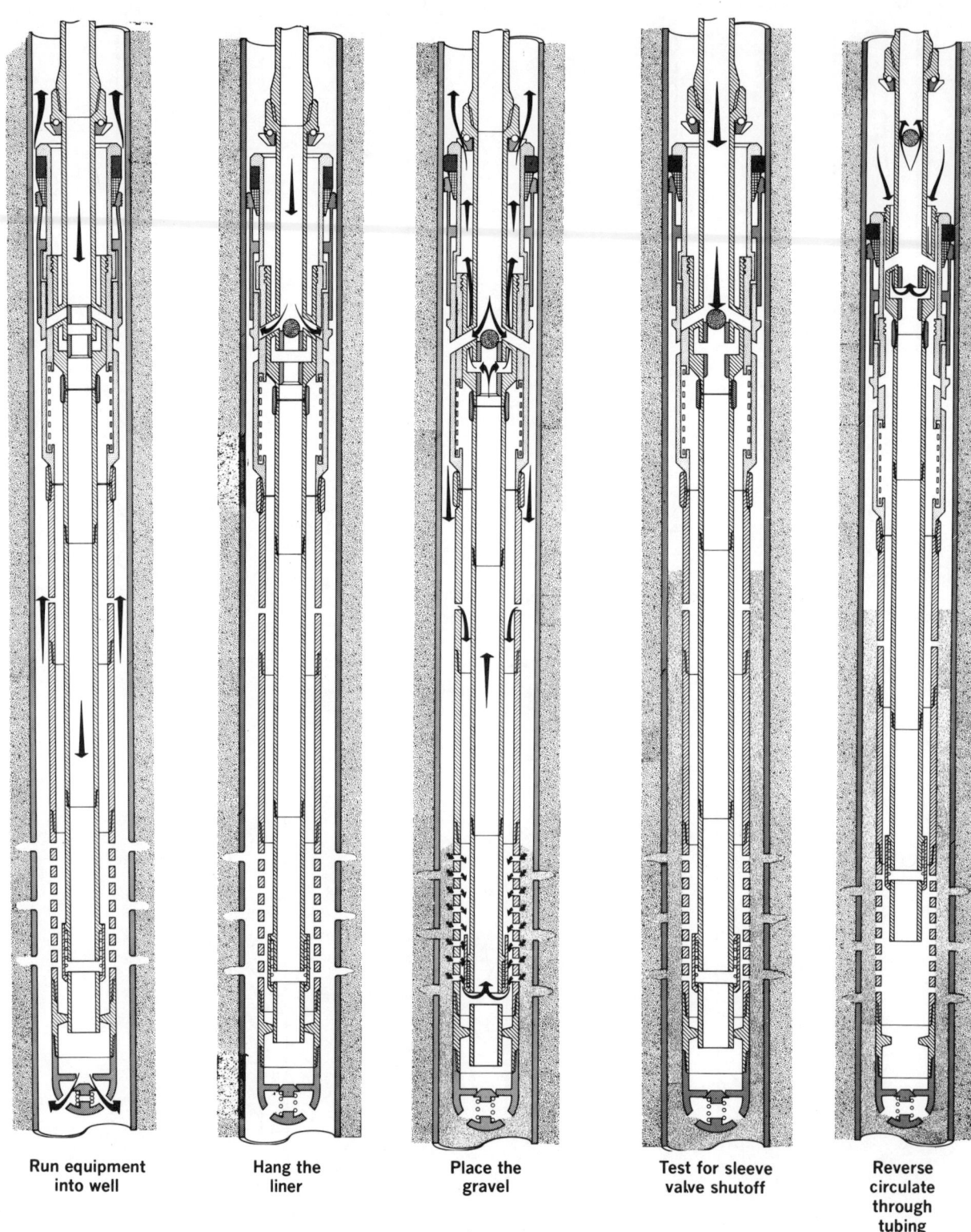

FIG. 4-17—*Inside-casing gravel-pack procedure. Courtesy of Brown Oil Tools.*

holes have reported numerous failures due to inability to fill the annular hole with gravel. Laboratory studies by Maly et al.[16] show that the problem becomes severe as the deviation angle approaches 60° from vertical, Figure 4-18. Apparently, gravel drops out near the top of the open hole section, bridging to prevent further downward movement.

Possibilities for overcoming this problem may include:

1. Use of a series of baffles inside the slotted liner to direct most of the fluid flow to the annular section, thus increasing drag forces moving the gravel toward the bottom.[16]

2. Use of high viscosity placement fluids to suspend the gravel and, again, to increase the drag forces moving the gravel toward the bottom.

3. Use of properly sized wash pipe to increase flow resistance in the wash pipe-liner annulus.[17]

Putting Well on Production is Critical Time—After gravel packing, the manner in which the well is put on production is quite important from the standpoint of formation damage as well as sand control. These points should be considered:

1. The formation pores around the well bore are "loaded" with fines carried in by fluids filtering into the formation or by inherent fines made movable by fluid filtrate effects.

2. These fines should be started back toward the well bore as soon as possible, but at a *low rate* to minimize plugging due to high velocity.

3. To promote bridging of formation sand on the gravel, gradually increasing production rates is also a step in the right direction.

So, put the well on production as soon as possible after the gravel is in place. Start with a low rate, gradually increasing to desired rate over a period of several weeks. Increased P.I.'s should result.

Life of the Gravel Pack—A gravel pack should be considered to have a definite life cycle. If the criteria of stopping all formation sand at the outer edge of the pack could be achieved, then the pack should last forever. In practice this is not done, thus sand gradually invades the pack, reducing pack permeability and increasing flow velocity in other sections of the pack. Failure is then a progressive plugging situation. Proper gravel sizing and placement are important factors to longer life. Production techniques are also im-

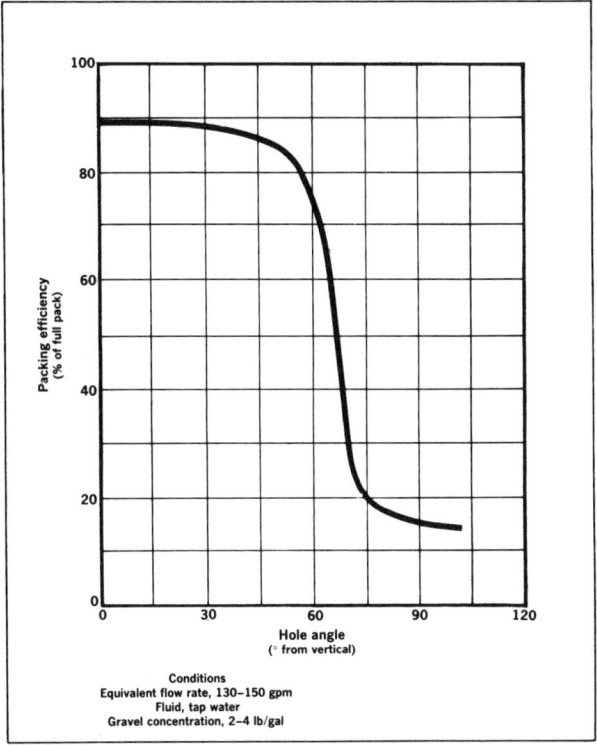

FIG. 4-18—*Effect of hole angle on gravel packing efficiency.[16] Permission to publish by The Society of Petroleum Engineers.*

portant. Fluctuating flow rates, for example, cause periodic breaking and reforming of bridges, and increase rate of sand penetration into the pack.

In Venezuela, a good open-hole gravel pack should last 8 to 10 years, an inside casing pack somewhat less.

The migration of fines into a gravel pack can be minimized by the use of clay stabilizing organic polymers (Cla-Sta) in the placement fluid. These polymers are strongly adsorbed by clays, feldspars and other negatively charged silicates to prevent their migration from the formation into the gravel pack. Thus, higher production rates can be maintained as well as extending gravel pack life.

Use of Screen or Liner Without Gravel

For low production rates per foot of section, use of properly sized screens or slotted liner can be a low cost means of controlling sand.

With non-uniform sand size distribution (C > 5), Coberly's bridging rule, slot size twice the 10 percentile sand size, can be used.

With uniform sand size distribution (C < 3), slot width probably should not be greater than the 10 percentile size.

Selection of a screen or slotted liner, once the slot size has been determined, depends on well conditions. Sawcut slots are cheaper. Wire-wrapped screen permits use of harder, more corrosion-resistant metal.

Screens set inside casing usually reduce productivity since fine sand moving through the perforations fills the annulus between the screen and casing. Use of largest diameter screen possible is good practice.

In open hole, screen should also be sized as large as possible to prevent caving of shale laminations. Underreaming is not desirable.

The effect of completion fluid cannot be overemphasized. Clean fluids are the key to success. A no-solids fluid should be used for drilling-in for open hole completions, and for perforating a cased completion.

Use of wash pipe permits washing outside the screen, but will not effectively remove mud cake from perforations or open hole.

Production should be initiated slowly, particularly with screens set inside casing to reduce erosion until a stable sand bridge is formed.

RESIN CONSOLIDATION METHODS OF SAND CONTROL
Theory of Resin Consolidation

The basic objective of resin sand consolidation is to increase the strength of the formation sand around the wellbore such that sand grains are not dislodged by the drag forces of the flowing fluids at the desired production rate.

Sand consolidation is accomplished by precipitating resin uniformly in the sand near the well-bore. Resin, attracted to the sand grain contacts, hardens to form a consolidated mass having a compressive strength on the order of 3,000 psi.

Permeability to oil is reduced because the resin occupies a portion of the original pore space, and also because the resin surface is oil-wet. Hopefully, formation strength is increased enough to prevent sand production even though pressure drawdown and drag forces are increased at a given production rate.

Resin Consolidation Advantages—Properly applied under the right conditions, resin consolidation has inherent advantages:

—Suitable for through tubing application
—Applicable in small diameter casing
—Leaves full open well bore
—Suitable for multiple reservoir completions
—Can be applied readily in abnormal pressure wells
—Works well in fine sands difficult to control with gravel packing

Problems Are Multiple—The basic problem is to increase the strength of the formation uniformly through the completion zone without excessive reduction in permeability.

Many practical problems evolve from the fact that all sand consolidation techniques utilize multistage processes in which several fluids, carefully formulated for the specific well conditions, must be uniformly injected sequentially into a perforated interval.

Uniform coverage of each perforation is a critical requirement. Fingering of one fluid through another must be controlled by low injection rates and by designing each fluid so that its viscosity is similar or slightly greater than the fluid it displaces.

Low fluid viscosities are desirable so that reasonable injection rates can be obtained at low injection pressures. Time allowed for injection is usually limited.

Most materials used are toxic and highly inflammable.

Resins are very expensive.

Two Types of Resin Consolidation Systems Used—Phase separation systems consist of 15 to 25% active resin which, properly attracted to the grain contact points, hardens to form the consolidation. The remaining inert material fills the center portion of the pore space to insure that permeability is retained. A curing agent or catalyst is added at the surface with the amount added dependent on formation temperature.

Very accurate control of displacement is required to place the resin through the perforations, but without over-displacement.

In an overflush system, the resin solution contains a high percentage of active material. Thus, in the initial step, active resin occupies most of the pore space. Permeability must be reestablished by displacing further into the formation all but a residual resin saturation. This, with proper sand wettability, remains at the sand grain contacts. Curing agent is usually contained in the overflush fluid, but can be added to the initial resin solution.

Accurate control of displacement is not as critical, but all sections not overflushed will be plugged.

Sands containing more than about 10% clay present

a problem to phase-separation systems, since these systems have only a small percentage of active material in the resin solution. The much higher surface area of dirty sands robs the limited amount of resin from the sand-grain contacts and reduces the strength of the consolidation.

Where sand production has created a cavity behind the perforations, the cavity must be filled with 40–60 or 20–40 mesh sand before either of the basic resin systems can be used.

Resin Sand-Pack System—For wells where sand production has occurred (or where a cavity can be created by washing through perforations), the resin sand injection system has application. In the present form, this system is much like gravel packing. Gravel sized to control the formation sand is coated with resin on the surface, then injected into the perforations at below frac pressure until a sand-out is achieved.

With an external catalyst system, the sand remaining inside the casing is then washed back with oil, and the resin coated sand remaining in the perforations and cavity is consolidated by injecting the catalyst.

With an internal catalyst system, the resin coated sand is allowed time to consolidate, then is drilled out of the casing.

A screen or slotted liner is not required, but is sometimes used as a supplemental control device.

Resin Processes

Many sand-consolidating resins are commercially available. *A Phase-separation system* is Eposand 9, an epoxy process developed by Shell (similar to Eposand 10, 20 & 30).

Overflush systems include: Sanset, a phenolic process developed by Exxon; Eposand 112, an epoxy process developed by Shell (internal catalyst); Sanfix, a furane process developed by Halliburton (external catalyst); and K-200, a modified phenolic process developed by Dowell (external catalyst).

Resin sand pack systems include: Conpac I, a furane resin system using low viscosity oil placing fluid and external catalyst (Halliburton); and Sandlock III, an epoxy or phenolic resin system using low viscosity oil carrying fluid and internal or external catalyst (Dowell). Sandlock IV is an internally catalyzed resin coated sand carried at high concentration (15 lb/gal) in a high viscosity oil fluid (50 cp @ bhc) (Dowell). Sandlock V is similar, but uses a water fluid. Conpac II is a furane resin system with internal catalyst using high viscosity carrying fluid and high sand concentration (Halliburton).

Table 4-5 shows some specifics of representative resin systems.

Placement Techniques

Although a number of resin systems are available and many improvements have been made since resin consolidation was first introduced, it should be recognized that well completion and workover methods, placement techniques, and job supervision and performance are usually more important than the specific resin system used.

Well Preparation—A good primary cement job is very important. If any possibility of mud filled channels exists, a low fluid-loss low-pressure squeeze cement job should be done. Perforating several holes above and below the zone to be consolidated, circulating fluid between these perforations, then using a low fluid-loss circulation squeeze cementing technique is ideal in this case.

The well must be perforated in clean, no-solids fluid. Shaley zones should not be perforated. Perforation density is significant—four holes per foot should be used to improve resin distribution and reduce drawdown. Figure 4-19 compares 1, 2 and 4 shot/ft densities in South Louisiana plasticizing experience.

Length of perforated zone should be short. Ten ft is the maximum zone that can be treated in one stage. (Resin sand-pack technique has application in longer zones 40–50 ft.)

Perforations should be cleaned by flow into the well-bore. Any debris left in the perforations will be "glued" in place by the resin. Back-surge technique appears to increase success ratio where perforating was done with differential pressure into formation.

The fluid system must be especially clean. Workover rig tanks must be scrubbed. Filter system should be used (2 micron filter) for injection of all fluids.

Work strings must be cleaned. Circulating sequestered HCl to bottom of tubing string, then reversing or flowing it back out is one method of removing rust and scale from tubing. Pipe dope should be used sparingly and applied only to the pin. It should not contain insoluble solids.

Service company tanks should be resin coated or stainless steel and should only be used for sand consolidation work.

TABLE 4-5
Specifics of Representative Resin Systems

Eposand 9 Process (Eposand 20 & 30 are similar)

—Catalyst is mixed with resin components on the surface. Formulation is dependent on bottom hole temperature.
—Contact between resin and formation fluids must be prevented by preflush. Formation oil is displaced with diesel, and formation water with Isopropyl Alcohol.
—Overflush is very undesirable (latch-in wiper plug desirable)
—Pumping time: 1 to 3 hours
—Formation temperature limits: 100° to 200°F.
—Permeability retention: 70%
—Well shut-in time: 8 hours before circulating
  24 hours before production

Sanfix process (overflush system-external catalyst)

—Diesel preflush with surfactant
—Resin injected into formation first, followed with the curing agent; mixing occurs in formation.
—Pumping time: No limit
—Formation temperature limit: 400°F.
—Permeability retention: 90% in clean sand
—Well shut-in time: 1 hour before circulating
  4 to 24 hours before production

Sandlock V process (resin sand pack system)

—Diesel preflush (50 gal/ft)
—Gravel (sized to control formation sand) is mixed in 10 bbl batches with a high viscosity water gel at concentration of 15 lb/gal. An internally catalyzed resin is added in low concentrations (1 gal per 100 lb) to coat the gravel.
—Viscous slurry is injected at low rates (1.0 bpm) until sand out. Additional batches may be required.
—Pumping time: 2–4 hours
—After hardening, material inside casing is drilled out.
—Temperature limit: 180°F.

Brine workover fluid (1% KCl) is usually satisfactory. A denser brine (9.5 lb/gal) placed below the perforations and a lighter brine or diesel placed above the perforations aid in preventing mixing of well fluids with resin materials (density 8.8 lb/gal) during placement.

Fluids containing starch should not be used with isopropyl alcohol since an insoluble precipitate is formed. Isopropyl alcohol may also precipitate salt out of highly concentrated brine.

An injectivity test showing less than 2 gal/min per perforation at below frac pressure indicates stimulation is needed.

In dirty sands, a preflush stimulation treatment designed to increase permeability may be required to improve uniformity or resin placement and to provide needed productivity following consolidation.

HCl-HF treatment is preferred. HCl may be sufficient in clean sands to clear perforating charge debris. Hydroxy aluminum or Cla-Sta may be used to hold clay particles in place.

Preflush—The objective of preflushing is (1) remove reservoir fluids which might contaminate the resin (water is particularly bad with epoxy resins), and (2) oil wet sand grain surfaces so resin will form continuous coating.

The type of preflush used will depend upon the resin system, and the recommended preflush must be used, or the consolidation could be a failure. Some systems use diesel oil containing a surfactant. Systems containing an epoxy resin require an additional preflush of isopropyl alcohol to remove all water from the formation to be consolidated. A non-aqueous spacer is used to prevent isopropyl alcohol contact with the resin solution.

Resin Mixing and Injection—Each resin system has its own peculiarities and requires expert service company supervision. (Detailed procedures for Eposand

Sand Control

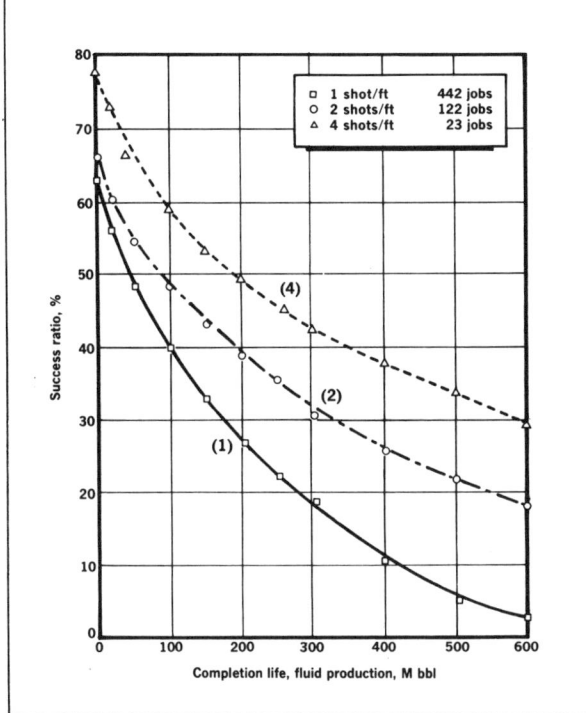

FIG. 4-19—*Plasticizing success vs. perforation density.*[15] *Permission to publish by The Society of Petroleum Engineers.*

so that fluid is displaced to, but not beyond, the perforations. With Eposand, the resin ideally is displaced to within 0.5 bbl of the top of the perforations. Under-displacement has apparently not caused problems. Calibrated displacement tanks must be used.

The overflush systems allow some margin for error and probably should be used where control of displacement is difficult.

After displacement is complete, the well must be shut in so that no fluid movement either way occurs through the perforations. A latch-in wiper plug is beneficial. Again, if it is not possible to prevent movement of wellbore fluid into the perforations (as in a low fluid level well) the overflush system should probably be used.

The details of the Eposand 112 system are shown in Appendix 4C.

COMPARISONS OF SAND CONTROL METHODS—SUMMARY

Reduction of Drag Forces—Flow rate per unit area, if applicable, should be given first consideration. Increase flow area if possible. Good well completion practices are paramount.

112 are shown in Appendix 4C.) Formation temperature at the time of injection must be known to select the basic resin material in some cases and the amount of accelerator in other cases. With internal catalysts, time is a very important factor; both injection time for the resin and (with the overflush systems) the time overflush operation must begin.

Injection of all solutions must be done at low rates (1.0 gal/min per perforation) to promote uniform coverage and at below fracture pressure. The volume of resin varies with the process and, perhaps, the uniformity of the sand. Figure 4-20 shows that South Louisiana "success" increased almost in direct proportion to volume of resin used. Generally volume is based on that required to fill reservoir pore volume 3 to 5 ft back from the well bore (90 to 150 gal/ft).

The resin sand-pack system is much more economical in the use of resin, requiring 20 to 30 gal/100 lb of sand with Conpac I or Sandlock III Systems—or only about 1.0 gal/100 lb of sand with Conpac II or Sandlock IV Systems.

Displacement—Careful control of displacement volume is required with the Phase Separation system

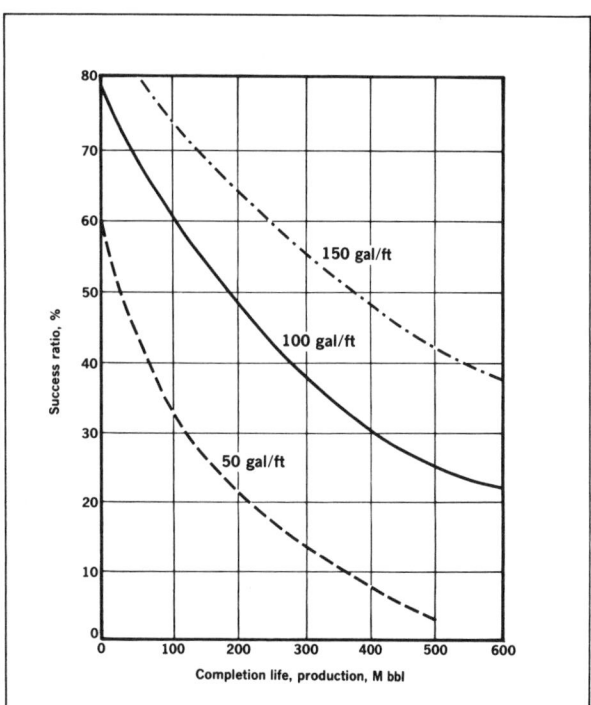

FIG. 4-20—*Plasticizing success vs. volume of resin.*[15] *Permission to publish by The Society of Petroleum Engineers.*

Gravel Packing—This offers the only practical sand control for long zones. Gravel packing may also be most practical for short zones—but remedial work, multiple completions, small hole diameters, and abnormal pressures increase difficulty and cost.

Open hole gravel pack should always be used on single completions where water or gas shut-off or other change of completion interval is not anticipated.

Inside casing, gravel pack restricts productivity—but productivity may be maximized by a sufficient number of large clean perforations and effective placement of the gravel, perhaps to the point of approaching the productivity of the OHGP.

Resin Consolidation—This is used in short zones where, for one reason or another, a gravel pack cannot be used. Some of the applications are: small pipe diameter, top zone of a dual completion, offshore or isolated location where tubing hoist is not available, and abnormal formation pressures make through tubing work advisable.

Resin Sand Pack—This has most of the same problems and advantages of the inside casing gravel pack.

REFERENCES

1. Coberly, C. J.: "Selection of Screen Openings for Unconsolidated Sands," API Drill. & Prod. Practice (1937).
2. Hill, K. E.: "Factors Affecting the Use of Gravel in Oil Wells," API Drill. & Prod. Practice (1941).
3. Sage, B. H. and Lacey, W. N.: "Effectiveness of Gravel Screens," AIME Trans. Vol. 146 (1942).
4. Tausch, G. H. and Corley, C. B.: "Sand Exclusion in Oil & Gas Wells," API, Houston (1958).
5. Rike, J. L.: "Review of Sand Consolidation Experience in South Louisiana," J. Pet. Tech., May 1966, p. 545.
6. Cirigliano, J. A. and Leibach, R. E.: "Gravel Packing in Venezuela," Seventh World Petroleum Congress, Mexico City, April 2–8, 1967.
7. Schwartz, David H.: "Successful Sand Control Design for High Rate Oil and Water Wells," J. Pet Tech., Sept. 1969.
8. Williams, B. B., Elliott, L. S., and Weaver, R. H.: "Productivity of Inside Casing Gravel Pack Completions," J. Pet. Tech., April 1972, p. 419.
9. Stein, N. and Hilchie, D. W.: "Estimation of Maximum Production Rates Possible From Friable Sandstones Without Using Sand Control Measures," J. Pet. Tech., Sept. 1972, p. 1, 156.
10. Saucier, R. J.: "Considerations in Gravel Pack Design," J. Pet Tech., Feb. 1974, p. 205.
11. Sparlin, Derry and Copeland, Travis: "Pressure Packing with Concentrated Gravel Slurry," SPE 4033, Oct. 1972.
12. Testing Screens and Their Uses, Handbook 53-1973, C. E. Tyler Division Combustion Engineering, Inc.
13. Tixier, M. P., Loveless, G. W., Anderson, R. A.: "Estimation of Formation Strength from the Mechanical Properties Log." SPE Paper 4532, Las Vegas, 1973.
14. Gulati, M. S., and Maly, G. P.: "Thin-Section and Permeability Studies Call for Smaller Gravels in Gravel Packing," J. Pet. Tech., Jan. 1975, p. 107.
15. Schroeder, R. H., and Tucker, M. J.: "Evaluation of Completion Practices for Improved Sand Control Success and Longevity," SPE 5027, Oct. 1974.
16. Maly, George P., Robinson, Joel P., and Laurie, A. M.: "New Gravel Pack Tool For Improving Pack Placement." J. Pet Tech., Jan 1974, p. 19.
17. Gruesbeck, C., Salathiel, W. M., and Echols, D. E.: "Design of Gravel Packs in Deviated Wells," SPE 6805, Oct. 1977.
18. Penberthy, W. L., and Cope, B. J: Design and Productivity of Gravel Packed Completions," SPE Paper No. 8428, Oct. 1979.
19. Recommended Practices for Testing Sand Used in Gravel Packing Operations. API RP. Dec. 1981.

Appendix 4A

Comparison Table of U. S. A., Tyler, Canadian, British, French, and German Standard Sieve Series

U.S.A. (1)		TYLER (2)	CANADIAN (3)		BRITISH (4)		FRENCH (5)		GERMAN (6)
*Standard	Alternate	Mesh Designation	Standard	Alternate	Nominal Aperture	Nominal Mesh No.	Opg. M.M.	No.	Opg.
125 mm	5″		125 mm	5″					
106 mm	4.24″		106 mm	4.24″					
100 mm	4″		100 mm	4″					
90 mm	3½″		90 mm	3½″					
75 mm	3″		75 mm	3″					
63 mm	2½″		63 mm	2½″					
53 mm	2.12″		53 mm	2.12″					
50 mm	2″		50 mm	2″					
45 mm	1¾″		45 mm	1¾″					
37.5 mm	1½″		37.5 mm	1½″					
31.5 mm	1¼″		31.5 mm	1¼″					
26.5 mm	1.06″	1.05″	26.5 mm	1.06″					
25.0 mm	1″		25.0 mm	1″					25.0 mm
22.4 mm	⅞″	.883″	22.4 mm	⅞″					
19.0 mm	¾″	.742″	19.0 mm	¾″					20.0 mm
									18.0 mm
16.0 mm	⅝″	.624″	16.0 mm	⅝″					16.0 mm
13.2 mm	.530″	.525″	13.2 mm	.530″					
12.5 mm	½″		12.5 mm	½″					12.5 mm
11.2 mm	⁷⁄₁₆″	.441″	11.2 mm	⁷⁄₁₆″					
									10.0 mm
9.5 mm	⅜″	.371″	9.5 mm	⅜″					
8.0 mm	⁵⁄₁₆″	2½	8.0 mm	⁵⁄₁₆″					8.0 mm
6.7 mm	.265″	3	6.7 mm	.265″					
6.3 mm	¼″		6.3 mm	¼″					6.3 mm
5.6 mm	No. 3½	3½	5.6 mm	No. 3½			5.000	38	5.0 mm
4.75 mm	4	4	4.75 mm	4			4.000	37	4.0 mm
4.00 mm	5	5	4.00 mm	5					
3.35 mm	6	6	3.35 mm	6	3.35 mm	5			
							3.150	36	3.15 mm
2.80 mm	7	7	2.80 mm	7	2.80 mm	6			
2.36 mm	8	8	2.36 mm	8	2.40 mm	7	2.500	35	2.5 mm
2.00 mm	10	9	2.00 mm	10	2.00 mm	8	2.000	34	2.0 mm
1.70 mm	12	10	1.70 mm	12	1.68 mm	10	1.600	33	1.6 mm
1.40 mm	14	12	1.40 mm	14	1.40 mm	12	1.250	32	1.25 mm
1.18 mm	16	14	1.18 mm	16	1.20 mm	14			
1.00 mm	18	16	1.00 mm	18	1.00 mm	16	1.000	31	1.0 mm
850 μm	20	20	850 μm	20	850 μm	18			
710 μm	25	24	710 μm	25	710 μm	22	.800	30	800 μm
							.630	29	630 μm
600 μm	30	28	600 μm	30	600 μm	25			
500 μm	35	32	500 μm	35	500 μm	30	.500	28	500 μm
425 μm	40	35	425 μm	40	420 μm	36			
							.400	27	400 μm
355 μm	45	42	355 μm	45	355 μm	44	.315	26	315 μm
300 μm	50	48	300 μm	50	300 μm	52			
250 μm	60	60	250 μm	60	250 μm	60	.250	25	250 μm
212 μm	70	65	212 μm	70	210 μm	72	.200	24	200 μm
180 μm	80	80	180 μm	80	180 μm	85	.160	23	160 μm
150 μm	100	100	150 μm	100	150 μm	100			
125 μm	120	115	125 μm	120	125 μm	120	.125	22	125 μm
106 μm	140	150	106 μm	140	105 μm	150	.100	21	100 μm
90 μm	170	170	90 μm	170	90 μm	170			90 μm
75 μm	200	200	75 μm	200	75 μm	200	.080	20	80 μm
63 μm	230	250	63 μm	230	63 μm	240	.063	19	71 μm
									63 μm
									56 μm
53 μm	270	270	53 μm	270	53 μm	300	.050	18	50 μm
45 μm	325	325	45 μm	325	45 μm	350	.040	17	45 μm
									40 μm
38 μm	400	400	38 μm	400					

(1) U.S.A. Sieve Series - ASTM Specification E-11-70
(2) Tyler Standard Screen Scale Sieve Series.
(3) Canadian Standard Sieve Series 8-GP-1d.
(4) British Standards Institution, London BS-410-62.
(5) French Standard Specifications, AFNOR X-11-501.
(6) German Standard Specification DIN 4188.

*These sieves correspond to those recommended by ISO (International Standards Organization) as an International Standard and this designation should be used when reporting sieve analysis intended for international publication.

Appendix 4B

U. S. A. SIEVE SERIES AND TYLER EQUIVALENTS
A.S.T.M.—E-11-70

Sieve Designation		Sieve Opening		Nominal Wire Diameter		Tyler Screen Scale Equivalent Designation
Standard(a)	Alternate	mm	in (approx. equivalents)	mm	in (approx. equivalents)	
125 mm	5 in.	125	5	8	.3150	.....
106 mm	4.24 in.	106	4.24	6.40	.2520	.....
100 mm	4 in.(b)	100	4.00	6.30	.2480	.....
90 mm	3½ in.	90	3.50	6.08	.2394	.....
75 mm	3 in.	75	3.00	5.80	.2283	.....
63 mm	2½ in.	63	2.50	5.50	.2165	.....
53 mm	2.12 in.	53	2.12	5.15	.2028	.....
50 mm	2 in.(b)	50	2.00	5.05	.1988	.....
45 mm	1¾ in.	45	1.75	4.85	.1909	.....
37.5 mm	1½ in.	37.5	1.50	4.59	.1807	.....
31.5 mm	1¼ in.	31.5	1.25	4.23	.1665	.....
26.5 mm	1.06 in.	26.5	1.06	3.90	.1535	1.050 in.
25.0 mm	1 in.(b)	25.0	1.00	3.80	.1496	.....
22.4 mm	⅞ in.	22.4	0.875	3.50	.1378	.883 in.
19.0 mm	¾ in.	19.0	0.750	3.30	.1299	.742 in.
16.0 mm	⅝ in.	16.0	0.625	3.00	.1181	.624 in.
13.2 mm	.530 in.	13.2	0.530	2.75	.1083	.525 in.
12.5 mm	½ in.(b)	12.5	0.500	2.67	.1051	.....
11.2 mm	7⁄16 in.	11.2	0.438	2.45	.0965	.441 in.
9.5 mm	⅜ in.	9.5	0.375	2.27	.0894	.371 in.
8.0 mm	5⁄16 in.	8.0	0.312	2.07	.0815	2½ mesh
6.7 mm	.265 in.	6.7	0.265	1.87	.0736	3 mesh
6.3 mm	¼ in.(b)	6.3	0.250	1.82	.0717	.....
5.6 mm	No. 3½(c)	5.6	0.223	1.68	.0661	3½ mesh
4.75 mm	No. 4	4.75	0.187	1.54	.0606	4 mesh
4.00 mm	No. 5	4.00	0.157	1.37	.0539	5 mesh
3.35 mm	No. 6	3.35	0.132	1.23	.0484	6 mesh
2.80 mm	No. 7	2.80	0.111	1.10	.0430	7 mesh
2.36 mm	No. 8	2.36	0.0937	1.00	.0394	8 mesh
2.00 mm	No. 10	2.00	0.0787	.900	.0354	9 mesh
1.70 mm	No. 12	1.70	0.0661	.810	.0319	10 mesh
1.40 mm	No. 14	1.40	0.0555	.725	.0285	12 mesh
1.18 mm	No. 16	1.18	0.0469	.650	.0256	14 mesh
1.00 mm	No. 18	1.00	0.0394	.580	.0228	16 mesh
850 μm	No. 20	0.850	0.0331	.510	.0201	20 mesh
710 μm	No. 25	0.710	0.0278	.450	.0177	24 mesh
600 μm	No. 30	0.600	0.0234	.390	.0154	28 mesh
500 μm	No. 35	0.500	0.0197	.340	.0134	32 mesh
425 μm	No. 40	0.425	0.0165	.290	.0114	35 mesh
355 μm	No. 45	0.355	0.0139	.247	.0097	42 mesh
300 μm	No. 50	0.300	0.0117	.215	.0085	48 mesh
250 μm	No. 60	0.250	0.0098	.180	.0071	60 mesh
212 μm	No. 70	0.212	0.0083	.152	.0060	65 mesh
180 μm	No. 80	0.180	0.0070	.131	.0052	80 mesh
150 μm	No. 100	0.150	0.0059	.110	.0043	100 mesh
125 μm	No. 120	0.125	0.0049	.091	.0036	115 mesh
106 μm	No. 140	0.106	0.0041	.076	.0030	150 mesh
90 μm	No. 170	0.090	0.0035	.064	.0025	170 mesh
75 μm	No. 200	0.075	0.0029	.053	.0021	200 mesh
63 μm	No. 230	0.063	0.0025	.044	.0017	250 mesh
53 μm	No. 270	0.053	0.0021	.037	.0015	270 mesh
45 μm	No. 325	0.045	0.0017	.030	.0012	325 mesh
38 μm	No. 400	0.038	0.0015	.025	.0010	400 mesh

(a) These standard designations correspond to the values for test sieves apertures recommended by the International Standards Organization Geneva, Switzerland.

(b) These sieves are not in the fourth root of 2 Series, but they have been included because they are in common usage.

(c) These numbers (3½ to 400) are the approximate number of openings per linear inch but it is preferred that the sieve be identified by the standard designation in millimeters or μm.

1000 μm = 1 mm.

Appendix 4C

Application Details for Eposand Resin 112 System for Sand Consolidation

Eposand Resin 112 is an overflush resin system for sand consolidation involving clean or dirty sands.

Sand consolidated by Eposand 112 exhibits a compressive strength of 3000^+ psi with indefinite retention of initial strength in the presence of hot (160°F) reservoir fluids. The retained absolute permeability will be at least 70%.

Eposand 112 contains three components: Eposand 112A resin, Eposand 112B curing agent, and when required, Eposand 112C accelerator. Temperature limits are 100°F to 220°F. Below 150°F, Eposand 112C accelerator should be added.

Eposand 112A—Eposand 112A contains the resin in a solvent. It is a clear, straw colored solution with a viscosity of 13 cp and a density of about 9.1 lb/gal. It is stable up to at least one year; however, contamination by water should be avoided. Safety goggles must be worn when handling.

Eposand 112B—Eposand 112B is an amine curing agent. It is a pale white, solid flake with melting point at 90°C. While only slightly irritating, contact with the eyes and skin should be avoided.

Eposand 112C—Eposand 112C is an organic catalyst used to accelerate polymerization at temperatures below 150°F. Contamination by water must be avoided.

The Eposand 112 Process

The Eposand 112 process involves: (1) three successive preflush fluids to prepare the sand, (2) the mixed Eposand 112 solution and (3) the overflush fluid. The overflush fluid displaces most of the resin solution and extracts the solvent, which speeds polymerization. The well is then shut in for a period dependent upon formation temperature to allow the resin to cure. For temperatures below 150°F, Eposand 112C may be added to accelerate the reaction and reduce the shut-in time required.

Mixtures Containing No Accelerator—The resin solution starts to polymerize as soon as mixed and, if left in the well or formation without being overflushed, it will solidify into an impermeable mass. Correct distribution of resin in the formation depends upon displacement efficiency of the overflush, thus the overflush fluid should have a viscosity near that of the resin solution. The resin solution must be overflushed before polymerization increases resin viscosity above 10 cp.

After combining Eposand 112A and 112B, the time at which resin viscosity reaches 10 cp is termed resin thickening time, Figure 4-21. After this time, resin viscosity begins to increase more rapidly. If no Eposand 112C accelerator is added, the resin components may be maintained in their mixed state at surface temperatures for several hours without gelling.

The important point is that the resin solution should not be left in the lower portion of the tubing or in the formation without overflushing beyond the resin thickening time.

Once overflushing has occurred, polymerization is accelerated and the well should be shut in until the resin has cured to a solid. The minimum shut in time after overflush versus bottom hole temperature is shown in Figure 4-22.

Mixtures Containing Eposand 112C Accelerator—Mixtures containing Eposand 112C accelerator polymerize more rapidly and must be handled with care, even at relatively low temperatures. For example, a mixture containing one-half package of Eposand 112C accelerator per unit at 80°F surface temperature will exhibit a temperature rise and begin

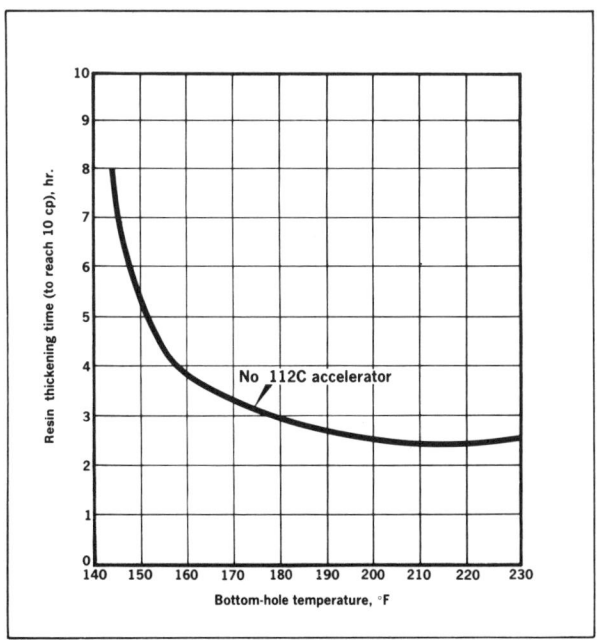

FIG. 4-21—*Resin-thickening time vs. bottom-hole temperature for Eposand 112.*

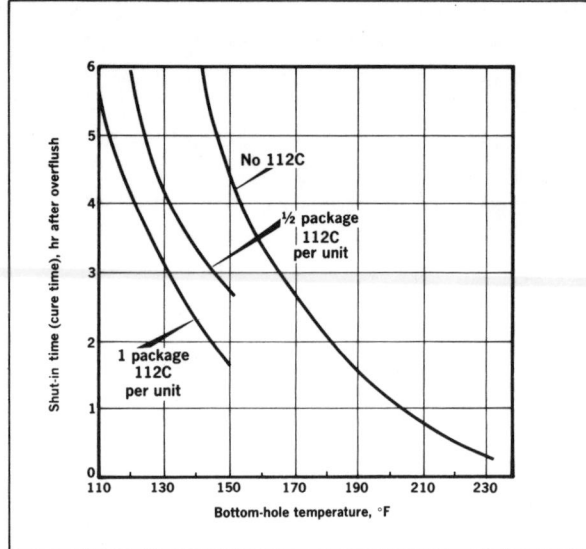

FIG. 4-22—*Shut-in time vs. bottom-hole temperature for Eposand 112.*

to thicken in surface mixing equipment approximately two hours after adding the accelerator.

If the solution is held in the surface equipment for an additional two hours, the resin solution will boil vigorously and be completely cured thirty minutes later. This reaction is proportionately faster when greater amounts of Eposand 112C are used. Difficulty can be avoided by pouring unused resin on the ground shortly after the resin thickening time has elapsed.

Figure 4-23 is a nomograph which can be used to determine maximum surface handling time for accelerated Eposand 112 mixtures. Maximum surface handling time is defined as the maximum time available for mixing and pumping the Eposand 112 components in surface equipment after Eposand 112C has been added. The procedure for determining maximum surface handling times is illustrated by the example on Figure 4-23.

To use the nomograph, begin by entering with the depth of the formation (4,000 ft), turn at the proper tubing size intersection, (2 in. nominal) and turn at the anticipated pump rate intersection (¼ bbl/minute) to find downhole pumping time (65 minutes). Re-enter the nomograph with the anticipated surface fluid temperature (80°F) and turn at the intersection of Eposand 112C proportion to be used (½ package per unit) to find resin thickening time (140 minutes). The maximum surface handling time (75 minutes) is found by connecting the two traces as shown.

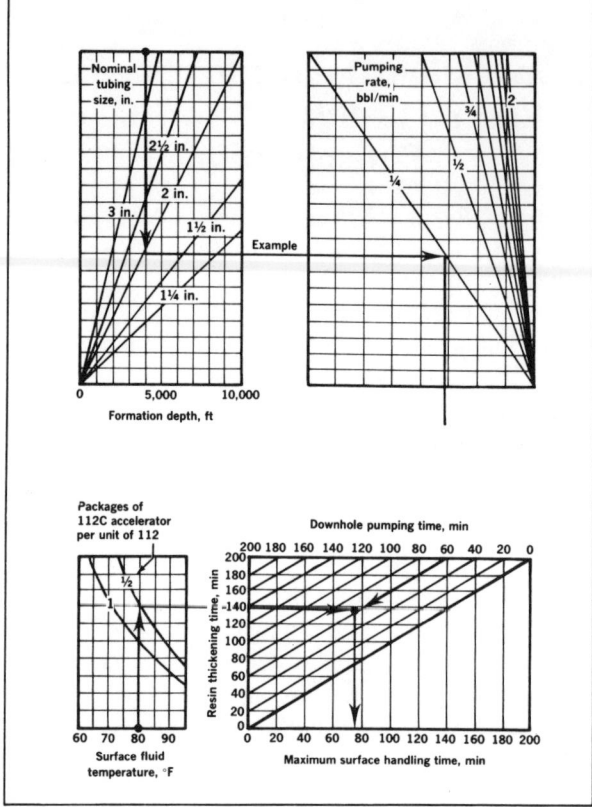

FIG. 4-23—*Maximum surface-handling time for accelerated Eposand 112.*

The maximum surface handling time of 75 minutes in the example means that, once Eposand 112C has been added to the mixture, the entire mixture must be pumped into the well before 75 minutes has elapsed. Any resin mixture remaining on the surface when time has elapsed should be discarded immediately and not pumped into the well.

If the resin mixture has all been pumped into the well-head before the 75 minutes has elapsed (say, within 50 minutes), the remaining time is slack. All or part of this slack (25 minutes, in this example) may be used before overflushing, if necessary, to interrupt pumping, to switch tanks, connections or perform minor equipment repairs. If no slack time remains, it will be necessary to continue pumping uninterrupted and displace overflush fluid into the formation immediately behind the resin solution.

If conditions such as pumping rate change as the job progresses, it is suggested that maximum surface handling time be redetermined from Figure 4-23 to ensure that no difficulty will be encountered.

Sand Control

After overflushing, the well should be shut in until the resin has cured to a solid. The minimum shut-in time for the accelerated system containing Eposand 112C is shown on Figure 4-22.

General Procedure

Resin volume is based on consolidating a zone of 3-ft radius around the wellbore with a length equal to the perforated interval plus 6 ft. Pore volume means the volume of this zone.

1. Spot heavy brine solution (density greater than 9.5 lb/gal) to fill the casing below the perforations. This prevents resin solution from displacing lighter fluids such as salt water weighing less than 8.8 lb/gal into the formation. Water entering the formation at this time would be detrimental to the consolidation.

2. Test injectivity with clean filtered salt water or diesel oil. Acidize if necessary to obtain at least a minimum practical injection rate.

3. The injection rate should be such that the resin mixture is in the formation well ahead of the time the resin reached 10 cp viscosity as shown on Figure 4-21. If a low temperature formation is being treated and the resin system will contain Eposand 112C accelerator, injection rate limitations will be determined from maximum surface handling time on Figure 4-23.

4. Inject preflush fluids, including:
   a. One pore volume of No. 2 diesel oil to clean the reservoir and to displace the crude oil or gas from the formation surrounding the wellbore.
   b. Two pore volumes of anhydrous (99.5%) isopropyl alcohol to remove the connate water. Elimination of all water is essential to obtain full consolidation strength.

5. Inject one pore volume of a spacer fluid, Shell medium aromatic oil, 84054, to physically separate the alcohol from the Eposand solution. The spacer prevents accelerated polymerization through mixing of the resin and alcohol and ensures maximum strength and proper usable life.

6. Inject resin:
   a. ⅔ of one pore volume of Eposand 112.
   b. Five pore volumes of an overflush solution containing Valvata Oil 79 and No. 2 diesel oil to disperse the resin and extract the solvent (1:1 ratio). Extraction of the solvent from the resin solution increases the rate of consolidation and thus speeds consolidation. Suggested overflush blends are given in Table 4-6.

The blends in Table 4-6 will assure that adequate overflush viscosity is maintained at bottomhole conditions during injection. Kerosene (No. 1 diesel) is an attractive substitute for No. 2 diesel oil, although it is somewhat more expensive and not as readily available in the oil field. Laboratory and field tests have shown that use of No. 1 diesel oil in the overflush solution blend gives somewhat higher and more consistent consolidation strengths.

Material Specifications

1. All liquids should be transparent and free from sediment and suspended matter. It is recommended that No. 2 diesel oil used in the preflush and overflush be filtered through a 2-micron filter element.

2. The isopropyl alcohol should be the anhydrous grade and have a water-clear color. The preferred range of specific gravity density at various temperatures is listed on Table 4-7.

3. The spacer fluid should be clear, pale amber with a density of 0.860–0.890 at 77°F.

TABLE 4-6
Suggested Overflush Blends for Eposand 112

Bottom-hole temperature range (°F)	Injection rate (bbl/min)	Suggested Overflush Blends* for EPOSAND 112 with:		
		No accelerator	½ pkg. 112C per unit	1 pkg. 112C per unit
Below 125	Any	67/33	67/33	67/33
125–160	Any	67/33	50/50	50/50
Above 160	Above 0.5	67/33	—	—
Above 160	Below 0.5	50/50	—	—

*67/33 refers to vol % No. 2 diesel oil and vol % Valvata Oil 79, respectively.

4. The Eposand 112A should be clear with no more than a trace of turbidity. The density at 77°F should be 1.10.

5. The Eposand 112B is a pale white, solid flake.

6. The Eposand 112C is a white, crystalline material.

7. The overflush has the appearance of unused or fresh motor oil.

Eposand 112 Unit—One unit of Eposand 112 is made up by blending one drum of Eposand 112A (55 gal weighing approximately 480 lbs net) and one fiber drum of 112B (weighing approximately 81 lb net). Eposand 112C accelerator is supplied in packages weighing 5.7 lb net each to be added as needed.

After combining Eposand 112A and 112B, one unit has a volume of approximately 64 gal and weighs 561 lb. The final Eposand 112 blend has an approximate density of 1.055 or 8.8 lb/gal at 70°F.

Equipment—It is recommended that clean, separate, and lined tanks be used to store the different components, and that the tank used to mix the Eposand 112 components be used for nothing else.

Should the equipment, tanks, lines, and pumps be used for any other material, they should be thoroughly cleaned of all foreign liquids and solids (with acetone, toluene, or similar solvents) using a high speed flush and dried before placing in Eposand service again. The Eposand 112 mixing tank should be rinsed with toluene followed by an alcohol rinse immediately after each batch of resin is mixed to prevent resin build-up.

TABLE 4-7
Suggested Specific Gravities for Isopropyl Alcohol Used as a Preflush Fluid

Fluid temperature (°F)	Specific gravity of isopropyl alcohol	
	Preferred range	Not acceptable
60	0.789–0.794	0.803 and above
70	0.784–0.790	0.797 and above
80	0.780–0.786	0.792 and above
90	0.775–0.783	0.789 and above

Inspection of equipment should include an examination to make certain no resin or other solids, residual solvent, and particularly, water is present. Equipment out of service should be sealed to avoid contamination before use.

Do's and Don't's for Using Eposand 112

1. In establishing the well pump-in rate, do not use fresh or dirty water. Do use filtered diesel oil or filtered salt water (sodium or potassium chloride) at a salt level which avoids swelling of clays which may be present in the formation to be treated.

2. Do check the specific gravity of isopropyl alcohol preflush to assure minimum water content as listed on Table 4-7.

3. Do keep a record of lot numbers of materials used.

4. The use of wiper plugs to prevent product mixing is optional but not required. If wiper plugs are used, they should be oil-resistant rubber. Do not use cement wiper plugs. It should be recognized however that wiper plugs loosens corrosion products and other potential plugging materials from tubing walls, and steps may be required to avoid any such injectivity impairment.

5. Do make sure that the workstring or tubing is clean, free from mud, scale, rust, sand or pipe dope. Apply pipe dope to pin end only. Such particulate matter as this may cause high injection pressures, production impairment or act as a diverting agent. This results in non-uniform treatment of the perforated interval.

6. Do flush out all lines from mixing tank to wellhead to avoid entrapment of foreign materials after each step.

7. Do not interchange solvents or revise the order in which injected.

8. Do not allow any water or mud in resin solution or preflush solvents.

9. Do not mix diesel oil with resin solution.

10. Do not blend aromatic oil with diesel oil.

Chapter 5 Formation Damage

Occurrence, significance of formation damage
Basic cause, damage mechanisms
Particle plugging within the matrix
Formation clay effects
Fluid viscosity effects
Diagnosis of formation damage

INTRODUCTION

All wells are susceptible to formation damage to some degree, from relatively minor loss of productivity to complete plugging of specific zones.

Occurrence of Formation Damage

Appendix 5-A details specific well operations where formation damage may occur. Practices employed in carrying out these operations have been developed over a long period of time and frequently have been optimized on the basis of minimizing current operating costs. Productivity was not always an overly important factor due to proration and other artificial restrictions which set a maximum limit on per-well producing rate.

Many of these practices were also developed in areas where wells were completed in single stringers or short sections. Thus the problem of insuring that each zone of a multi-zone completion contributed proportionately to the total production was not relevant. These same practices, perhaps acceptable in single thin-zone completions, do not give adequate consideration to the problem of proportionate production from each section in wells where many sections are open to the wellbore.

Significance of Formation Damage

Flow surveys almost invariably show that a high percentage of the zone open to the wellbore is not contributing to the total flow. With the inherent barriers to vertical flow present in most zones, formation damage can restrict or prevent effective depletion. Thus reserves may remain trapped in a high percentage of the potentially productive zone.

Ultimate economics usually favor control of formation damage rather than stimulation to overcome limited productivity. For many situations, complete restoration of productivity is not possible.

Formation damage means reduced current production.

Example: A 6-in diameter well completed in 10-ft formation having undamaged permeability of 100 md and containing 0.5 cp oil, produces 100 bopd. If this well is damaged to the extent that permeability is reduced to 1.0 md over a radius of 2 ft, then, with the same pressure drawdown, it will produce only 5 bopd.

Figure 5-1 relates productivity loss to degree and depth of damage. The important point is that with radial flow, the *critical area* is the first few feet away from the well bore.

BASIC CAUSES OF DAMAGE

Contact with a foreign fluid is the basic cause of formation damage. This foreign fluid may be a drilling mud, a clean completion or workover fluid, a stimulation or well-treating fluid, or even the reservoir fluid itself if the original characteristics are altered.

Most oilfield fluids consist of two phases—liquid and solids. Either can cause significant formation damage through one of several possible mechanisms.

Plugging Associated with Solids

Plugging by solids occurs on the formation face, in the perforation, or in the formation. Solids may be

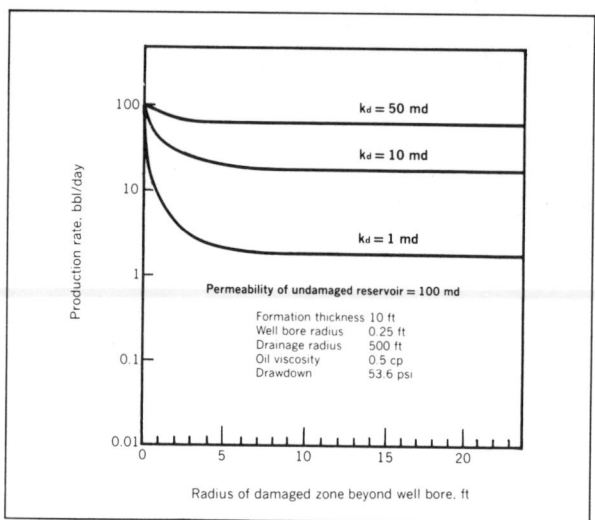

FIG. 5-1—*Effect of formation damage on well productivity.*

weighting materials, clays, viscosity builders, fluid-loss-control materials, lost-circulation materials, drilled solids, cement particles, perforating charge debris, rust and mill scale, pipe dope, undissolved salt, gravel pack or frac sand fines, precipitated scales, or paraffin or asphaltenes.

Large Solids—Large solids cause formation damage by plating out on the face of a filter media or a short distance behind the face of the filter media. Thus, when the pressure differential is into the formation, a permeable producing or injection zone serves as very effective filter for any contacting fluid. Plugging may occur in a perforation tunnel, on the face of an open hole zone, on the face of a natural or created fracture, or in a fracture channel.

Sometimes, depending on the size relationships and the manner of deposition, the plated-out solids may be removed by reverse flow. However, many times it is not possible to achieve sufficient differential pressure at the right point. Initially, a small part of the zone is opened; then, with the influx of fluid into the wellbore from the first part of the zone opened, there may not be sufficient differential pressure to complete the damage removal process. With the existence of inherent barriers to vertical flow, many zones may remain partially or completely sealed for the life of the well.

Small Solids—Very small solids such as iron oxides, clays, or other silicate particles may be carried for some distance into the pores of relatively permeable formations to create serious plugging. In frac jobs or gravel packing operations, fines are frequently carried in or created by the treating fluids, then held in place by the frac sand or gravel to reduce fracture flow capacity or cause an internal plug in the gravel pack.

Solids Precipitation—Solids may also be precipitated within the formation. For example, scale often precipitates due to mixing of incompatible waters; asphaltene or paraffin may be precipitated due to changing equilibrium conditions.

Plugging Associated with Fluid Filtrate

The liquid filtrate may be water containing varying types and concentrations of positive and negative ions and surfactants. It may be a hydrocarbon carrying various surfactants. The liquid is forced into porous zones by differential pressure, displacing or commingling with a portion of the virgin reservoir fluids. This may create blockage due to one or more of several mechanisms that may reduce the absolute permeability of the pore, or restrict flow due to relative permeability or viscosity effects.

Particle migration effects include hydration or dehydration of clays; dispersion or flocculation of highly swellable or slightly swellable clays and formation particles; or dissolution of cementing materials allowing fines, clays or other particles to move within the pore constrictions.

Increased water saturation causes waterblocking or reduced relative permeability to oil or gas.

Liquid filtrate may create a viscous emulsion with the virgin reservoir oil or water or may tend to oil-wet the rock, reducing relative permeability to oil. Stable emulsions within a formation appear to be associated with partially oil-wet systems.

Viscosity effects include emulsions, but also plugging by a high-viscosity treating fluid, which for some reason does not "break" or is not sufficiently diluted to readily return to the wellbore under the influence of the available differential pressure.

CLASSIFICATION OF DAMAGE MECHANISMS

The numerous mechanisms that result in formation damage may be generally classified as to the manner by which they decrease production:

—Reduced absolute permeability of formation—results from plugging of pore channels by induced or inherent particles.

—Reduced relative permeability to oil—results from

an increase in water saturation or oil-wetting of the rock.
—Increased viscosity of reservoir fluid—results from emulsions or high-viscosity treating fluids.

In a radial flow system, any reduction in permeability around the wellbore results in serious reduction of productivity or injectivity. In a linear flow situation, some plugging of the face of the fracture can be tolerated due to the large area represented by the faces of the fracture. However, plugging of the fracture itself results in serious reduction in productivity or injectivity.

REDUCED ABSOLUTE PERMEABILITY

This results from plugging of pore channels by induced or inherent particles.

Particle Plugging Within the Formation

The pore system provides a tortuous path to the wellbore. Recent laboratory techniques, including the scanning electron microscope, have provided a clearer picture of particle plugging within the formation pore system, both by particles inherent to the formation and by particles carried into the formation by various fluid filtrates.

Figure 5-2 shows use of the scanning electron microscope in detecting formation damage in Berea sandstone. Damage occurred after injecting 4 pore volumes of dilute clay-base mud. Permeability before injection was 562 md; after injection, it was 2.3 md.

A sandstone formation contains many interconnected pores. Typical pore diameters range from perhaps 10 to 100 microns. One square inch of Berea sandstone contains over 3,000 pores. Figure 5-3 shows size relationships among pore channels, formation fines, formation particles, and fluid-loss additives.

Fluids moving through such pore caverns are subject to frequent changes in direction and velocity. As fluid nears the well bore in a radial flow system, velocities increase and, in the smaller pores, may reach the turbulent range.

A sandstone formation is, in effect, an excellent filter, utilizing three basic filter mechanisms: screening (pore openings); adsorption (high surface area); and sedimentation (pore depth). It can be classed as a depth filter, and, as such, is highly sensitive to flow rate and pressure differential. Depth filters plug rapidly when their design flow rate is exceeded.

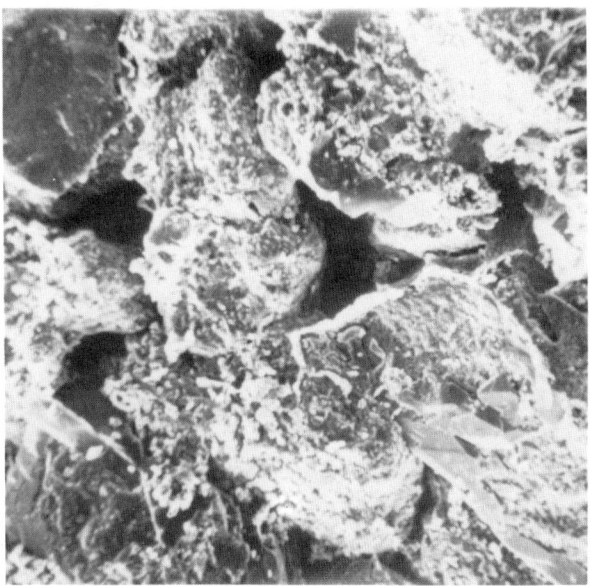

FIG. 5-2—*Scanning electron microscope shows damage in Berea sandstone after injecting only 4 pore volumes dilute clay-base mud (10% mud, 90% clear water). Permeability before, 562 md; after, 2.3 md.*[14] Permission to publish by Society of Petroleum Engineers.

Particles can move through the pore system. Electron microscope pictures of sandstones show that even clean sands contain a relatively large amount of small particulate materials. These particles, clays, feldspar, and other minerals, appear to be stuck to the rock matrix; however, lab tests indicate that if the flow velocity reaches a high enough level, these particles can be picked up and moved from one pore cavern to another. If the next pore cavern is larger and flow velocity drops, the particle may settle out.

If several particles moving through the caverns meet pore restrictions having an opening less than about three times the particle size, they will bridge. Such bridging, causing partial or complete plugging, will force fluids to seek other paths to the well-bore.

Differences in permeability with direction of flow in back-flush tests provide laboratory evidence of this particle movement and bridging within a core.

Particle movement is affected by wettability and by the fluid phases in the pore system. In a normal producing situation, an oil zone pore system contains water and oil. With a water-wet system, water is in contact with the rock matrix, and oil flows through the center portion of the pore cavern.

Where the clays and other fines are water-wet,

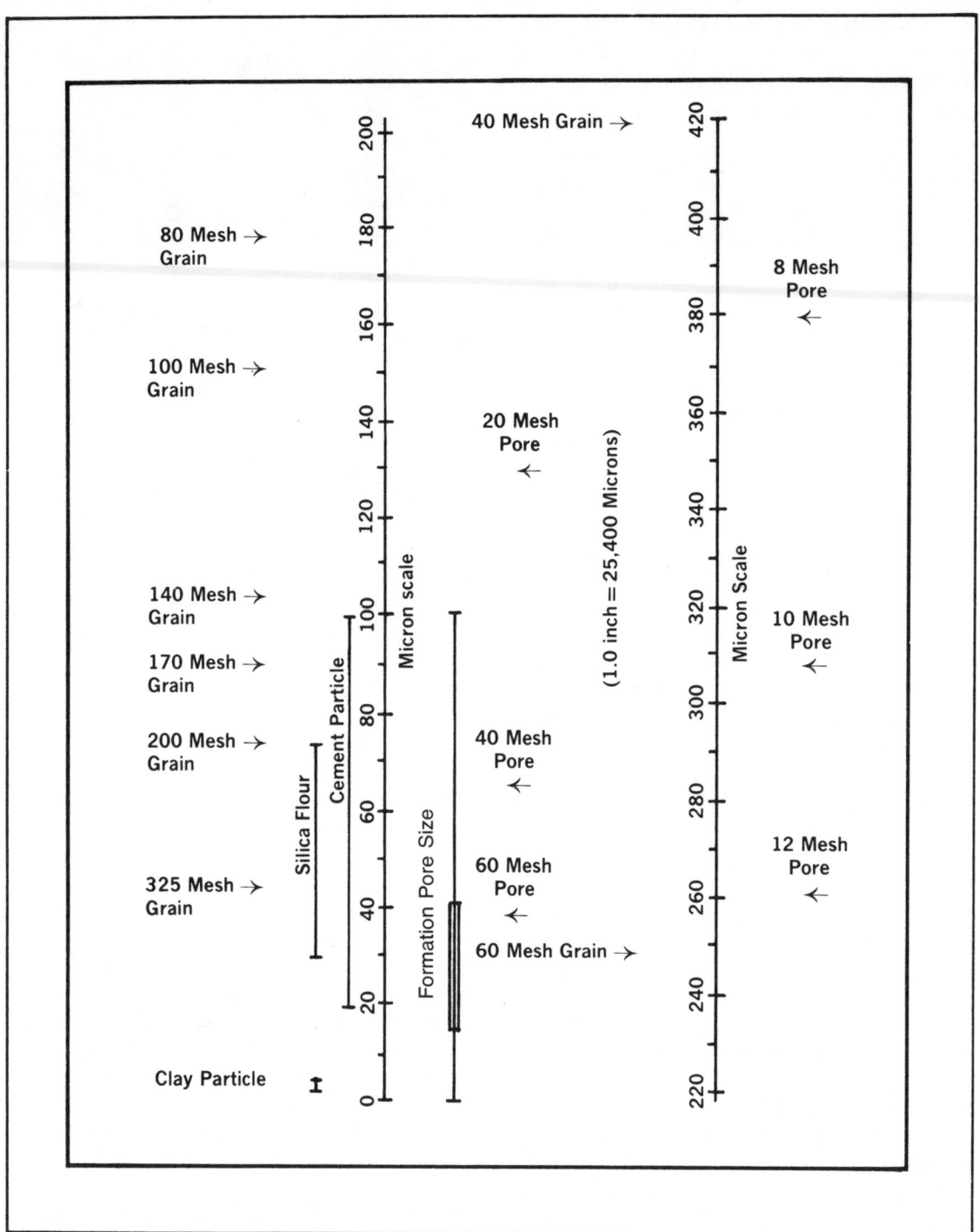

FIG. 5-3—*Formation size relationships.*

these particles are attracted to and are immersed in the water envelope. Thus, their movement and plugging effect primarily occurs with the flow of water.

At low water saturations where little flow of water occurs, these particles may not cause problems.

However, if these particles become oil-wet or partially oil-wet due to some outside influence, they tend to move with the oil, and the resulting plugging may be much more severe.

Loss of filtrate may produce single-phase flow on cleanup. Following completion or workover operations where the wellbore has been filled with a kill fluid, the formation pore system contains a very high saturation of the kill-fluid filtrate. The initial flow through the pore caverns near the well-bore occurs essentially as a single-phase flow.

A water filtrate produced back at a high rate could cause severe plugging due to bridging of inherent formation particles which, under normal producing conditions, might not be free to move. More important, the filtrate will have carried into the pore system thousands of foreign particles.

Thus, as the well is put on production, the pore system around the wellbore will be loaded with moveable inherent and induced particles.

Plugging by particles is rate sensitive. Work by Krueger et al has shown that the extent of permeability reduction due to particle-bridging at pore constrictions depends on flow rate. At high rates, randomly-dispersed particles apparently tend to interfere with each other as they approach pore constrictions and finally bridge.

At low flow rates, however, particles are in more gentle movement and may either (1) gradually align themselves so that one by one they can work their way through the constriction without bridging, or (2) are displaced in the water envelope to a nonblocking position out of the main flow stream.

In core tests, gradual increases in flow rates provided higher ultimate permeabilities than did one-step increases to the same flow rate. This effect is shown in Figure 5-4.

Once a bridge is formed by high flow rates, efforts to displace it by backflow have been only sporadically successful.

Damage Reduction—An important consideration in reducing formation damage due to particle plugging in the formation is to eliminate all possible sources of particles extraneous to the formation. Well-killing fluids are an obvious source of extraneous particles. Usually, cation content is adjusted to prevent disturbance of formation clays, but, with little fluid loss control, a large amount of fine particles are carried into the formation. Although this cannot be prevented entirely, these steps are in the right direction:

1. Surface fluid tanks and workover tubulars must

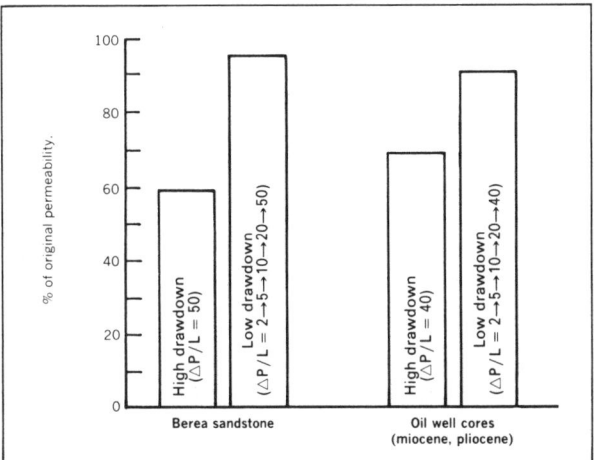

FIG. 5-4—*Effect of drawdown pressure during cleanup on permeability regain of damaged sandstone cores.[9] Permission to publish by The Society of Petroleum Engineers.*

be clean. Apply pipe dope sparingly to the pin with a small brush.

2. Filter all fluids through a 2-micron filter at the surface.

3. Add oxygen scavenger to flow system to prevent formation of iron oxide particles downhole. A sequesterant should be used to prevent formation of iron hydroxide.

4. Reduce hydrostatic pressure of wellbore fluid to a near balanced or even under balanced condition relative to formation pressure.

These precautions will reduce permeability reduction even if large amounts of the clean brine are lost to the formation. If fluid loss control is necessary, powdered calcium carbonate sized for bridging can be used. The resulting "mud cake" can be at least partially removed with hydrochloric acid. However, if large amounts of calcium carbonate fines enter the formation pore system, complete removal is limited due to bypassing and fingering of the acid.

It is also important that fine particles as well as the rock surfaces be left in a water-wet condition after completion, workover, or well treatment.

The fact that bridging by particles at pore constrictions is rate sensitive suggests that cleaning of a well after completion or workover using high flow rates should be avoided. Laboratory data, backed up by field experimentation, show that particles free to move in the pore channels around the wellbore can be best removed by initiating production slowly and by gradually increasing production rate until the desired producing rate is reached.

Formation Clays (Inherent Particles)

Occurrence of Clays—Nearly all oil-producing sandstones contain some clays occurring as a coating on individual sand grains and/or discrete particles mixed with the sand. Carbonate rocks may also contain clays. Frequently, however, these clays are encapsulated in the rock matrix and are not seriously affected by invading fluids. A sand that contains 1 to 5% clay is usually termed a "clean" sand. A dirty sand may contain 5 to greater than 20% clay.

Clays most frequently found in hydrocarbon zones are montmorillonite (bentonite), illite, mixed-layer clays (primarily illite-montmorillonite), kaolinite, and chlorite. The name, *Smectite,* seen in current literature refers to *Montmorillonite*.

Along the United States Gulf Coast, montmorillonite and mixed-layer (montmorillonite-illite) clay predominate at shallow to medium depths. At deeper depths (9000–10,000 ft.) montmorillonite percentage decreases and illite, kaolinite and chlorite increase. Some kaolinite and chlorite are found at all depths.

Clay packets are frequently concentrated at junctions of sand grains and are more concentrated near shale lenses. Kaolinite is usually found in isolated spots; montmorillonite in smears throughout the area where it is present; illite in streaks close to shale lenses; and mixed-layer clay in irregular streaks throughout the sand.[13]

Clay Migration—All clay types are capable of migrating when contacted by a foreign water which alters the ionic environment. Examples of foreign waters are filtrate loss from drilling fluids, cement, completion fluids, workover fluids, and stimulation fluids. In the case of montmorillonite and mixed-layer clays, a change in size due to swelling or water retention enhances their probability of migrating.

Figure 5-5 illustrates clay particles in a sandstone system (1) in a stable, flocculated (unexpanded) condition, and (2) in an unstable, deflocculated expanded condition.

A dehydrated clay particle has a diameter of about 4 microns compared to a pore size in an average sandstone of 10 to 100 microns.

It should be remembered that high flow rate alone (even with no change in environmental conditions) is sufficient to cause particle migration.

Thus, anytime a clay (or other fine particle) is pres-

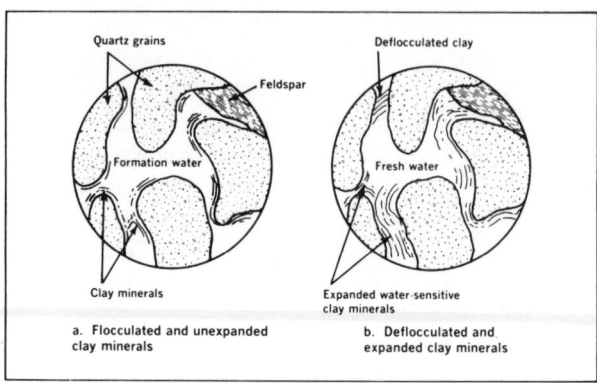

FIG. 5-5—*Occurrence of clay in sandstone.*[5] *Permission to publish by The Petroleum Engineer.*

ent, it can be assumed that permeability damage may occur. The degree of damage will depend on the type and concentration of clays or particles present, their relative position in the rock, the severity of the ionic environmental change, and, to an extent, the rate of fluid flow.

Clay Structure—Clay crystals are thin platelets which under normal conditions of deposition are oriented in deck-of-cards stacks or packets. Surface area of clay packets may be very high. Dispersed montmorillonite may have a surface area of 750 sq m/gm, whereas sand grains have a surface area of 150–200 sq cm/gm.

Crystals consist of various combinations of two building units, the silica tetrahedral sheet and the alumina octahedral sheet. These crystals extend indefinitely in two directions but have a definite thickness ranging from 7 to about 17 angstrom units (one angstrom unit equals one 10 millionth of a millimeter). See Figures 5-6 and 5-7.

Montmorillonite is a three-layer 2:1 lattice structure composed of a central alumina octahedral sheet with a silica tetrahedral sheet on each side (Figure 5-8). The tips of the tetrahedrals are oriented inwardly to contact the octahedrals, producing a chemical bond between adjacent oxygen molecules.

Substitution of Fe^{++} or Mg^{++} for Al^{+++} in the octahedral sheet and Al^{+++} for Si^{++++} in the tetrahedral sheet, results in a charge unbalance of minus 1 per 1.5 unit crystals. This unbalance is neutralized by absorption of Ca^{++}, Mg^{++}, H^+, K^+, and Na^+ on the outside surface of the clay crystals.

The weak bond between crystals permits water molecules to enter and expand the distance between clay crystals. Thus, the width of the basic montmo-

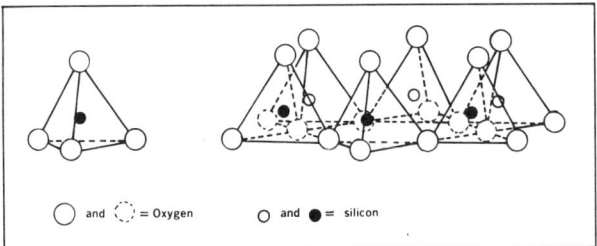

FIG. 5-6—*Single-silica tetrahedron and sheet structure of silica tetrahedrons arranged in hexagonal network.*[8] *Permission to publish by McGraw-Hill Book Co.*

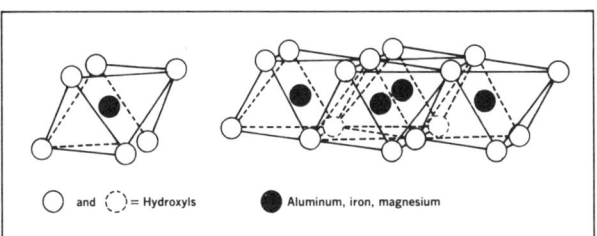

FIG. 5-7—*Single octahedral unit and sheet structure of octahedral units.*[8] *Permission to publish by McGraw-Hill Book Co.*

rillonite crystal, when dry, is about 9.5 angstroms but, exposed to formation waters, can vary from 15.5 to about 17 angstroms depending on the exchangeable cation adsorbed on the structure.

The *Illite* crystal is a 2:1 lattice structure composed of two silica tetrahedral sheets with a central octahedral sheet similar to montmorillonite. Substitution of Al^{+++} for Si^{++++} in the tetrahedral sheet; and Fe^{++}, Fe^{+++}, and Mg^{++} for Al^{+++} in the octahedral sheet results in a charge unbalance of minus 1 per unit crystal. This unbalance is greater than montmorillonite and also occurs mainly in the outer tetrahedral sheets.

The unbalance is neutralized by potassium ions, which position themselves so as to tightly bind successive illite crystal units together and prevent expansion due to entry of water molecules. The thickness of the illite unit is 10 angstroms. However, if the structure is altered by leaching of the potassium ions, it can change to a clay that will expand in contact with water.

Kaolinite is composed of one silica tetrahedral sheet and one alumina octahedral sheet (gibbsite) in a two-layer 1:1 lattice structure. The crystals are held tightly together by hydrogen bonds which prevent expansion of the lattice structure due to penetration of water into the lattice. The kaolinite crystal has a thickness of 7.2 anstroms.

Mixed-layer clays are composed of more than one clay mineral. Irregular mixed-layer clays usually contain montmorillonite and illite, and show definite swelling characteristics. The weak montmorillonite crystal may allow the mixed-layer conglomerate to break apart in large clumps which then have enhanced plugging capabilities. Another example of a mixed layer clay is the montmorillonite-chlorite intergrade. However, this type is not nearly as prevalent as the montmorillonite-illite mixed layer clay.

Effect of Water—Montmorillonite is the only clay that swells by adsorbing ordered water layers between crystals. Mixed-layer clay, which contains montmorillonite, will also swell, but the illite portion of this clay is only slightly swellable. Kaolinite, chlorite, and illite may be classed as slightly swellable clays. Their crystals tend to remain in packets instead of being dispersed like montmorillonite crystals; however, they do adsorb some water.[7]

1. Cation exchange capacity. All clays are negatively charged. With montmorillonite, these charges are predominant on the faces of the clay crystal, while

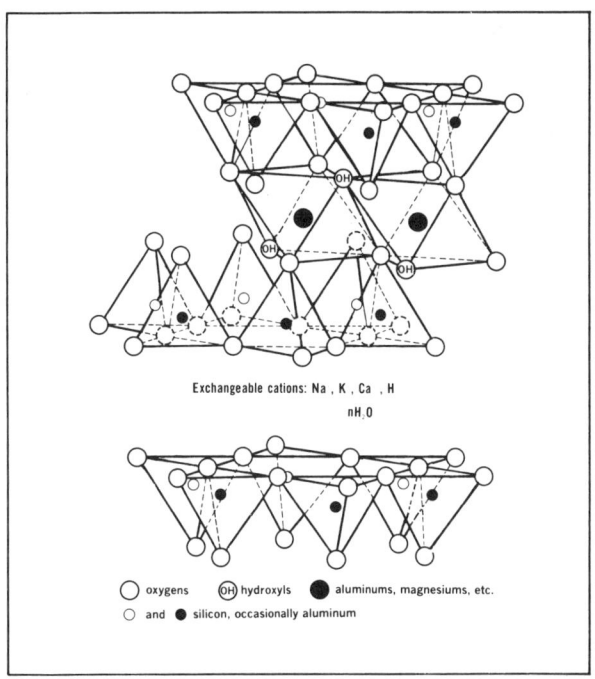

FIG. 5-8—*Structure of montmorillonite according to Hofman, Endell, and Wilm, Marshall, and Hendricks.*[8] *Permission to publish by McGraw-Hill Book Co.*

the edges of the crystal are positively charged. The density of the negative charges on a clay crystal is known as the cation exchange capacity of the clay. Montmorillonite has far greater cation exchange capacity than other types of clay (Table 5-1).

2. Hydration of cations. Swelling of clays in contact with water is due to the hydration of the cations attached to the clay and hydrogen bonding. The degree of swelling depends on the cation adsorbed on the clay and the amount of salts dissolved in the water contacting the clay. When calcium is the exchangeable cation, the clay will adsorb a well-ordered water layer only a few molecules thick. However, when sodium is the exchangeable cation, the clay will adsorb a poorly-oriented water layer on top of the well-ordered water, which is much thicker.

3. Effect of cation type and concentration. When a montmorillonite clay in equilibrium with a particular formation brine is contacted by water having different salts, a cation exchange may occur. Calcium, a divalent ion, is quite effective in replacing monovalent ions, particularly sodium. Potassium is more effective than sodium in replacing calcium.

This cation exchange may cause the size of the clay particle to change. A calcium montmorillonite may increase in size (deflocculate) when contacted with water containing a low concentration of sodium ions. Conversely, flocculation may occur when a sodium clay is contacted by water containing a sufficient concentration of the calcium ions. A lower concentration of the divalent calcium ion is required to cause flocculation than the monovalent sodium or potassium ions. Figure 5-9 shows the degree of swelling of montmorillonite under various salt conditions at atmospheric pressure. Calcium montmorillonite has four layers of water under all conditions. Potassium montmorillonite undergoes very little swelling when contacted with water containing at least 0.4% potassium chloride.

A much higher concentration of sodium chloride is required to minimize the size of the sodium montmorillonite particle than with either calcium or potassium chloride. It should be noted that recent work shows pressure is a significant factor in limiting the water absorptions of montmorillonite.

4. Clay dispersion affected by pH. Scanning electron microscope studies have confirmed the effect of pH on clay particle disturbance. Clay particles in core pore spaces were significantly disturbed and thus made mobile in contact with 8 pH fluid. This effect was more noticeable when contacted with 10 pH fluid. Virtually no disturbance was noted when similar core samples were contacted with a 4 pH fluid.[15]

Control of Clay Damage—When clay particles in sandstones are rearranged or disturbed in any manner, it is usually impossible to restore the original permeability; thus, formation damage due to clays must be prevented rather than cured.

X-ray diffraction tests can easily determine the type and amount of clay in a particular sandstone. These tests can be used to indicate which formations warrant particular measures to avoid formation damage.

The position of clays in the rock is also important. This can be quickly determined by the use of dyes which exhibit characteristic colors when absorbed by different types of clays. Where detailed studies are warranted, the scanning electron microscope permits

TABLE 5-1
Cation-Exchange Capacity of Several Clays

Clay	Range of cation-exchange capacity
Montmorillonite	80 to 150
Illite	10 to 40
Kaolinite	3 to 15
Chlorite	10 to 40

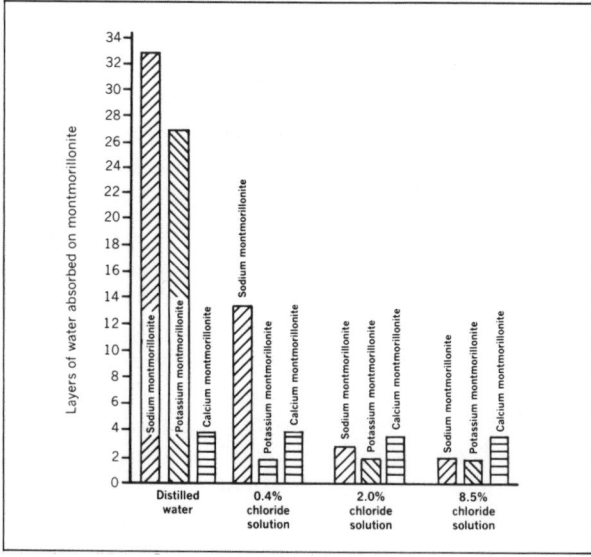

FIG. 5-9—*Degree of swelling exhibited by montmorillonite clay in various salt solutions. Courtesy of Halliburton Services.*

us to actually "see" the effect of certain actions on clays.

In virgin formation, the degree of swelling of a clay particle is in equilibrium with the type and concentration of salts in the connate water. Thus, clean formation water used in workover operations should not disturb this balance. If clean formation water is not available, it is necessary to synthesize an economical substitute.

Laboratory work has shown that, for many sandstone formations, a solution of 2.0% potassium chloride appears to be an ideal fluid from an economic point of view since this concentration of KCl is sufficient to minimize clay disturbance under most conditions.

Cement Clays in Place with Polymers—Clay migration can be effectively controlled by a process using the hydroxy-aluminum cation. This large cation effectively replaces all other cations and forms a polymer coating the surface of the clay. This inorganic coating then binds the clay particles together. The polymer is stable in the presence of brines, but will be removed if contacted with acid.

Zirconium oxychloride is used similarly. Zirconium hydrates in a manner that forms poly-nucleated cations which bond strongly to clay particles. This system is stable in the presence of HCl but is removed by contact with HF.

Multi-nuclei organic polymers (Halliburton CLA-STA and Dowell L53) have additional advantages in that they are not removed by HCl and are stable for at least one hour in 3% HF-12% HCl. They can be placed in any aqueous or alcohol fluid, HCl, or sea water. Both leave the clay and sand grains water-wet. In addition, there are no known temperature limitations for their use. They have been effective when placed with 700°F steam.

Asphaltene Plugging

Temperature and pressure reductions accompanying flow of crude oil and containing appreciable quantities of asphaltic or paraffinic material may result in deposition of these materials in the formation.

Deposition may reduce formation permeability by blocking pore spaces or by causing the formation to become oil-wet.

Plugging from asphaltenes and paraffins can also result from injecting stock tank crude into the formation.

Diagnostic procedures include lab tests on bottom-hole crude samples. Flow tests on cores using stock tank oil can show if particular crude should not be used as frac fluid or workover fluid.

REDUCED RELATIVE PERMEABILITY
Increased Water Saturation

Increased water saturation near the wellbore results from filtrate invasion or fingering or coning of formation water. Filtrate invasion is normally termed "water blockage." The extent of oil productivity reduction depends on the degree of water saturation and the radius of the affected area.

Figure 5-10 shows a typical oil-water relative permeability relationship. Permeability of this particular core to a single fluid is 214 md. Relative permeability to the non-wetting phase (oil) is shown by the curve labeled k_{ro}. Relative permeability to the wetting phase (water) is shown by the curve labeled k_{rw}. With an oil-water system (water being the wetting phase), an increase in water saturation from 30% to 50% reduces oil permeability from 135 md to 28 md in this core.

Oil Wetting

Oil wetting can result from surface-active materials carried in drilling or workover fluid or in various well treating fluids. Most cationic surfactants and certain nonionic surfactants cause silicate rock surfaces to become oil-wet.

Using the relative permeability relation of Figure 5-10 with a water saturation of 35% changing from a water-wet to an oil-wet condition, would reduce permeability to oil from 100 md (water-wet) to 40 md (oil-wet). Again, the effect of this change on well productivity would depend on the radius of the affected area as well as the reduction in permeability to oil. Percentagewise, greater reduction in permeability occurs in lower permeability rocks.

Corrective and Preventive Measures

A simple water block is largely self-correcting as the water-oil ratio will increase immediately after the block is formed, then decline as water is produced. Surfactant treatment may speed removal of blockage by reducing interfacial tension between the water and oil.

Coning or fingering can be distinguished from a water block by noting that water-oil ratio declines

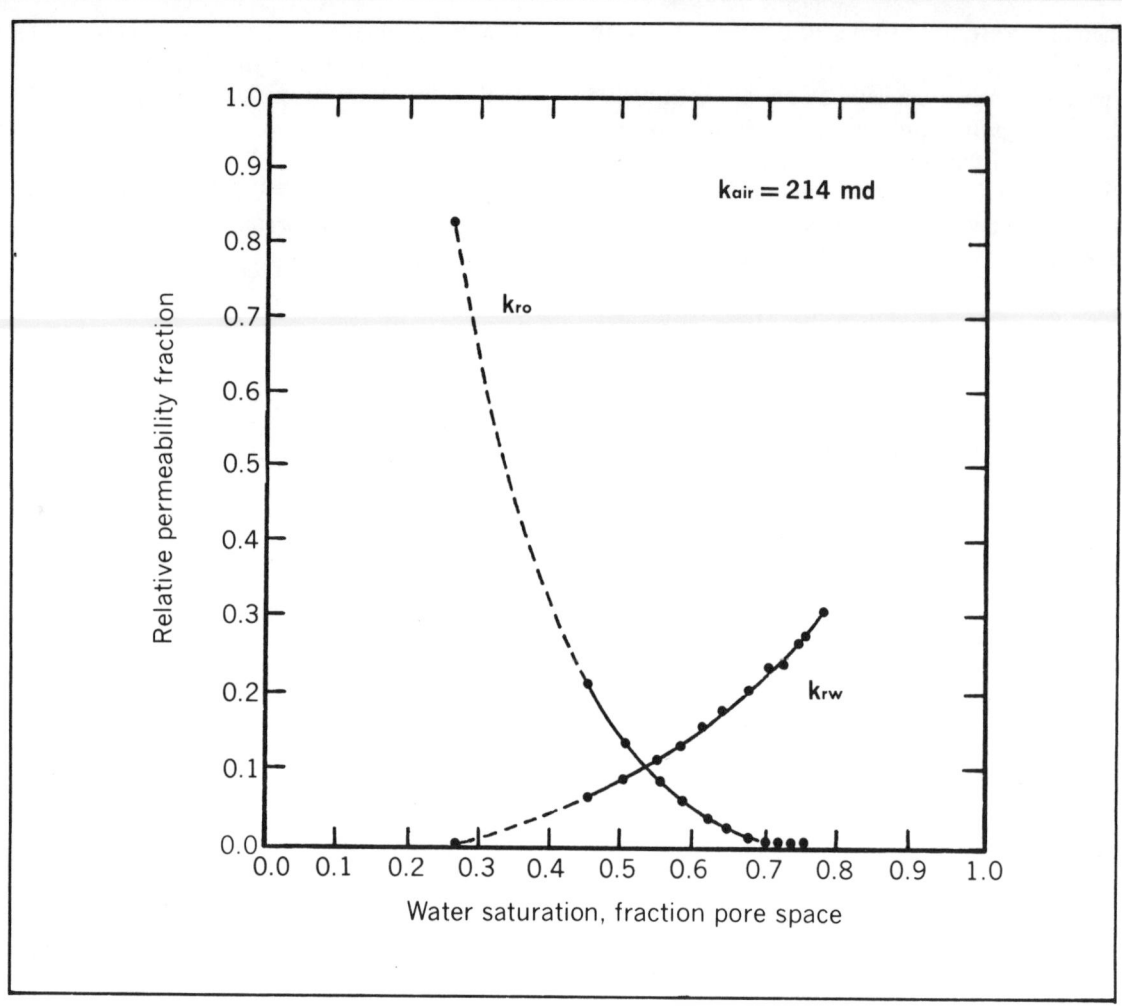

FIG. 5-10—*Oil-water relative permeability (water-wet core).*[10] Permission to publish by The Society of Petroleum Engineers.

with a water block, whereas the ratio remains constant or increases with coning or fingering.

Coning or fingering is usually rate sensitive; thus, reduced production rates should reduce percentage of water production.

Oil wetting must usually be corrected by surfactant treatment. The surfactant should be selected on the basis of laboratory tests or previous experience. The treatment procedure must be designed to provide effective coverage of the zone. (See Chapter 6, Surfactants.)

Preventive Measures Are the Key—Since sand grains, once oil-wet, are sometimes difficult to restore to a water-wet condition, the most effective measures are preventive rather than corrective. Some cationic oil-wetting surfactants are almost impossible to remove from silica surfaces by any corrective treatment.

The best preventive measure is to run laboratory or even simple field tests to determine the wetting characteristics of prospective well-treating fluids, particularly when these fluids contain such surfactant materials as corrosion inhibitors, cationic emulsion breakers or oil-treating chemicals, bactericides, or scale inhibitors. These tests are described in detail in Chapter 6, Surfactants.

If a particular oil-wetting surfactant must be used, then perhaps the formation sand grains can be protected by a pre-treatment of a strong water-wetting surfactant or by a fluid such as mutual solvent, which apparently plates out on silica surfaces to prevent subsequent contact by surfactants.

INCREASED FLUID VISCOSITY

Plugging of the formation may occur due to the presence of emulsions in the pores of the formation.

In a radial flow situation, the extent of productivity reduction depends on viscosity of the emulsion and the radius of the affected area. Water-in-oil emulsions generally exhibit viscosities many times higher than the viscosities of oil-in-water emulsions.

Present evidence indicates that rarely does indigenous oil and water create emulsion blocks. Emulsion blocks probably occur when oil injected into formation becomes emulsified with formation water or when extraneous water enters the formation and becomes mixed with the oil phase. Usually, energy is required to form an emulsion, and a stabilizing mechanism is required to maintain the emulsion. Required energy exists in the restricted flow path areas around the wellbore where flow from all directions converges to move in toward a perforation.

Emulsions are stabilized by surface active materials and by small solid particles such as formation fines, drilling or completion fluid clays, or solid hydrocarbon particles.

Cationic surfactants (corrosion or scale inhibitors, biocides, and even surface emulsion breakers) often tend to stabilize water-in-oil emulsions.

The presence and the character of "fines" contribute significantly to emulsion stability. These fines may occur because of the character of the formation—or may be released as a result of a stimulation treatment or contact with a foreign fluid. Generally, fine-particle wettability is an important factor in emulsion stability and in determining the continuous phase of the emulsion. Strongly water-wet fines tend to reduce emulsion stability.

Wettability of the formation is a significant factor in emulsion stability. Emusions exhibit much greater stability and viscosity in strongly oil-wet formations. In lab experiments, certain lower gravity crude oils are capable of forming emulsions with brines in capillaries, even where no agitation or mixing of fluids occurs. The same crudes, normally those with significant asphaltene content, can also develop a stiff organic film between the crude oils and brines or acid solutions. Both the spontaneous emulsion and the organic film can cause production blockage. Such blockage can occur in formation pore spaces when an aqueous solution enters the rock, and subsequently, the fluid remains in a static condition for a period of time.

Emulsion blocks exhibit a "check valve" effect which can be detected by comparing injectivity and productivity tests.

Emulsion blockage is normally treated with surfactants. Emulsion formation can usually be prevented by including a carefully selected surfactant in well treatments. (See Chapter 6, Surfactants.)

DIAGNOSIS OF FORMATION DAMAGE

It is usually possible to determine that formation damage or "skin effect" exists in a particular well. This can be done through well tests such as injectivity or productivity tests. Analysis of pressure buildup or fall-off tests may indicate the relative magnitude of the damage or skin effect. Production logging surveys may show zones not contributing to the total flow stream.

Comparison of productivity of the subject well with productivities of surrounding wells can provide clues. It is, of course, first necessary to rule out mechanical problems such as sand accumulation in the wellbore or artificial lift difficulties.

Diagnosis of the specific cause of formation damage is many times hampered by lack of sufficiently detailed information on the characteristics of the reservoir rock and reservoir fluids and the characteristics of well control or treating fluids.

Careful examination of well completion reports or workover reports is sometimes helpful. Liberal reading between the lines is usually necessary to tie down significant clues. This can be done effectively only if the "detective" is well grounded in well operations in the field as well as being knowledgeable on formation damage mechanisms.

Production history of the specific well is often a key diagnostic tool. The following production history examples, Figures 5-11, 5-12, and 5-13, have been prepared to show well behavior as influenced by specific types of damage problems.

REFERENCES

1. Glenn, E. E.; Slusser, M. L.; and Huitt, J. L.: "Factors Affecting Well Productivity—Parts I & II," J. Pet. Tech, May 1957, p. 126.

2. Monaghan, P. H., Salathiel, Morgan, and Kaiser: "Laboratory Studies of Formation Damage in Sands Containing Clays," SPE TP 8076, Fall meeting, 1958.

3. Graham, J. W.; Monaghan, P. H.; and Osoba, J. S.: "Influence of Propping Sand Wettability on Productivity of Hydraulically Fractured Oil Wells," Petroleum Transactions, AIME, Vol. 216, 1959, p. 324.

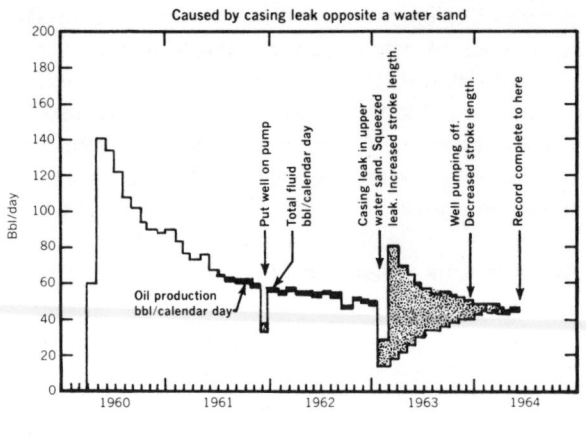
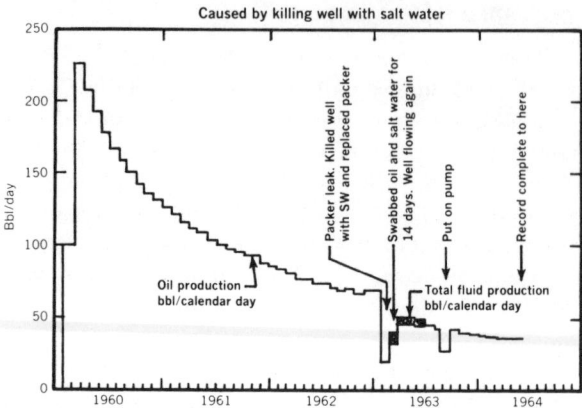
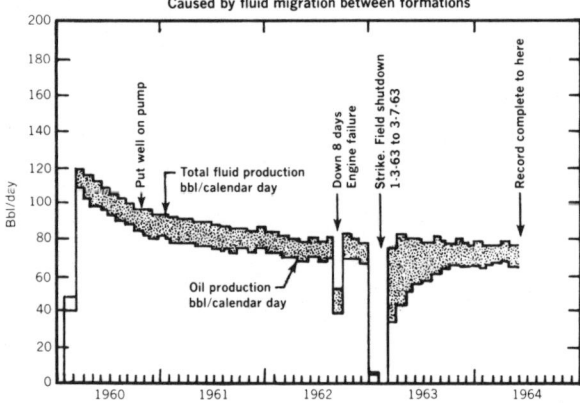
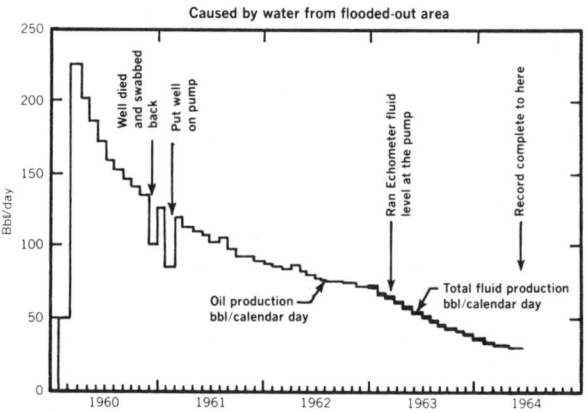

FIG. 5-11—*Simple water blocks.*

FIG. 5-12—*Emulsion blocks.*

4. "Influence of Chemical Composition of Water on Clay Blocking of Permeability," J. Pet. Tech., April 1964.

5. Moore, J. E.: "Clay Mineralogy Problems in Oil Recovery," The Petroleum Engineer, Feb. 1960.

6. Moore, J. E.: "How to Combat Swelling Clays," The Petroleum Engineer, March 1960.

7. Black, H. N.; and Hower, W. F.: "Advantageous Use of Potassium Chloride Water for Fracturing Water-Sensitive Formations," API Paper 850-39-F, 1965.

8. Grim, R. E.: *Clay Mineralogy*, McGraw-Hill, 1953.

9. Krueger, R. F.; Vogel, L. C.; and Fischer, P. W.: "Effect of Pressure Drawdown on the Clean-Up of Clay or Silt-Blocked Sandstone," J. Pet. Tech., March 1967, p. 397.

10. Morgan, J. T.; and Gordon, D. T.: "Influence of Pore Geometry on Water-Oil Relative Permeability," J. Pet. Tech., Oct. 1970, p. 1199.

11. Gidley, J. L.: "Stimulation of Sandstone Formations with the Acid-Mutual Solvent Method," J. Pet. Tech., May 1971, p. 1551.

12. Allen, T. O.: "Creative Task Force Attack on Profit Loss Due to Formation Damage," SPE Paper 4658, Oct. 1973.

13. Hower, W. F.: "Influence of Clays on the Production

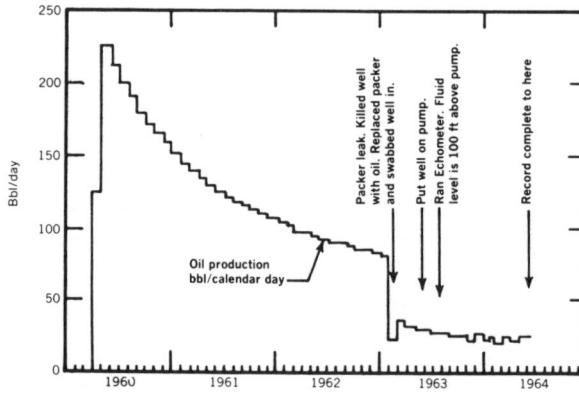

FIG. 5-13—*Insolubles in crude oil.*

of Hydrocarbons," SPE Paper 4785, New Orleans, Feb. 1974.

14. Holub, R. W.; Maly, G. P.; Noel, R. P.; and Weinbrandt, R. M.: "Scanning Electron Microscope Pictures of Reservoir Rocks Reveal Ways to Increase Oil Production," SPE Paper 4787, New Orleans, Feb. 1974.

15. Simon, D. E.; McDaniel, B. W.; and Coon, R. M.;

"Evaluation of Fluid pH Effects on Low Permeabiity Sandstones," SPE Paper 6010, New Orleans, Oct. 1976.

16. Reed, M. G.: "Formation Permeability Damage of Mica Alteration and Carbonate Dissolution," SPE 6009, New Orleans, Oct. 1976.

17. Neasham, J. W.: "The Morphology of Dispersed Clay in Sandstone Reservoirs and its Effect on Sandstone Shaleness, Pore Space, and Fluid Flow Properties," SPE No. 6558, October 1977.

18. Muecke, T. W.: "Formation Fines and Factors Controlling Their Movement in Porous Media," SPE Paper 7007, Third Annual Formation Damage Symposium, Lafayette, La., February, 1978.

19. Hill, Donald G.: "Clay Stabilization—Criteria for Best Performance," SPE Paper No. 10656, Fifth SPE Symposium on Formation Damage Control, March, 1982.

20. Gabriel, G. A., and Inamadar, G. R.: "An Experimental Investigation of Fines Migration in Porous Media," SPE 12168, Oct., 1983.

Appendix

Formation Damage During Specific Well Operations

1. *Damage during drilling of oil and gas zones in wildcat or development wells.*
   a. Mud solids may block pores, vugs, and natural or induced fractures.
   b. Mud filtrate invasion into oil or gas zones may oil-wet the formation and cause water or emulsion blocks. The filtrate may also cause the clays or other fines to flocculate, disperse, swell, shrink, or move, and block the formation.
   c. Pores or fractures near the wellbore may be sealed by the trowelling action of the bit, drill collars and drill pipe.
2. *Damage during casing and cementing*
   a. Cement or mud solids may plug large pores, vugs, and natural or induced fractures.
   b. Chemical flushes used to scour hole ahead of cement may cause changes in clays in the producing formation.
   c. Filtrate from high fluid loss cement slurries may bring about changes in the producing formation.
3. *Damage during completion*
   a. Damage During Perforating
      (1) Perforations may be plugged with shaped charge debris and solids from perforating fluids.
      (2) The formation around the perforation is crushed and compacted by perforating process. If this essentially zero permeability zone is held in place by solids in perforation tunnel, the perforations may be completely blocked.
   b. Damage While Running Tubing and Packer
      (1) If returns are lost while running tubing, solids in the well fluid may plug any fracture system near the wellbore.
      (2) Perforations may be plugged if solids are forced into perforations by the hydrostatic differential pressure into the formation.
   c. Damage During Production Initiation
      (1) Damage may be caused by incompatible circulation fluids and by loss of clays or other fines into perforations, formation pores, vugs, and fractures with high fluid loss fluids including oil or water.
      (2) Damage may result from depositing of mill scale, clay, or excess thread dope from tubing collars in perforations when circulating to clean a well.
      (3) Completion fluids containing blown asphalt may cause damage by oil-wetting the formation and by plugging perforations and formation.
      (4) Clean-up of a well at high rates can result in severe plugging within the formation by particles which, for one reason or another, are free to move.
4. *Damage during well stimulation*
   a. Perforations, formation pores, and fractures may be plugged with solids while killing or circulating a well with mud or with unfiltered oil or water. Even filtered fluids may result in plugging due to solids scoured from the tubing, open hole, or casing.
   b. Filtrate from circulating fluids may cause damage. (See 1-b)
   c. Breaking down or fracturing the formation with acid may shrink the mud cake between the sand face and cement or may affect mud channel in the annulus allowing vertical communication of unwanted fluids.
   d. Acidizing sandstone with hydrofluoric acid may leave insoluble precipitates in formation. Properly designed treatment minimizes this effect.
   e. Hydraulic Fracturing
      (1) Propped fractures may be plugged with

frac fluids, solids or frac sand fines.
   (2) Inadequate breakers for high viscosity frac fluids may cause blocking of propped fractures.
   (3) Fluid loss or diverting agents may cause plugging of the perforations, formation pores, or propped fractures.
  f. Fracture Acidizing of Carbonates
   (1) Failure to employ clean compatible fluids may cause plugging of etched-fracture flow channels and adjacent formation matrix.
   (2) Paraffin, asphalt, scale, thread dope, silt, or other solids in the tubing or wellbore may result in plugged perforations, formation, or etched fractures.

5. *Damage caused by cleaning of paraffin or asphalt from the tubing, casing, or wellbore*
  a. When cleaning paraffin or asphalt from a well with hot oil or hot water, the formation and perforations will be plugged unless the melted asphalt or paraffin is swabbed, pumped, or flowed from the well before the wax cools.
  b. While cutting paraffin or asphalt from the tubing, if removed particles are circulated down the tubing and up the annulus, a portion of the scraped material will be pumped into perforations and into pores, vugs, or fractures adjacent the wellbore.

6. *Damage during well servicing or workover*
  a. Essentially all of the same types of damage associated with initial completion (See item 3) may occur during well servicing or workover.
  b. Perforations, formation pores, vugs, or fractures may be plugged with solids when killing or circulating a well with mud or unfiltered oil or water.
  c. Filtrate invasion by incompatible water, oil, or other chemicals may cause water blocks, emulsion blocks, oil wetting the formation, or changes in formation clays.
  d. If necessary to pull a packer, mud or other damaging fluids above the packer may be dumped on the producing zones.
  e. If a well has been previously hydraulically fractured and propped, any solids entering the fractures will tend to bridge between the sand grains or other proppants and cause permanent reduction of fracture flow-capacity. Previously acid-etched fractures in carbonate rock may also be plugged by the introduction of clays, barite, or other debris into the fractures.

7. *Damage during producing phase*
  a. Corrosion inhibitors, scale inhibitors, or paraffin inhibitors usually cause some permeability reduction if allowed to contact the producing or injection zone.
  b. Precipitated scale may plug the wellbore, perforations, and formation if an oil or gas well produces water from the normal producing zone, a channel, or a casing leak.
  c. Asphalt may be deposited around the wellbore in wells producing relatively high viscosity asphaltic oil. Asphalt deposition will cause oil-wetting and as a result, emulsions may form around the wellbore.
  d. Wells in reservoirs nearing pressure depletion are more susceptible than high pressure wells to plugging with paraffin or asphalt.
  e. The wellbore opposite the producing interval in both carbonate and sandstone wells may become plugged with silt, shale, mud, frac sand, or other types of fill.
  f. Screens or gravel packs may become plugged with silt, clay, mud, scale, or other debris.
  g. Sand-consolidated wells may become plugged with silt, mud, or other debris. In addition, the sand consolidating material will reduce formation permeability to varying degrees.

8. *Damage during water injection*
  a. Oil-wetting surfactants in water obtained from stock tanks or heater-treaters may oil-wet the formation around the wellbore. Under these conditions, emulsions may occur in the formation adjacent the wellbore.
  b. The tubing, casing, perforations, screen, gravel packs, formation face, or fractures may be plugged with mud, silt, clay, paraffin, asphalt, emulsions, rust, mill scale, thread dope, scale, scale inhibitors, corrosion inhibitors, or bactericides.

9. *Damage during gas injection*
  a. The wellbore, perforations, formation fractures, vugs, and pores may be plugged with mill scale, thread dope, or other solids scoured by injected gas from injection lines or tubing.
  b. Lubricating oil from the gas compressors may build up an oil saturation around the wellbore, oil-wet the injection zone, and cause an emulsion to form in the formation.
  c. The injection of corrosion inhibitors into gas zones will usually reduce well injectivity or productivity.

Chapter 6 Surfactants for Well Treatments

Surfactant characteristics
Action of surfactant types
Damage susceptible to surfactant treatment
Surfactant selection
Well-treatment techniques

Surfactants, or surface-active agents, are chemicals that can favorably or unfavorably affect the flow of fluids near the wellbore. The use of surfactants should be considered for all well completion, well killing, workover, and well stimulation. To appreciate the role of surfactants, it is necessary to understand the operation of liquids.

In the bulk volume of a liquid, molecules exert a mutual attraction for each other. This force, a combination of Van der Waals' forces and electrostatic forces, is balanced within the bulk of a liquid but exerts "tension" at the surface of the liquid. Similar effects take place between two immiscible liquids, or between a liquid and a rock or metal surface.

CHARACTERISTICS OF SURFACTANTS

A surface-active agent, or surfactant, can be defined as a molecule that seeks out an interface and has the ability to alter prevailing conditions. Chemically, a surfactant has an affinity for both water and oil. The surfactant molecule has two parts—one part that is soluble in oil and another part that is soluble in water. The molecule is thus partially soluble in both water and oil. This promotes the surfactant accumulation at the interface between two liquids, between a liquid and a gas, and between a liquid and a solid. A surfactant with stronger affinity for oil is usually classed as oil soluble, and one with a stronger attraction for water is classed as water soluble. Some surfactants are classed as water or oil dispersible.

Surfactants can bring about the following changes in reservoir fluids and reservoir rocks:

1. Raise or lower surface and interfacial tension
2. Make, break, weaken, or strengthen an emulsion
3. Change the wettability of reservoir rocks and casing, tubing, or flowline
4. Disperse or flocculate clays and other fines

Surfactants have the ability to lower the surface tension of a liquid that is in contact with a gas by adsorbing at the interface between the liquid and gas. Surfactants can also reduce interfacial tension between two immiscible liquids by adsorbing at the interfaces between the liquids, and can reduce interfacial tension and change contact angles by adsorbing at interfaces between a liquid and a solid.

Because the primary action of most surfactants is due to electrostatic forces, a surfactant is classified by the ionic nature of the molecule's water-soluble group. Schematically, the water-soluble part of the molecule is represented by a circle and the oil-soluble part as a bar, as shown in Figure 6-1.

Anionic surfactants are organic molecules where the water-soluble group is negatively charged. A model of an anionic is shown in Figure 6-2, where M^+ represents a positive ion such as Na^+.

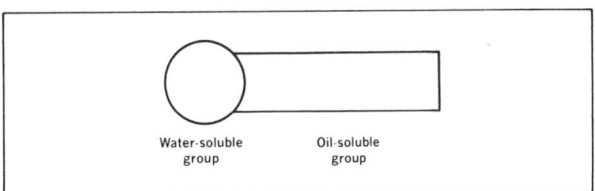

FIG. 6-1—*Surfactant molecule.*

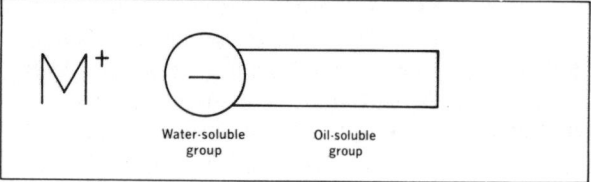

FIG. 6-2—Anionic surfactant.

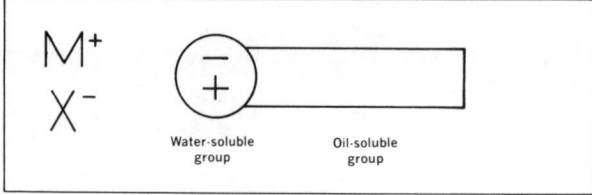

FIG. 6-5—Amphoteric surfactant.

Examples of anionics include sulfates represented as $R - OSO_3$; sulfonates as $R - SO_3$; phosphates as $R - OPO_3$; and phosphonates as $R - PO_3$, where "R" represents the oil-soluble group. The most common anionics are sulfates and sulfonates.

Cationic surfactants are organic molecules where the water-soluble group is positively charged. A model of a cationic is shown in Figure 6-3, where X^- represents a negative ion such as Cl^-. Most cationics are amine compounds such as:

$$\text{Quaternary Ammonium Chloride} \begin{bmatrix} & R_2 & \\ & | & \\ R_1 - & N - & R_3 \\ & | & \\ & R_4 & \end{bmatrix}^+ Cl^-$$

Nonionic Surfactants are organic molecules that do not ionize and, therefore, remain uncharged. A model of a nonionic is shown in Figure 6-4.

Most nonionic surfactants contain water-soluble groups that are polymers of either ethylene oxide or propylene oxide. Examples are polyethylene oxide as $R - O - (CH_2CH_2O)_x H$ and polypropylene oxide as $R - O - [CH_2CH(CH_3)O]_x H$, where "R" represents an oil-soluble group.

Amphoteric Surfactants are organic molecules where the water-soluble group can be either positively charged, negatively charged, or uncharged. The charge on an amphoteric surfactant is dependent upon the pH of the system. A model of an amphoteric is shown in Figure 6-5.

Wettability

Wettability is a descriptive term used to indicate whether a rock or metal surface has the capacity to be preferentially coated with a film of oil or a film of water. Surfactants may adsorb at the interface between the liquid and rock or metal surface and may change the electrical charge on the rock or metal, thereby altering the wettability. Although the surface of a solid can have varying degrees of wettability under normal reservoir conditions, these conditions usually exist:

—Sand and clay are water-wet and have a negative surface charge.

—Limestone and dolomite are water-wet and have a positive surface charge in the pH range of 0 to 8.

Mechanics of Emulsions

Emulsions can occur between two immiscible liquids and may be stable depending on effects that occur at the interface. Energy is required to create the emulsion, and stabilizers must collect at the interface between the liquids to keep the emulsion from breaking. The most significant stabilizers of emulsions are:

1. Fine particles of clay or other materials
2. Asphaltenes
3. Surfactants

Surfactants have the ability to break an emulsion by acting on the stabilizing materials in such a way as to remove them from the interfacial film surrounding an emulsion droplet.

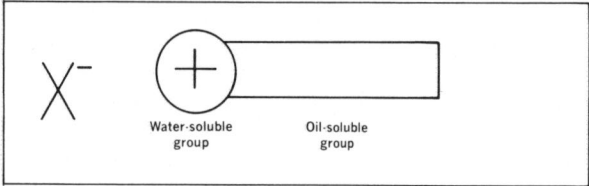

FIG. 6-3—Cationic surfactant.

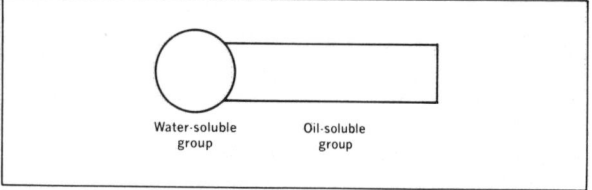

FIG. 6-4—Nonionic surfactant.

USE AND ACTION OF SURFACTANTS

Well-treating surfactants are usually a combination of anionic and nonionic surfactants. Anionic and cationic surfactants should not be used together because the combination may produce an insoluble precipitate. Surfactants may be adsorbed on solids to replace surfactants previously adsorbed and give the solids the wetting characteristic of the stronger surfactant.

Action of Anionic Surfactants—Anionics will normally:

—Water-wet negatively-charged sand, shale, or clay
—Oil-wet limestone or dolomite up to a pH of 8
—Water-wet limestone or dolomite if the pH is 9.5 or above
—Break water-in-oil emulsions
—Emulsify oil in water
—Disperse clays or fines in water

Action of Cationic Surfactants—Cationics will normally:

—Oil-wet sand, shale, or clay
—Water-wet limestone or dolomite up to a pH of 8
—Oil-wet limestone or dolomite if the pH is 9.5 or above
—Break oil-in-water emulsions
—Emulsify water in oil
—Disperse clays or fines in oil
—Flocculate clays in water

Action of Nonionic Surfactants—Nonionics are probably the most versatile of all surfactants for well stimulation because these molecules do not ionize. In combination with other chemicals, nonionics can add such features as high tolerance to hard water and acid pH.

Most nonionics are derivatives of ethylene oxide or propylene oxide-ethylene oxide blends. The water solubility of nonionics is caused by hydrogen bonding or attraction of water for the oxygen of the ethylene oxide. This attraction is reduced at elevated temperatures and/or high salt concentrations, causing most nonionic surfactants to separate from solution.

There is an increasing use of nonionics to lower surface tension and to break emulsions. This is accomplished without changing the static charge of the formation rocks involved.

Action of Amphoteric Surfactants—These are molecules containing both acidic and basic groups. In acidic pH, the basic part of the molecule becomes ionized and gives surface activity to the molecule. In basic pH, the acidic part of the molecule is "self-neutralized" and usually has less surface activity than at other pH values. There is limited use of amphoterics; however, some are being used as corrosion inhibitors.

Summary of Wetting Action by Anionic and Cationic Surfactants

Although sophisticated chemical research can bring about radical changes in the nature and action of surfactants, the usual action by surfactants of each class may be accepted with reasonable confidence.

Because anionics and cationics are the primary surfactants used to change the electrostatic forces involved in the association of liquids and solids, the usual wetting action of these two classes of surfactants in the normal pH range is summarized for quick reference in Figure 6-6.

FORMATION DAMAGE SUSCEPTIBLE TO SURFACTANT TREATMENT

A number of types of formation damage can be prevented or alleviated with surfactants. The most effective approach is to use surfactants to prevent damage that might otherwise occur during nearly all phases of well operations including drilling, well completion, well killing, workover, and well stimulation. However, extreme care must be exercised in the selection and use of surfactants. A specific surfactant may prevent or alleviate one type of damage and create another type of damage. Types of damage which may be prevented, alleviated, or aggravated by surfactants are:

1. Oil-wetting of formation rock
2. Water blocks
3. Viscous emulsion blocks
4. Interfacial film or membrane blocks

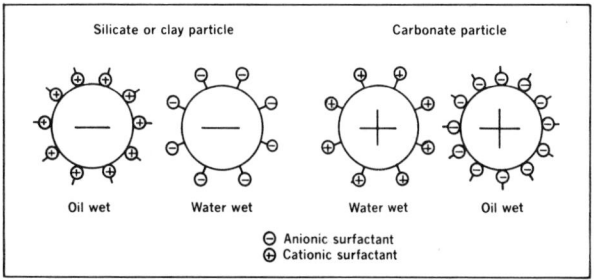

FIG. 6-6—*Usual wetting action in normal pH range for anionic and cationic surfactants.*

5. Particle blocks caused by dispersion, flocculation or movement of clays or other fines

6. Flow restriction caused by high surface or interfacial tension of liquid

Oil-Wetting

Oil-wetting a normally water-wet formation can reduce permeability to oil by 15 to 85% with an average reduction of about 40%. When the formation near the wellbore becomes oil-wet, oil is preferentially attracted to the surface of reservoir rock. This appreciably increases the thickness of the film coating the reservoir rock and reduces the size of flow paths from the reservoir as well as decreasing the relative permeability to oil.

Gas wells are also adversely affected by oil-wetting the formation. Oil-wetting a reservoir rock can result in severe water or emulsion blocking.

Sources of oil-wetting in oil and gas wells are:

— Surfactants in some drilling mud filtrates and workover and well stimulation fluids may oil-wet the formation.
— Corrosion inhibitors and bactericides are usually cationic surfactants, which will oil-wet sandstone and clay.
— Many stock tank or heater-treater emulsion breakers are cationics, which will oil-wet sand and clay.
— Oil-base mud containing blown asphalt will oil-wet sandstone, clays, or carbonates. Oil emulsion muds usually contain considerable cationic surfactants and may oil-wet sandstones and clays. Laboratory tests indicate that some available oil emulsion muds exhibit strong oil-wetting characteristics, whereas, others appear to be less oil-wetting. Other drilling or completion fluids may oil-wet reservoir rock under specific pH conditions.

A strong water-wetting surfactant may convert some oil-wetted surfaces to water-wet surfaces. This will enlarge flow paths to oil and restore oil permeability to that of the undamaged water-wet matrix around the wellbore. However, cationic surfactants are extremely difficult to remove from sandstone and clay. The best approach is to avoid contact of formation sands and clay with cationic surfactants.

Water Blocks[10]

When large quantities of water are lost to a partially oil-wet formation, the return of original oil or gas productivity may be slow, especially in partially pressure-depleted reservoirs. This problem is caused by a temporary reduction in relative permeability near the wellbore to oil or gas. It is usually self-correcting but may persist for months or years.

Water blocking can usually be prevented by adding to all injected well fluids about 0.1 to 0.2% by volume of surfactant selected to lower surface and interfacial tension. The surfactant should also water-wet the formation and prevent emulsions.

Cleanup of a water-blocked well can be accelerated by injecting into the formation a solution of 1% to 3% by volume of a selected surfactant in clean water or oil. The surfactant should lower surface and interfacial tension and leave the formation in a water-wet condition. Removing damage requires many times the volume of surfactant required to prevent damage.

Emulsion Blocks

Viscous emulsions of oil and water in the formation near the wellbore can drastically reduce the productivity of oil or gas wells. There is some question as to whether emulsions can form in sandstone formations unless the formation is oil-wet or unless emulsifying surfactants are present. In carbonates, emulsions are usually associated with fracture acidizing.

Emulsions in the formation can be broken by injecting demulsifying surfactants into the formation, provided intimate contact is made between the surfactant and each emulsion droplet. To break the emulsion, the surfactant must be absorbed on the surface of the emulsion droplets and lower interfacial tension so that emulsion droplets can coalesce.

Breaking an emulsion in the formation usually requires the injection of 2% to 3% by volume of demulsifying surfactant in clean water or clean oil. The treatment volumes should be at least equal to or greater than the volume of the damaging fluid previously lost to the formation. The amount of surfactant required to remove an emulsion block will usually be 20 to 30 times the volume of surfactant required to prevent the block.

Diagnosis of emulsion block—It has been proved that if an emulsion block exists, the calculated average well permeability as determined by injectivity tests will be manyfold higher than the average permeability determined from production tests. This provides a reliable way to predict emulsion blocks and is frequently called the "check valve" effect. If an emulsion-blocked well is producing water, increasing or decreasing producing rates will not appreciably change water percentages.

Interfacial Films or Membranes

Film-forming material, including surfactants, can be adsorbed at the oil-water interface and cause formation plugging. Interfacial films are intimately related to oil-wetting and emulsion properties of crudes.

Fines, clays, and asphaltenes appreciably increase film strength. An increase in percent salt in a solution increases film strength. Oil exposed to air may form tough films. Specific surfactants may increase film strength in a particular oil-water system.

Surfactants may sometimes cause films to resolubilize in oil and thus reduce formation blocking. The use of solvents as a carrier for surfactants is usually beneficial in removing tough films.

Particle Blocks

As a rule, it is desirable to maintain formation clays in the original condition in the reservoir. However, an oil or gas productive formation may be blocked by transmission of clays into the formation in water or oil or mud filtrate. Dispersing, flocculating, breaking loose, or moving clays probably causes more damage to wells than the swelling of clays. Careful analysis should be made of any suspected clay block. For example, if properly diagnosed as a clay-dispersion problem, a nonionic surfactant could be selected to flocculate the clays and reduce the block. However, if improperly diagnosed and the problem were actually clay-flocculation, a flocculating surfactant would probably cause the block to be more severe.

Dispersion of Clays—Clay dispersion is a frequent cause of formation damage:

—Anionic surfactants will disperse clays in acid solutions.
—High pH fluids tend to disperse clays.
—If surfactants are employed to disperse clays in mud-plugged perforations, high concentrations of these same surfactants may damage the formation by dispersing clays in the formation.

Flocculation of Clays—Clay flocculation may sometimes reduce or increase formation damage:

—Specific nonionic surfactants may be used to flocculate clays.

—Acid and other low pH fluids tend to flocculate clays.

Change of Particle Size Affects Formation Damage

Oil-wetting of clays with cationic surfactants greatly increases the size of clay particles, and thereby increases the severity of clay-blocking. Because cationic surfactants are so difficult to remove from clays and sand, the use of cationics in injection or circulation fluids should be avoided in sandstone wells. However, a mutual solvent preflush may prevent cationic adsorption on clay and sand.

Hydrated sodium clays may be reduced in size by an HCl acid treatment. When a hydrated sodium clay reacts with acid, the hydrogen ions (H^+) will replace the sodium ions (Na^+) by ion exchange. Since the hydrogen clay is a smaller particle than the original hydrated sodium clay, treating with HCl tends to increase formation permeability. However, the HCl may have caused flocculation of clays, thus creating damage. An anionic surfactant should usually be used when acidizing sandstone with HCl to prevent clay flocculation as well as to prevent emulsions and to increase the water-wettability of the formation.

High Surface and Interfacial Tension

High surface tension of liquids near the wellbore will reduce the flow of oil and gas into a well and increase well cleanup time. Selected surfactants may be added to well completion, workover, or well stimulation fluid to maintain low surface and interfacial tension in fluids around the wellbore.

Lowering surface and interfacial tension will aid in preventing emulsion and water blocks and will accelerate well cleanup.

Key to Well Treating

A correct surfactant, which is designed for specific well conditions, can lower surface and interfacial tension, favorably change wettability, break or prevent emulsions, prevent or remove water blocks, and cause clays to disperse, flocculate, or remain in place as desired. In summary, the proper use of surfactants during well completion, workover, or well stimulation can prevent or remove many types of damage and result in increased productivity or injectivity.

Susceptibility to Surfactant-Related Damage

Sandstone wells are usually more susceptible to damage caused by oil-wetting, emulsion blocks, water blocks, and changes in clays than limestone wells.

1. Since most cationic surfactants will oil-wet clay and sandstone and stabilize water-in-oil emulsions, caution should be exercised in the use of cationics in sandstone reservoirs. This precaution applies to sandstone acidizing and to all other fluid injection or circulation activities.
2. Organic corrosion inhibitors and bactericides are usually cationic surfactants. Before squeezing any cationic corrosion inhibitor into a sandstone formation, laboratory tests should be run on formation cores to determine the effect of the specific corrosion inhibitor on formation permeability.
3. Caution should be exercised in using saltwater or oil from treaters or field stock tanks treated with cationic emulsion breakers if emulsion or water-blocking appears to be a problem. The damaging effects of the cationic surfactants may sometimes be overcome by adding carefully selected surfactants and/or mutual solvents.

Asphaltenes and high-molecular-weight hydrocarbons in crude oil are major factors in promoting formation damage. Resins show only weakly-damaging properties. The remaining oil fraction has essentially no damaging properties. Wells producing crude containing more than one percent asphaltenes are the most likely to be associated with formation damage that is amenable to surfactant and solvent stimulation. Low API gravity crude usually contains a high percentage of asphaltenes.

Sandstone wells producing low gravity asphaltic crude are more susceptible to damage from oil-wetting, emulsion-blocking, and water-blocking. A 24 to 72-hr solvent-surfactant soak followed by an HF acid treatment is usually the best treatment for wells that contain a high percentage of asphaltenes. Above 37° API, the asphaltene content in crude is usually too low to promote formation damage.

Emulsion-blocking and water-blocking are usually not a problem in limestone and dolomite wells except during acidizing. A nonemulsifying surfactant should be used as a standard additive in all carbonate acidizing to prevent emulsions. When it is anticipated that appreciable undissolved fines will be released in fracture acidizing of carbonates, it may be desirable to use a suspending agent. In stimulation or well killing operations, most suspending agents are surfactants or polymers.

PREVENTING OR REMOVING DAMAGE

Emphasis in the use of surfactants should be aimed at preventing damage. As previously noted, acidizing can cause emulsions, water blocks, and high surface tension problems in carbonates. Sandstones are even more prone to damage during acidizing caused by water and emulsion blocking, high surface tension, oil-wetting, and clay dispersion or flocculation. A surfactant should be employed on all acidizing jobs and should be selected through tests outlined in API RP 42, revised 1977, "Recommended Practice for Laboratory Testing of Surface Active Agents for Well Stimulation."

Because sandstone wells are more susceptible to damage, all fluids and chemicals that are injected or circulated into sandstone wells during well servicing, workover, well completion, and stimulation should be tested for compatibility with formation fluids.

If laboratory tests show potential damage caused by fluid circulation or injection into a well, surfactants should be selected through laboratory tests to prevent damage.

Mud filtrate on drilling wells should be checked for compatibility with formation fluids. Also, the effect of filtrates on formation clays in sandstone wells should be analyzed.

Fluids compatibility tests, damage prevention tests, and damage removal tests with surfactants are similar to those described in API RP 42 for use during acidizing.

Workover fluid compatibility test—This test is based on API RP 42. To illustrate the required test procedure, assume that reservoir saltwater obtained from nearby field storage tanks will be used as a well-killing fluid for an oil well.

The subsequent steps should be followed in workover fluid compatibility tests:

1. The test equipment required is a high-speed stirrer such as Hamilton-Beach Model 936 with standard disc head or a Sargeant-Welch agitator S-76695; a 400 ml tall form beaker; 100 ml graduated cylinders; a stop watch or timer; and one milliliter graduated syringe.
2. Obtain samples of water to be injected into the well and samples of oil produced from the reservoir. The oil should be free of treating chemicals. Obtain

samples of surfactants for consideration and a small quantity of formation fines or silica flour and untreated bentonite. Under no circumstances should treated bentonite be employed in these tests.

3. Pour 25 ml of saltwater into the 400 ml beaker and then disperse into the water 2.5 grams of pulverized formation fines or 2.5 grams of a 50-50 mixture of silica flour and bentonite.

4. Add 75 ml of produced crude oil to the saltwater and dispersed solids. Stir the solution with a mixer at 14,000 to 18,000 rpm for 30 seconds. Pour the emulsion immediately into a 100 ml graduated cylinder and record the volumes of water breakout in ten minutes and in one hour.

5. If a clean breakout of water is not obtained in one hour, a surfactant is usually required in well killing fluid to prevent damage.

Surfactant-Selection Procedure to Prevent an Emulsion

If a surfactant is needed, tests must be run to determine the best surfactant for job. Tests for the selection of surfactants are similar to the fluid compatibility test previously described. The only difference between the two is that a prospective surfactant, usually 0.1% to 0.2% by volume, is added to oil or water prior to stirring the oil and water with the high speed blender. The following test procedure should be run on several surfactants to determine the best surfactant for the job.

1. The test equipment required is a high-speed stirrer such as Hamilton-Beach Model 936 with standard disc head or a Sargeant-Welch agitator S-76695; a 400 ml tall form beaker for stirring; 100 ml graduated cylinders; a stop watch or timer; and a one milliliter graduated syringe.

2. Obtain samples of water to be injected and samples of oil produced from the reservoir. The oil should be free of treating chemicals. Obtain samples of surfactants considered for use and a small quantity of formation fines or silica flour and untreated bentonite.

3. Pour 25 ml of saltwater into the 400 ml beaker and then disperse into the water 2.5 grams of pulverized formation fines or 2.5 grams of a 50-50 mixture of silica flour and bentonite.

4. Add 75 ml of produced crude oil to the saltwater and dispersed solids. Add the selected surfactant, usually 0.1 to 0.2% by volume. Stir the solution with a mixer at 14,000 to 18,000 rpm for 30 seconds. Pour the emulsion immediately into a 100 ml graduated cylinder and record the volumes of water breakout at various time intervals.

5. Tests should be repeated, using several surfactants and varying the percentage of surfactants to determine the most effective surfactant at the lowest cost. If the surfactant is effective, most emulsions will break within a few minutes after stirring is stopped.

Selection of an Emulsion Breaking Surfactant

If an emulsion block is indicated on a completed workover or well completion, emulsion breaking tests should be made using selected surfactants and samples of produced emulsion. If samples of the emulsion are not available, the alternative is to reconstruct, in the laboratory, a similar emulsion using the formation fines, fluids, and chemicals that caused the downhole emulsion.

The emulsion breaking tests involve adding a surfactant, usually 2 or 3% by volume, to the emulsion and stirring it with a high speed blender for 30 seconds. Pour into a graduated cylinder and record the percent of water breakout in one hour and in 24 hours. It is usually advisable to run several emulsion breaking tests using different surfactants to select the most effective emulsion breaker. Table 6-1 shows results of typical, emulsion breaking tests with seven different surfactants.

Systems that will not form stable emulsions usually will not require surfactants in the treating solutions. Conversely, if reconstructed systems involved in previous well treatments show stable emulsions, well damage may be due to emulsion blocking of the formation.

Wettability Tests Based on API RP 42

Various wettability measurements are described in API RP-42; however, only the visual wettability test is described here. This test is applicable for field use.

Equipment and materials required for test are:

—A 4 oz wide mouth bottle or 150 ml beakers
—Kerosene and/or crude oils to be tested
—Aqueous test fluid (water, brine, or acid)
—Clean sand and/or limestone particles, 40–60 mesh.

The procedure for testing oil soluble or oil dispersible surfactants is:

1. Place 50 ml of oil containing a surfactant at a desired concentration (usually one percent or less for

TABLE 6-1
Emulsion-Breaking Tests on a Low Gravity Asphaltic Crude and Brine

	% Emulsion broken					
	Well no. 1		Well no. 2		Well no. 3	
Agent	1 hour	24 hours	1 hour	24 hours	1 hour	24 hours
A	80	100	100	100	100	100
B	70	100	100	100	80	100
C	40	60	80	100	50	100
D	20	20	20	50	10	40
E	20	20	20	40	10	40
F	20	20	20	40	10	30
G	10	20	0	10	0	20

use in well treatments) into a bottle and add 10 cc of test sand.

2. After 30 minutes, slowly pour 50 ml of water into the bottle, taking care to prevent excessive mixing and emulsification.

3. Observe the relative dispersibility of the particles and their tendency to form clumps in both aqueous and oil phases by lifting a small quantity of sand with a semi-micro spoon spatula into the oil phase and allowing the sand to fall back into the water.

The procedure for testing water soluble or water dispersible surfactants is:

1. Place 50 ml of a water solution containing the surfactant at the desired concentration in a bottle and add 10 cc of test sand.

2. After 30 minutes, decant the solution into another bottle and carefully add 50 ml of oil on top of the solution.

3. Sift the treated sand slowly into the bottle, allowing it to fall through the oil and water.

4. Observe the relative dispersibility or tendency to form clumps in both aqueous and oil phase.

The procedure for wettability testing of surfactants used in acid solutions is:

1. Place 50 ml of acid containing the surfactant and corrosion inhibitor to be tested in a bottle and add 10 cc of sand. Carry out the remainder of the test in the same manner as described above for water-soluble surfactants.

2. Observe the appearance of sand grains in the acid, decant the acid and rinse the sand with formation or synthetic brine. Cover the sand with 50 ml of brine and 50 ml of oil. Again observe the condition of sand grains to determine whether the grains are coated with a water or an oil film.

Interpretation of Results:

—Strongly water-wet clays or other fines disperse readily in aqueous phase but agglomerate or clump in oil phase.

—Strongly oil-wet particles disperse readily in oil but agglomerate or clump in water phase.

—Since wettability exists in different degrees between the extremes of being either strongly water-wet or strongly oil-wet, observations of intermediate systems are difficult to distinguish and describe.

—Other factors should be considered. For example, when a dark-colored crude oil is used, oil-wet sands should approach the color of the crude. If the crude tends to form an emulsion spontaneously on contact with the aqueous surfactant solutions, the sand may have the same appearance as if it were oil-wet.

Requirements For Well Treating Surfactants

A surfactant used to prevent or remove damage should:

—Reduce surface and interfacial tension

—Prevent the formation of emulsions and break emulsions previously formed

—Water-wet the reservoir rock, considering salinity and pH of water involved

—Should not swell, shrink, or disturb formation clays

—Maintain surface activity at reservoir conditions

Surfactants for Well Treatments

Many commercial surfactants of all four classes appear to lose much of their surface activity above 50,000 ppm salt. To overcome this difficulty, it is sometimes desirable to pump a preflush of solvent or relatively low salinity water, such as 1% KCl, ahead of the surfactant treatment. The use of a solvent preflush may also reduce water production immediately following treatment. However, a solvent preflush should not be used in dry gas wells.

—Have solubility in the carrier or treating fluid at reservoir temperature. Some satisfactory surfactants are dispersed in their carrying fluid.

—Have tolerance for formation brine or produced fluids. Some anionic and cationic surfactants may be "salted out" of solution by high salt concentrations but are usually more soluble than nonionics at high temperatures.

Table 6-2 describes characteristics of three widely-used surfactants. All three are water-wetting for sandstone, clay, and carbonates in fresh water and salt water through 200,000 ppm NaCl. All are generally applicable for lowering surface and interfacial tension, preventing emulsions, and breaking emulsions. All three utilize specific chemical bonding to maintain water–wetting for carbonates, clays, and sandstones through a wide range of salinities.

TABLE 6-2
Comparable Data on Some Commercially-Available Surfactants

Property	Exxon Corexit 7652	Dowell W-52	Halliburton Morflo II
Ionic Nature	Nonionic & Anionic	Nonionic	Anionic
Specific Gravity	1.0321	0.920	1.058
Flash Point, F	121	69	70
Interfacial Tension			
w/0% Surfactant	39.8	39.8	39.8
w/.01% Surfactant	18.1	0.2	8.4
w/.1% Surfactant	—	0.1	<1
w/.2% Surfactant	—	0.1	—
Soluble In:			
Isopropyl Alcohol	Yes	Yes	No
Aromatics	Yes	Yes	No
Diesel Oil	Yes	No	No
Kerosene	Yes	No	No
Crude Oil	Yes	No	No
Water	No	Yes	Yes
Dispersible In:			
Oil	No	Yes	Yes
Saltwater	Yes	Yes	—
Fresh Water	Yes	Yes	—
Wettability of Sand:			
Fresh Water	Water Wet	Water Wet	Water Wet
w/NaCl— 50,000 ppm	Water Wet	Water Wet	Water Wet
75,000 ppm	Water Wet	Water Wet	Water Wet
100,000 ppm	Water Wet	Water Wet	Water Wet
150,000 ppm	Water Wet	Water Wet	Water Wet
200,000 ppm	Water Wet	Water Wet	Water Wet
Wettability of Carbonates			
Fresh Water	Water Wet	Water Wet	Water Wet
w/NaCl— 50,000 ppm	Water Wet	Water Wet	Water Wet
75,000 ppm	Water Wet	Water Wet	Water Wet
100,000 ppm	Water Wet	Water Wet	Water Wet
150,000 ppm	Water Wet	Water Wet	Water Wet
200,000 ppm	Water Wet	Water Wet	Water Wet

WELL STIMULATION WITH SURFACTANTS

The primary purpose of surfactants in well completion, workover, and well stimulation should be to prevent damage. The real problem in emulsion removal from sandstone formations with surfactants is the near impossibility of getting the surfactant in intimate contact with emulsion droplets in sandstone. Water blocking is relatively easy to treat. The objective is to increase relative permeability to oil and decrease interfacial tension.

Emulsion blocks can be treated; however, surfactant stimulation treatments tend to finger, or channel, through a viscous emulsion. If most of the emulsion is not broken during surfactant stimulation, the emulsion usually migrates back to the area immediately around the wellbore and restores the blocking condition.

If the damage problem is oil-wetting, this may be alleviated by injecting a strong water-wetting surfactant into the formation. However, if oil-wetting of a sandstone is caused by cationic surfactants, the cationics are very difficult to remove. The best approach is to avoid treating sandstone wells with cationic surfactants.

As a rule, it is quite difficult to positively diagnose well damage. However, assuming the problem has been diagnosed as amenable to surfactant treatments, the next step is to plan the job of removing existing damage without causing additional damage. The stimulation plan should include practical provisions to provide a clean carrying fluid for the surfactant, including a clean handling, mixing, and circulation system.

Prior to the surfactant treatment, it may be necessary to clean the tubing, wellbore, and perforations of rust, scale, paraffin, asphaltenes, sand, silt, and other debris. To aid in injecting the surfactant into all zones, reperforating may be desirable. Stimulation is usually carried out with a dilute solution of surfactant, usually 2% or 3% in filtered oil or filtered saltwater, which is free of extraneous chemicals.

Treatment Fluids Used

In stimulation treatments using oil as the surfactant carrying fluid, refined oil such as diesel oil, xylene, heavy aromatic naphtha, kerosene, or aviation hydroformate, is usually employed along with 2 or 3% surfactant, either miscible or dispersible in oil.

Clean, filtered crude oil may be used, but it should contain no materials such as corrosion inhibitors, dehydration agents, other extraneous chemicals, or suspended solids. It is difficult to remove suspended solids consisting of asphalt, paraffin, or fines from crude oil. Peco filters are the most satisfactory oil filters used in field operations.

For stimulation treatments using water as the carrying fluid, clean 2% KCl or clean saltwater is used along with 2% or 3% surfactant, either miscible or dispersible in water.

Treatment size should be equal or greater than the volume of fluid that damaged the formation. However, the exact volume of fluid lost or injected into the formation is frequently not known. An average treatment is about 100 gallons of 2% or 3% surfactant solution per foot of interval treated and is designed to contact a radius of three to five feet from the wellbore. The average surfactant treatment may be 4,000 to 5,000 gallons of a 2% surfactant solution.

Fluid Placement—The surfactant treatment should be planned to insure injection into all permeable zones that are open to the wellbore. In long zones, isolation techniques should be employed to insure that the treated interval does not exceed approximately 50 feet.

A straddle type of perforation washer may aid in directing the surfactant into all open intervals. A solvent preflush is sometimes useful in reducing water production following a surfactant treatment.

After squeezing surfactant into the formation at below frac pressure, the well should be shut in for about 24 hours to insure proper surfaction response.

INCREASING EFFECTIVENESS OF ROD PUMPING

Chevron Research[4] reported the successful use of surfactants to invert a water-in-oil emulsion to an oil-in-water emulsion downhole to improve production from wells pumping a very viscous emulsion of asphaltic crude.

In one field a 11°API asphaltic crude was being produced as a water-in-oil emulsion having a viscosity of 30,000 to 42,000 cp at 80°F. After inverting the emulsion downhole to an oil-in-water emulsion, the viscosity of the emulsion produced at the surface was about 5 cp.

In one three-well group, production was increased by 34% as a result of the reduction of viscosity from the 30,000 cp range to 5 cp. This viscosity reduction eliminated the rod drop problem in pumping wells.

Surfactant cost was 12 cents per barrel of additional oil produced. However, the cost of electricity for pumping the less viscous emulsion was reduced by an amount equivalent to the surfactant cost.

PREVENTION OF WELL DAMAGE

The best method of handling well damage from emulsion or water blocks and oil-wetting is to prevent the damage.

The addition of 0.1% to 0.2% by volume of a properly selected surfactant to water or oil used in well killing, well workover, or well servicing will usually prevent well damage caused by interfacial changes at low cost. Surfactant selection should be based on API RP 42, 1977. This type of treatment has reduced swabbing time by two-thirds in one major area where 300 to 500 barrels of oil or water is lost per workover. It should be standard practice to use surfactants in connection with all acid or frac jobs.

It appears that surfactants are used on far too few well completion, workover, and well stimulation jobs, and the correct surfactants are probably used on an even smaller number of jobs.

Perhaps a better understanding of the function of surfactants by field engineering and field supervisory personnel, plus more use of the 1977 revision of API RP 42, "Recommended Practice for Laboratory Testing of Surface Active Agents for Well Stimulation," will bring about a much greater and more effective use of surfactants in field operations.

REFERENCES

1. Somasundaran, P. and Agar, G. E.: "The Zero Point Charge of Calcite," *Journal of Colloid and Interface Science*, 24, 1967.

2. Strassner, J. E.: "Effect of pH on Interfacial Films and Stability of Crude Oil-Water Emulsions," *J. Pet. Tech.* March, 1968.

3. Hower, Wayne F. and Stegelman, J., Jr.: "Increased primary Oil Production by the Wise Use of Surface Active Agents," API Division of Production, Paper 826-27-A, April 26, 1956.

4. Simon, R. and Poynter, W. G.: "Downhole Emulsification for Improving Viscous Crude Production," *J. Pet. Tech.*, Dec. 1968, p. 1, 349.

5. Bobeck, J. E., Mattax, C. C., Denekas, M. O.: "Reservoir Rock Wettability—Its Significance and Evaluation," SPE of AIME Paper No. 895-G, 1957.

6. Jeffries-Harris, M. J. and Coppel, C. P.: "Solvent Stimulation in Low Gravity Oil Reservoirs," *J. Pet. Tech.*, Feb. 1969, p. 167.

7. API RP 42, Revised 1977, "Recommended Practice for Laboratory Testing of Surface Active Agents for Well Stimulation."

8. Hall, B. E., and Lasater, R. M.: "Surfactants for Well Stimulation." Presented before the Div. of Petroleum Chemistry, American Chemical Society, Feb. 1970.

9. Reese, D. D.: "Demulsifier Squeeze Treatments Elk Basin Embar Tensleep Reservoir," SPE 7575, Oct. 1978.

10. Ribe, Kitt. "Production Behavior of a Water-Blocked Oil Well," SPE Well Completion Reprint Series No. 5. 1970.

11. McClaflin, G. D., Clark, C. R., and Sifferman, Thomas R.: "The Replacement of Hydrocarbon Diluent with Surfactant and Water for the Production of Heavy, Viscous Crude Oil," SPE 10094, Oct., 1981.

12. Hjelmeland, O. S. and Larrondo, L. E.: "Experimental Investigation of the Effect of Temperature, Pressure, and Crude Oil Composition on Interfacial Properties," SPE 12124, Oct., 1983.

13. Adkins, J. D.: "Field Results of Adding Surfactant to Cyclic Steam Wells," SPE 12007, Oct., 1983.

Chapter 7 Acidizing

Acid types, characteristics
Application of additives
Reaction rates, retardation methods
Carbonate acidizing techniques
Sandstone acidizing techniques

Acid may be used to reduce damage near the wellbore in all types of formations. Inorganic, organic, and combinations of these acids, along with surfactants, are used in a variety of well stimulation treatments. In carbonate formations, acid may be used to create linear flow systems by acid fracturing. Acid fracturing is not applicable to sandstone wells.

The two basic types of acidizing are characterized through injection rates and pressures. Injection rates *below* fracture pressure are termed matrix acidizing, while those *above* fracture pressure are termed fracture acidizing.

Matrix acidizing is applied primarily to remove skin damage caused by drilling, completion, workover or well-killing fluids, and by precipitation of deposits from produced water. Due to the extremely large surface area contacted by acid in a matrix treatment, spending time is very short. Therefore, it is difficult to affect formation more than a few feet from the wellbore.

Removal of severe plugging in sandstone, limestone, or dolomite can result in a very large increase in well productivity. If there is no skin damage, a matrix treatment in limestone or dolomite could stimulate natural production no more than one and one-half times. Matrix treatments tend to leave zone barriers intact if pressures are maintained below frac pressures.

One of the problems in matrix acidizing is that fracture pressure is not always known. Because breakdown or fracture pressure may decrease with a decrease in reservoir pressure, it is frequently necessary to run "breakdown" tests to determine fracture pressure of a specific zone or reservoir. Figure 7-1 illustrates pressure behavior, during a test to determine fracture pressure.

The test procedure is to start pumping water or clean oil into the formation at a very slow rate, perhaps $\frac{1}{4}$–$\frac{1}{2}$ bbl/min for short zones, and measure pump in pressure. Then, increase pump rate by steps and read injection pressure until the injection rate-pressure curve breaks as indicated in point B. If desired matrix acidizing pressure is reached before "breakdown," the acidizing may be carried out at that pressure or slightly lower pressure.

Fracture acidizing is an alternative to hydraulic fracturing and propping in carbonate reservoirs. In fracture acidizing, the reservoir is hydraulically fractured and then the fracture faces are etched with acid to provide linear flow channels to the wellbore.

Acid fracturing of relatively homogenous carbonates will produce smooth fracture faces that will retain little fracture flow capacity when treating pressure is released. (Hydraulic fracturing and propping is preferred for homogeneous carbonates.) Fracture acidizing of heterogeneous carbonates can develop nonuniform etching of the fracture face. The area that is not etched acts as a support for the etched areas, thus providing flow channels in the fracture and significant increases in well productivity. However, it is suggested that laboratory tests be conducted on cores to determine the etching characteristics of the rock before treating a well. In some instances, the nonetched area of a heterogeneous carbonate will be softened by acid and will not support the fracture when treating pressure is released. The only alternative would be hydraulic fracturing and propping with sand.

Two other problems can exist in fracture acidizing. Undissolved fines can significantly reduce fracture flow capacity if not removed with spent acid. Suspending agents, usually surfactants or polymers, will materially aid in the removal of these fines. Emul-

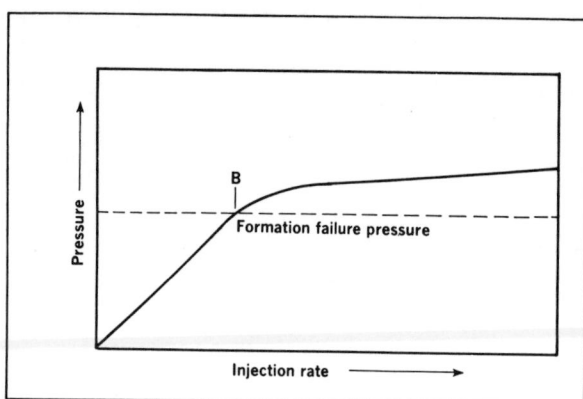

FIG. 7-1—*Tests to determine fracture pressure. Courtesy of Dowell.*

sions can block the etched fracture. API RP 42 tests should be performed to select an emulsion preventing surfactant for the acid treatment.

Fracture acidizing has no application in sandstone wells. "Breakdown" of a sandstone well with acid at fracture pressures may break down natural vertical permeability barriers to adjacent unwanted zones.

"Breaking-down" with acid may also tend to open channels between the cement and the formation, even though formation fracturing pressure is never reached. If breakdown of mud-plugged perforations is necessary, a less hazardous procedure from the standpoint of maintaining zone isolation is to break down perforations with clean water or oil and a suitable surfactant using ball sealers or other diverting agents. Short fractures created by water or oil breakdown of the formation will usually close without permanent damage to adjacent zone barriers.

ACIDS USED IN WELL STIMULATION

The basic types of acid used are: Hydrochloric, Hydrochloric-Hydrofluoric, Acetic, Formic, and Sulfamic. Also, various combinations of these acids are employed in specific applications.

Hydrochloric acid (HCl) used in the field is normally 15% by weight HCl; however, acid concentration may vary between 5% and about 35%. The freezing point of 15% acid is −27°F, less than −70°F for 20 to 29% acid, and −36°F for 35% acid. HCl will dissolve limestone, dolomite, and other carbonates.

A thousand gallons of 15% HCl will dissolve 1.840 lb. or 10.5 cu ft of zero porosity limestone ($CaCO_3$). This reaction will produce 2,050 lb of calcium chloride ($CaCl_2$), 812 lb of carbon dioxide (CO_2) or 6,600 cu ft of CO_2 gas at standard conditions of temperature and pressure, and 333 lb of water, in addition to the 7,600 lb of water injected as a carrier for HCl acid. In practice, after spending in limestone, 1,000 gal of 15% HCl becomes 1,020 gal of 20.5% solution of calcium chloride, weighing 9.79 lb/gal.

A thousand gallons of 15% HCl will dissolve 1,710 lb, or 9.6 cu ft of zero porosity dolomite, $CaMg(CO_3)_2$. The spending of 1,000 gal of 15% HCl in dolomite will produce 1,020 gal of a mixture of 10.5% calcium chloride and 9% magnesium chloride solution, weighing 9.7 lb/gal.

Acetic acid (HAc) is a weakly-ionized, slow-reacting organic acid. A thousand gallons of 10% acetic acid will dissolve about 704 lb of limestone. The cost of dissolving a given weight of limestone is greater with acetic acid than with HCl.

Acetic acid is relatively easy to inhibit against corrosion and can usually be left in contact with tubing or casing for days without danger of serious corrosion. Because of this characteristic, acetic acid is frequently used as a perforating fluid in limestone wells.

Other advantages of acetic acid in comparison to HCl are:

1. Acetic acid is naturally sequestered against iron precipitation.
2. It does not cause embrittlement or stress cracking of high strength steels.
3. It will not corrode aluminum.
4. It will not attack chrome plating up to 200°F. Therefore, acetic acid should be considered when acidizing a well with an alloy pump in the hole.

Formic acid is a weakly-ionized, slow reacting organic acid. It has somewhat similar properties to acetic acid. However, formic acid is more difficult to inhibit against corrosion at higher temperatures and does not have the widespread acceptance and use of acetic acid.

Hydrofluoric acid used in oil, gas, or service wells is normally 3% HF acid plus 12% HCl. It is employed exclusively in sandstone matrix acidizing to dissolve formation clays or clays which have migrated into the formation. One thousand gal of 4.2% HF acid will dissolve 700 lb of clay. Fast reaction time and precipitants make HF acid undesirable in carbonate-containing sands having more than 20% solubility in HCl. HF acid should never be used in carbonate formations.

Sulfamic acid, a granular-powdered material, reacts about as fast as HCl. The primary advantage of sulfamic acid is that it can be hauled to the location as a dry powder and then mixed with water. Unless sulfamic acid is modified, it will not dissolve iron oxides or other iron scales. Because of its molecular weight, the amount of calcium carbonate dissolved by one pound of sulfamic acid is only about one-third that dissolved by an equal weight of HCl. Acidizing with sulfamic acid is usually much more expensive than with HCl.

Sulfamic acid is not recommended for temperatures above 180°F because it hydrolizes to form sulfuric acid (H_2SO_4). When H_2SO_4 reacts with limestone or $CaCO_3$ scale, calcium sulfate ($CaSO_4$) can be precipitated.

ACID ADDITIVES

Acidizing can cause a number of well problems. Acid may (1) release fines, (2) create precipitants, (3) form emulsions, (4) create sludge, and (5) corrode steel. Additives are available to correct these and a number of other problems.

Surfactants should be used on all acid jobs to reduce surface and interfacial tension, to prevent emulsions, to water-wet the formation, and to safeguard against other associated problems. Surfactant type and concentration should be selected on the basis of tests outlined in API RP 42[24] dated 1977.

Surface tension of 15% HCl is 72 dynes/cm and can be reduced to about 30 dynes/cm by the addition of an effective surfactant. Swabbing and clean-up time after acidizing oil, gas, and service wells can be reduced by lowering surface tension.

Suspending Agents—Most carbonate formations contain insolubles which can cause blocking in formation pores or fractures if fines released by acid are allowed to settle and bridge. Suspension should be differentiated from dispersion. Dispersed particles usually settle in a short time. A suspending surfactant, such as Halliburton's HC-2, in concentrations of about five gallons per 1,000 gallons of acid may suspend fines for more than 24 hours, and possibly as long as seven days. Suspending agents are usually polymers or surfactants.

Clean-up after fracture acidizing can be accelerated by use of a suspending agent. Some suspending surfactants may also prevent emulsions, provide lower surface tension in both raw and spent acid, and water-wet the formation.

Sequestering agents act to complex ions of iron and other metallic salts to inhibit precipitation as hydrochloric acid spends. During acidizing if ferric hydroxide is *not* prevented from precipitating, this insoluble iron compound may be redeposited near the well-bore and cause permanent plugging. The effectiveness of these acid solutions is dependent on two chemicals that can act synergistically to keep iron in solution over a long period of time. Acetic acid is used to maintain a low pH as HCl spends. Citric acid acts as a chelating agent and is particularly useful when higher iron concentrations are present.

Sequestered HCl is most often used in the treatment of disposal and injection wells. Tubular goods are often coated with iron corrosion products which are soluble in HCl. If iron is in the oxidized condition, it will precipitate when HCl spends in the formation and cause plugging of the rock pores. Other sources of iron that could cause plugging in producing, waterflood, and disposal wells are iron sulfide and iron carbonate (Siderite).

Acid concentrations used in a well treatment are dependent on the amount of iron that may be dissolved and the formation temperature. HCl concentration may be increased from the normal 15% to as high as 25% where higher concentrations of iron oxide scales are in the well system. A normal acetic acid concentration of 10% is suggested for most applications; however, a higher concentration is required above 200°F. Most sequestering solutions also contain 50 lb of citric acid/1,000 gal of HCl. The use of additional citric acid is also possible, but should not exceed 150–200 lb/1,000 gal of HCl because calcium citrate may precipitate. Higher than normal concentrations of acetic and citric acid are recommended where appreciable quantities of iron carbonate are present in the formation. Lactic acid is also an effective sequestering agent but is not usually recommended where the temperature is higher than 200°F.

All sequestered acid solutions require a corrosion inhibitor to minimize acid reaction on tubular goods. Adequate protection to 450°F can be obtained with the correct inhibitor.

Well conditions should be thoroughly analyzed to determine the most effective sequestered acid solution. Iron precipitation can be prevented for as long as fifteen days. The addition of a compatible surfactant to the acid solutions will improve acid contact with iron compounds. A sequestering agent should probably be used as a precautionary measure in most older wells because of the probability of a corrosion product in the casing or tubing.

Anti-Sludge Agents—Some crudes, particularly heavy asphaltic crudes, form an insoluble sludge when contacted with acid. The primary ingredients of a sludge are usually asphaltenes; sludges may also contain resins and paraffin waxes, high-molecular weight hydrocarbons, and formation fines or clays.

The addition of certain surfactants can prevent the formation of sludge by keeping colloidal material dispersed. Also these sludge-preventing surfactants usually function as emulsion preventers. Sludge is more of a problem with high strength acids.

Corrosion inhibitors temporarily slow down the reaction of acid on metal. Corrosion inhibition time varies with temperature, acid concentration, type of steel, and inhibitor concentration. The effect of temperature and types of steel on corrosion rate is illustrated in Figures 7-2 and 7-3.

Both organic and inorganic corrosion inhibitors have application in acidizing. Some organic inhibitors are effective up to the 300°F range. Extenders have been developed to increase the effective range to 400°F. Inorganic arsenic inhibitor can be used up to at least 450°F. Table 7-1 shows effectiveness of some organic and inorganic inhibitors at high temperatures.

Arsenic is more effective than organic inhibitors at all temperatures. However, the use of arsenic in oil well treatments has been banned in many areas because even small percentages of arsenic acts as a poison to refinery catalysts. Arsenic has proved to be very effective as a long-term corrosion inhibitor in some very corrosive gas wells. Also arsenic can be a safety problem since deadly arsine gas can be librated as a by-product of corrosion.

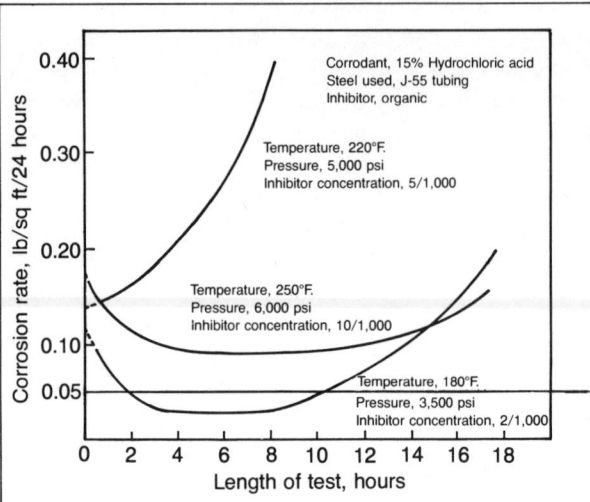

FIG. 7-3—*Effect of temperature on acid corrosion. Courtesy of Halliburton Services.*

Alcohol, normally methyl or isopropyl alcohol in concentrations of 5% to 30% by volume of acid, may be mixed with acid to lower surface tension. The use of alcohol in acid may accelerate the rate of well clean-up, particularly in dry gas wells. Disadvantages are increased inhibitor problems, possible salt precipitation, and increased costs.

Fluid Loss Control Agents may be required to reduce acid leak-off in fracture acidizing. The preferred method of selecting fluid loss control agents is to run fluid loss tests on cores from the formation to be acidized. Benefits and possible damage from specific fluid loss control agents should be determined from

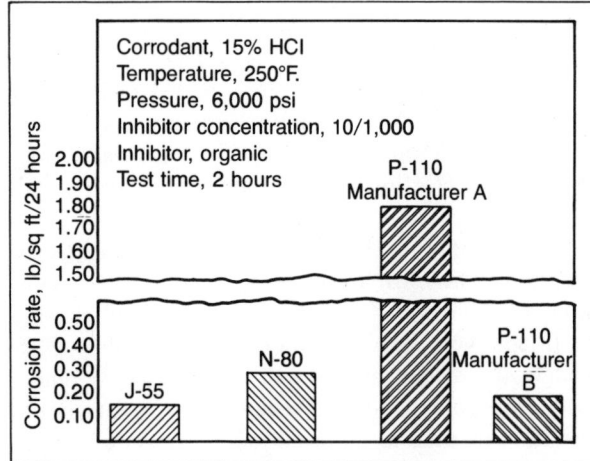

FIG. 7-2—*Effect of hydrochloric acid on different types of steel. Courtesy of Halliburton Services.*

TABLE 7-1
Effectiveness of Corrosion Inhibitors at High Temperatures in 15% HCl

Type inhibitor	Inhibitor concentration %	Temperature °F	Protective time* Hrs.
Organic	0.6	200	24
	1.0	250	10
	2.0	300	2
Inorganic	0.4	200	24
	1.2	250	24
	2.0	300	12

*Time required for 15% HCl to remove 0.05 lb metal per sq ft of exposed metal area.

these tests. Fluid loss agents should be degradable or slowly soluble in the treating solution, produced oil, or produced water. An example of this type of material is Halliburton's Matriseal I, an oil-soluble resin plus a limited swelling natural gum.

Diverting or Bridging Agents—As illustrated in Figure 7-4, acid will follow the path of least resistance, usually the lesser damaged intervals, unless diverting or bridging agents are employed to allow relatively uniform acidizing of all porous zones open to the wellbore.

Dowell's J-237, Halliburton's Matriseal O, and Western's ASP-530 are finely divided oil-soluble resins used for diverting in matrix acidizing oil wells.

Temporary Bridging Agents—These materials may be used as temporary bridging agents:

—Benzoic acid is slowly soluble in water or oil and is available as finely divided particles and as flakes. It is marketed by Halliburton as TLC-80; by Dowell as J-227; by BJ Hughes as Divert II; and by Western as Westblock III-X.

—Graded rock salt is available in a range of one-fourth inch through 125 mesh. For perforated completions, use 10 to 15 lb per perforation in gelled salt water.

—Rock salt and Benzoic acid are sometimes used on a 50-50 basis.

—Union's Unabeads are wax beads and can be selected for desired melting temperature. They are available in one-fourth inch size and graded small particles.

—Ball Sealers are effective.

CARBONATE ACIDIZING

The objective of acidizing limestone and dolomite wells is to remove damage near the wellbore or to create linear flow channels by fracturing and etching. Acid may also be used in sandstone wells to dissolve carbonates in the form of cementing materials, discrete particles, and scale. The time required for a specified volume and concentration of HCl acid to spend to about 3.2% in a selected formation under given conditions is defined as Acid Reaction Time.

A major problem in fracture acidizing of carbonate formations is that acids tend to react too fast with carbonates and spend near the wellbore.

Factors Controlling Acid Reaction Rate

Factors controlling the reaction rate of acid are: area of contact per unit volume of acid; formation temperature; pressure; acid concentration; acid type; physical and chemical properties of formation rock; and flow velocity of acid.

Effect of Area-Volume Ratio on Acid Spending Time—Reaction time of a given acid is indirectly proportional to the surface area of limestone or dolomite in contact with a given volume of acid. Extremely high area-volume ratios are the general rule in matrix acidizing. Therefore, it is very difficult to obtain a significant acid penetration before spending during matrix treatments.

Figure 7-5 shows the comparative effect of area-volume ratio on acid spending time in a 6-in. wellbore, a 0.1-in. fracture and the matrix of a limestone formation.

Effect of Temperature on Acid Spending Rate—As temperature increases, acid spends faster on carbonates, as shown in Figure 7-6. It is often necessary to increase pumping rate during acid fracturing to place acid effectively before it is spent.

Effect of Pressure on Acid Spending Time—An increase in pressure up to 500 psi will increase spending time for HCl. Above this pressure, only a very small increase in spending time can be expected with increases in pressure.

Effect of Acid Strength on Spending Time—As concentration of HCl increases, acid spending time in-

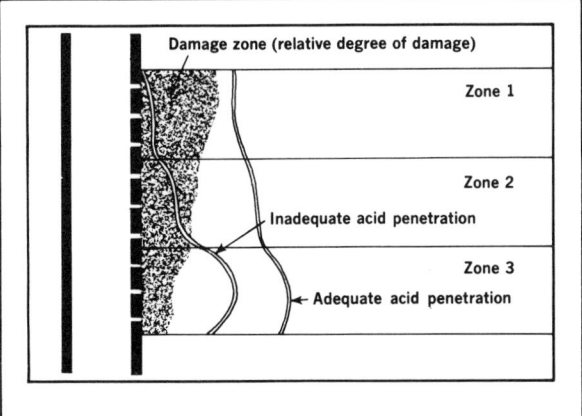

FIG. 7-4—*Injected acid follows the path of least resistance. Courtesy of Dowell.*

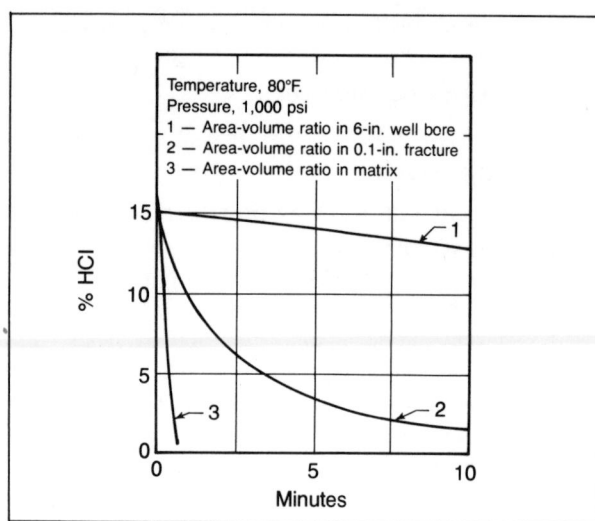

FIG. 7-5—*Effect of area-volume ratio on HCl-CaCO₃ reactions. Courtesy of Halliburton Services.*

creases because the higher strength acid dissolves a greater volume of carbonate rock. This reaction releases greater volumes of CaCl$_2$ and CO$_2$ which further retards HCl. Figure 7-7 illustrates the effect on reaction rate of increasing acid concentration.

Effect of Formation Composition on Acid Spending Time—Physical and chemical composition of the formation rock is a major factor in determining spending time. Generally, the reaction rate of limestone is more than twice that of dolomite; however, at high temperatures reaction rates tend to be nearly equal.

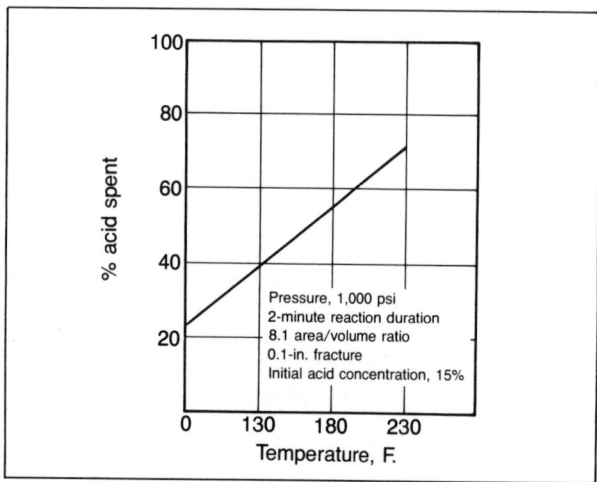

FIG. 7-6—*Effect of temperature on HCl-CaCO₃ reactions. Courtesy of Halliburton Services.*

Effect of Acid Velocity on Spending Time—The effect of acid velocity on reaction rate appears to be slight, except possibly with high strength HCl. In fracture acidizing, an increase in pumping rate increases fracture width. This decreases area-volume ratio, thereby increasing acid reaction time.

Retardation of Acid

To achieve deeper penetration in fracture acidizing, it is often desirable to retard acid reaction rate. This can be done by emulsifying, gelling, or chemically retarding the acid. Also, HCl can be retarded by adding CaCl$_2$, CO$_2$, or HAc. Another approach is to use naturally retarded acetic or formic acid. Comparative results from the principal retardation systems are shown in Figure 7-8 for various temperatures. For retarded high strength acids, the relation illustrated in Figure 7-8 may *not* be correct because high acid concentrations may break down the retarding systems of gelled, emulsified, or surfactant-retarded acid. If a fluid-loss additive is used, or if an acid is gelled, emulsified or chemically retarded, spending time is increased because a lesser area of formation rock is contacted per unit of time with raw acid.

Emulsification is the most used technique in fracture acidizing to retard reaction rate of HCl on limestone and dolomite within the temperature range of 80°F to 300°F. Emulsified acid usually produces the longest spending time of any retarded acid. It may also serve as a diverting agent between stages of conventional acid. Because of its excellent suspending qualities, oil or water-soluble bridging agents can be pumped with emulsified acid to divert acid to less permeable zones. Because of high viscosity and high friction loss, it is not normally used for matrix acidizing except as a diverting agent.

Gelled acid provides minor retardation in the temperature range from 80°F to 200°F. Gels usually have high viscosity and low friction loss and provide some fluid loss reduction. Their primary application is as a diverting agent. Some gels will suspend fines while other gels flocculate fines.

Chemically-Retarded Acid—Retardation of HCl is obtained by the addition of a unique surfactant to the acid which causes oil-wet and water-wet spots on the faces of a fracture in limestone or dolomite having impurities. Oil, injected ahead of or with the acid, adheres to the oil-wet spots and reduces acid reaction on the oil-wet areas of the fracture faces. The result

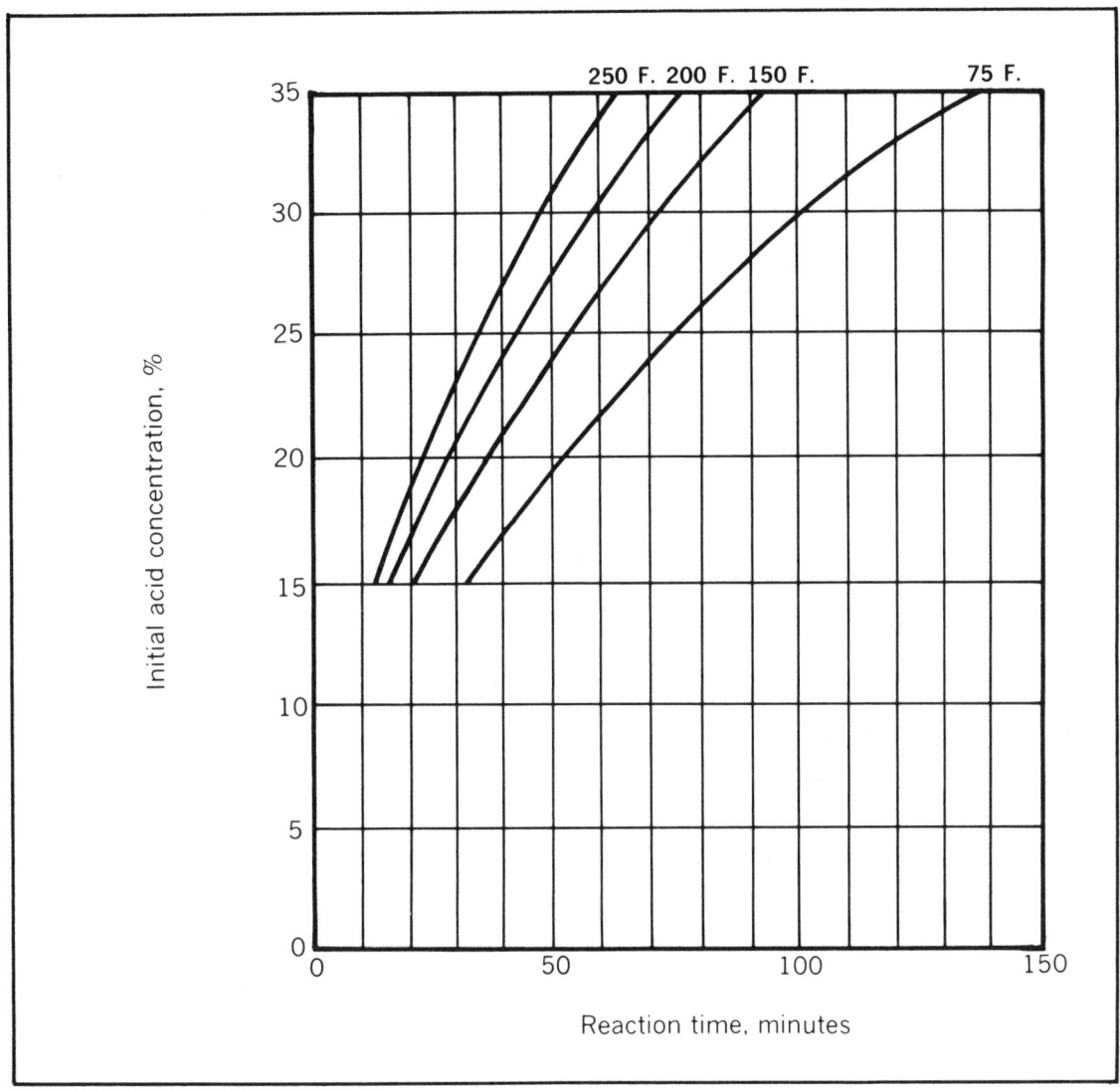

FIG. 7-7—*Effect of acid concentration on reaction time at various temperatures. Courtesy of Halliburton Services.*

is rock pillars between the fracture faces to provide permanent flow channels.

Chemically-retarded acid can be modified by adding non-emulsifiers, suspending agents, friction reducers, sequestering agents, and fluid-loss additives. Chemically-retarded acid is marketed by Halliburton as CRA; by Dowell as Retarded Acid; by Western as Reactrol; and by BJ Hughes as SRA.

Retardation of HCl with Calcium Chloride ($CaCl_2$)—Calcium chloride is beneficial when acidizing formations containing anhydrite, because $CaCl_2$ decreases the solubility of anhydrite ($CaSO_4$) in HCl. As a result, the quantity of anhydrite reprecipitated when the acid spends will be reduced. Except for this use, $CaCl_2$ is not generally used as a retarder, because it is not competitive with other forms of retardation on a cost basis.

Retardation of HCl with CO_2—Carbon dioxide retards HCl by cooling and by changing the kinetics of reaction. CO_2 expands and provides additional cleanup following acidizing, especially in low-pressure wells.

Retardation of HCl with Acetic Acid—Acetic acid reacts with limestone to form calcium acetate $Ca(C_2H_3O_2)_2$, which acts as a buffer to HCl. The CO_2, released by reaction of HCl and acetic acid on lime-

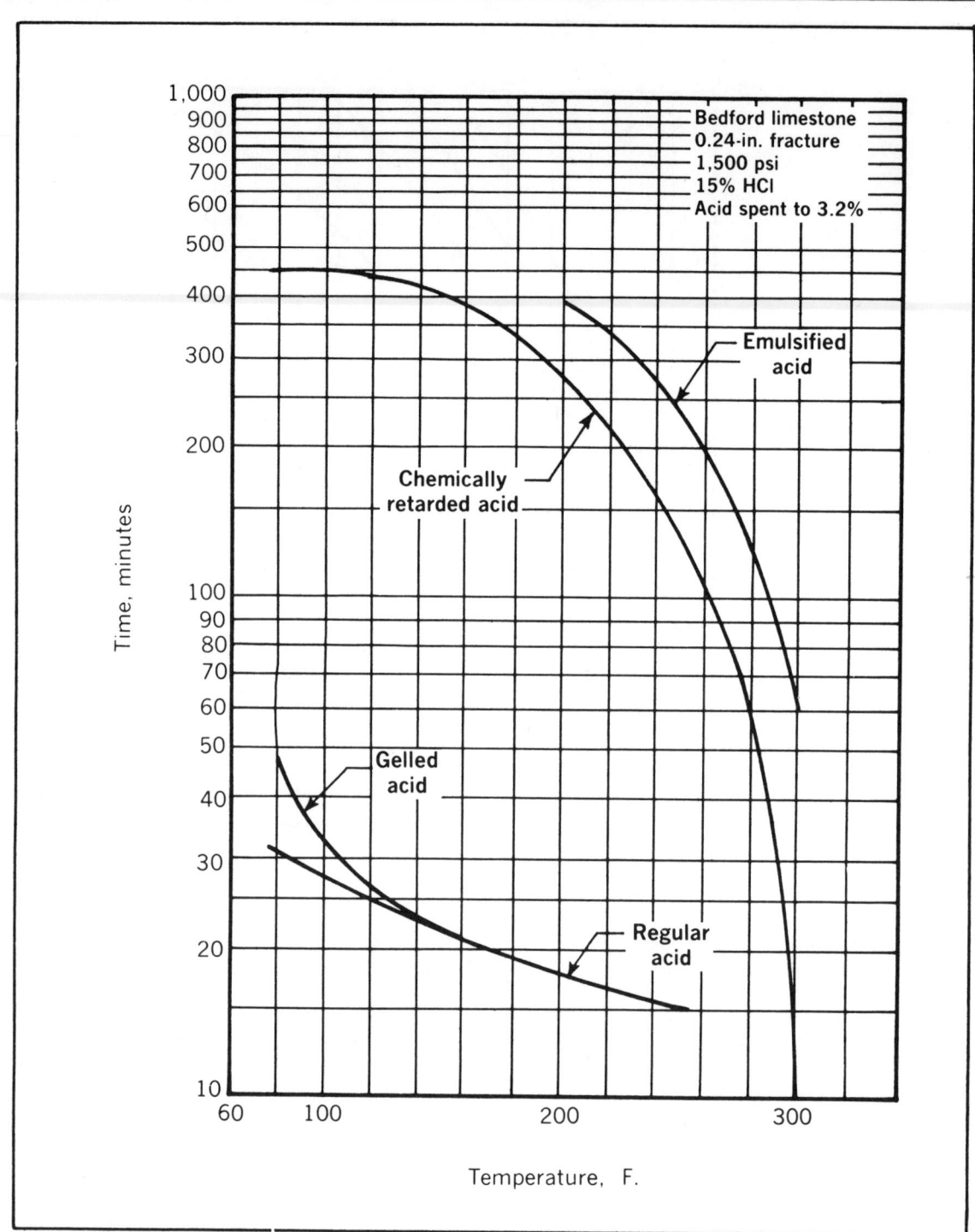

FIG. 7-8—*Comparative retardation of HCl. Courtesy of Halliburton Services.*

stone, retards the rate of reaction of both the HCl and acetic acid.

Naturally-Retarded Acetic and Formic Acid—The reaction product, carbon dioxide (CO_2), is released from limestone or dolomite and reduces the reaction rate of acetic acid until the reaction almost ceases when about 42% of acid is spent. After CO_2 bleeds off into the formation, or is absorbed by oil, water, or hydrocarbon gas, reaction of acetic acid is resumed. Because of its higher ionization constant, formic acid continues to react until 86% is spent. Formic acid reacts faster than acetic acid.

ACIDIZING TECHNIQUES FOR CARBONATE FORMATIONS

Carbonate acidizing may be divided into three types: acidizing to remove or bypass formation damage including acid-soluble scales, matrix acidizing, and fracture acidizing.

In actual practice, acidizing is usually divided into two categories based on pressure. Acid jobs performed below fracture pressure are called matrix acidizing and are usually aimed at damage removal. Fracture acidizing involves fracturing with water and then etching with acid to form pillars and posts to hold open the etched fracture. These jobs are usually performed to open new linear flow channels to the wellbore; however, many jobs are performed to bypass damage relatively near the wellbore.

Matrix Acidizing Carbonate Formations

The primary purpose of matrix acidizing is to remove or bypass damage due to scale, mud, clay, or hydrocarbon deposits, and to restore natural formation permeability. Matrix treatments are usually performed by soaking, jetting or agitation, or circulation below fracture pressure.

Fifteen percent HCl is normally used, with modifications as required. Acetic acid or acetic-HCl mixtures are being employed to a greater extent for specific types of applications. Acetic acid should be considered for temperatures above 250°F because of effectiveness of inhibitors at high temperature.

Since the depth of damage is seldom more than a few feet, the volume of acid needed is relatively small. With 15% porosity limestone, one bbl of acid per ft of section is required to reach a distance of three feet from wellbore. If there is no damage around the wellbore, removal of all the reservoir rock to a radius of five feet would increase production only 65%. If the matrix around a wellbore is 90% plugged, restoration of permeability to a radius of five feet may increase production 350% or more.

Additives Required in Most Matrix Acidizing Jobs—A surfactant should be selected to prevent emulsions of spent acid and to lower interfacial and surface tension. A corrosion inhibitor should be selected on the basis of treating temperature and grade of steel tubing and casing. Fluid loss control agents, diverting agents, or temporary bridging agents may be required to promote uniform injection into long sections. Sequestering agents may be employed to prevent the precipitation of $FE(OH)_2$, which is insoluble in spent acid. Suspending agents are beneficial when considerable insoluble fines are released by acidizing.

Fracture Acidizing Carbonate Formations

Fracturing of limestone or dolomite wells is designed to open linear flow paths, usually near vertical, from the wellbore to some point within the reservoir. In acid fracturing, the objective is to develop permanent flow channels by etching the faces of the hydraulically-created fractures. The alternate to fracture acidizing is to prop open the fracture faces with sand or glass beads.

The choice between fracture acidizing and conventional hydraulic fracturing is often a difficult decision. If both systems appear equally feasible to obtain desired fracture flow capacity, then the decision may be based on comparative costs.

The major problems in obtaining fracture flow capacity in fracture acidizing are usually inadequate flow paths resulting from the etching process, and plugging of fracture channel with undissolved fines. Etching flow channels in the fracture faces is usually feasible in heterogeneous carbonate rocks. However, in homogeneous rocks, the fracture faces tend to etch uniformly, resulting in fracture closure soon after completion of the acidizing job.

Figure 7-9 illustrates the results of a simple acid etching test used to estimate whether the etched fracture faces will provide sufficient pillars and posts to hold open the fracture following an acid fracturing job. In this case, good fracture support can be expected. However, in many relatively homogeneous carbonates such as those with oolitic or pin-point porosity, normal acid etching techniques may not provide adequate flow paths through the etched fractures.

The effectiveness of a fracture is a function of both its conductivity and penetration. There is no absolute means for determining the amount of formation rock that must be dissolved or the area that must be affected to achieve a desired improvement in production; however, dynamic laboratory etching tests are very useful in estimating the probability of appreciably increasing productivity of a particular formation by fracture acidizing.

Created fracture area is proportional to fluid vol-

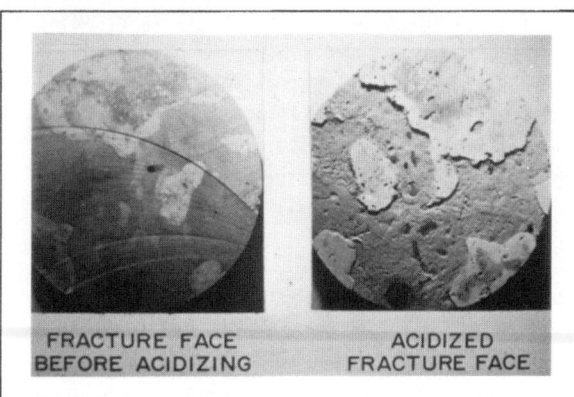

FIG. 7-9—*Acid-etching test.*

ume and inversely proportional to fluid loss coefficient. Viscous fluids tend to provide wider fractures. Longer fractures are created by higher pumping rates, larger volumes, fluid loss control, and retardation of acid. Above 5 to 10 md permeability, adequate fluid loss control may *not* be practical. For further details of fracture initiation and propagation, refer to the Hydraulic Fracturing chapter of this volume.

In some formations, regular or high strength acid fracturing may not provide sustained production increases if large quantities of fines are released in the fracturing by the acid treatment, thus blocking the fracture. Also high strength acid may over-etch some very soluble carbonates, resulting in crushing of the pillars and posts after fracturing pressure is released or at a later date when reservoir pressure declines. Thus, the fracture would close, blocking linear flow paths to the wellbore.

Fracture Etching in Homogeneous Carbonates—Many relatively homogeneous limestone or dolomite formations etch uniformly so that inadequate flow capacity is obtained following fracture acidizing. Productivity of fracture-acidized wells may decline rapidly due to the fracture closure in cases where inadequate channels are formed in the fracture faces, or fractures may close over a period of time due to crushing of support pillars as the reservoir pressure declines.

To combat the fracture closure problem in acidizing relatively homogeneous carbonates, the following techniques have been developed over the past 15 years.

Viscous Preflush in Fracture Acidizing—As illustrated in Figure 7-10, one method of developing fracture flow channels in a relatively homogeneous carbonate is to inject a high viscosity pad, normally a high viscosity gel or emulsion, ahead of the etching acid. Fingering of low viscosity acid through a viscous pad develops a valley and ridge-type etching pattern which aids in maintaining linear flow channels.

The viscous pad also reduces acid leak-off thus resulting in greater fracture length and width. The etching of channels, as opposed to etching the entire fracture faces, allows deeper acid penetration for a given volume of acid.

One of the initial viscous pads used as a preflush has an apparent viscosity up to 20,000 cp. However, viscosity may be tailored to break back to near 1 cp after the acid job is complete. Various viscous gels are currently being used as a preflush in fracture acidizing.

Chemically Retarded Acid[3] for Selective Etching—The development of a partial oil-wetting and partial water-wetting surfactant acidizing system allows the etching of permanent pillars to hold open the fractures and provide increased fracture flow capacity. For this system to operate, it appears that there must be a given amount of impurities in the fracture face. Some of the best results have been in treating limestone formations having joints or incipient fractures at relatively low pump rates of 3–5 bbl/min.

Although chemical retardation with this combination oil-wetting/water-wetting surfactant has merit, etching of pillars and valleys to form permanent flow channels in the fracture faces of homogeneous limestone or dolomite is the more significant function of this acid.

With this system, about five barrels of oil are pumped into a fracture, followed by chemically re-

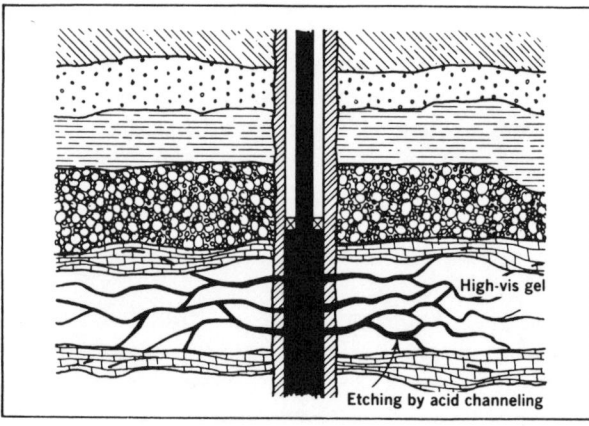

FIG. 7-10—*Viscous preflush causes irregular etching. Courtesy of Halliburton Services.*

tarded etching acid, and then five barrels of oil are injected behind each 1,000 gallons of CRA acid. An alternate approach is to mix at least one barrel of oil in each 1,000 gallons of acid in addition to the five barrels of oil injected ahead of the treatment. Halliburton's current version of this development is CRA-22 for 15% HCl, and CRA-35 for 20% to 35% HCl. Dowell's Retarded Acid V, using 15% HCl, BJ Hughes' SRA-1, and Western's Reactrol 70 and 72 are offered as competitive acids.

Combination Density and Viscosity Controlled Fracture Acidizing—Research has shown that density and viscosity of the preflush, the acid, and the afterflush can be controlled, so as to reduce the probability of etching flow channels either below or above the desired stimulation zone.[22] A density difference of 0.84 lb/gal is usually sufficient to maintain fluid layering within the fracture during the acid treatment. However, a 1.68 lb/gal density difference should be considered when extreme overriding or underriding of fluid is desired.

A lighter weight preflush tends to prevent the acid from etching above the desired interval and a heavier preflush tends to prevent the acid from etching below the desired zone. Increasing the viscosity of the preflush up to about 100 cp may assist in preventing the density-controlled acid from etching above or below the desired stimulation interval. If the viscosity of the preflush is above 100 cp, the acid tends to finger through the preflush.

To maintain desired layering of fluids in the fracture, it is necessary to inject alternate slugs of different density fluids. In one series of tests[22], alternate 2,000 to 3,000 gallon slugs of high density preflush followed by lighter weight retarded acid were injected to keep the acid from etching the fracture faces into the lower water zone.

Although viscosity difference plays a lesser role than density difference to prevent etching out of zone, viscosity difference can often be used along with density difference to aid the process.

Acetic-HCl Mixture for Matrix and Fracture Acidizing

Various acid combinations are employed as indicated by dynamic tests on cores. However, a frequent combination is 15% HCl and 10% glacial acetic acid. Outstanding results have also been obtained in many areas with mixtures of high strength HCl and high strength (10% or greater) acetic acid.

Advantages of Acetic-HCl mixture are:

1. For some limestone and dolomite, higher fracture flow capacity is obtained.
2. Acetic acid sequesters iron and prevents formation of ferric hydroxide, an insoluble precipitate.
3. It reduces sludging and emulsification with spent acid.
4. It provides a higher acid strength with a lower corrosivity than an equivalent strength of HCl.
5. Acetic acid with HCl maintains a low pH, thus minimizing swelling of clays.

Halliburton markets Acetic-HCl acid as MOD 101, MOD 202, and MOD 303, with higher numbers referring to higher strength HCl. Dowell sells the Acetic-HCl mixture as Retarded Acid IV and VIII, Western as A-20.2 and BJ Hughes as DR Acid.

Foamed Acidizing

Foam is becoming widely used for acid fracturing low permeability liquid-sensitive carbonate formations, especially gas wells. Foams employed are a dispersion of gas, normally nitrogen, in acid with a small amount of surfactant foaming agent. Volumetric gas content (foam quality) is usually between 65 and 85% with 15 to 35% liquid. Surfactant content is commonly 0.5 to 1.0% of acid volume along with a compatable corrosion inhibitor. Most jobs have been with HCl; however, HCl and organic acid combinations have been employed. Also, gelled water pads have sometimes been used ahead of the foamed acid in highly fractured reservoirs to reduce fluid loss of acid. One hundred mesh sand has been used for fluid loss control in some fractured reservoirs but may result in blocking of etched fractures. Regular foamed acid or foamed gelled acid usually provides fluid loss control as well as good foam stability.

Apparent advantages of foamed acid over conventional acid fracturing are:

1. The low fluid loss inherent in foamed acidizing makes more acid available for etching longer fractures.
2. There is less formation damage and wells clean up quicker.
   a. Most foamed acid jobs are performed without the addition of fine solids for fluid loss control.
   b. Additional foam stability and fluid loss control is obtained by foaming gelled acid.

c. Very stable foamed spent acid, usually removed from the well as soon as practicable after treatment, will carry out a large quantity of fines, thus reducing the possiblity of blocking the fractures with fines.
   d. The relatively high apparent viscosity of foamed acid results in wider fractures and increases acid spending time because of lower area/volume ratio during acidizing. Higher viscosity improves pumpability.
   e. The built-in gas assist of 65 to 85% nitrogen in the spent acid provides rapid cleanup, particularly in low pressure reservoirs.

Specialty Acids

Oil Soluble Acid (Halliburton) is a nonaqueous solution of acetic acid in hydrocarbon fluids. When this solution contacts water in the formation, the acetic acid is extracted and becomes active. OSA is noncorrosive until it contacts water, and even then, it is only slightly corrosive. Its main uses are (1) as a well completion fluid, (2) as a perforating fluid, and (3) for stimulation of low pressure oil wells. A water-wetting surfactant is normally present in the mixture.

Dowell Acid Dispersion (DAD) and Halliburton's Paragon Acid Dispersion (PAD) are a blend of aromatic hydrocarbons and any type of aqueous acid. These mixtures were developed to treat wells where production declines were caused by a combination of organic (paraffin and asphaltenes) and inorganic scales. Also, such a mixture has been used to (1) clean injection and disposal wells, (2) convert producing wells to injection wells, (3) clean liners and gravel packed wells, and (4) preflush for gypsum scale removal treatment.

Turflo Acid

Turflo, a product developed by Turbo Resources for well stimulation, is a concentrated phosphoric acid containing a corrosion inhibitor, surfactant, clay stabilizer, sequestering agent, and a crystal modifier. Turflo does not dissolve silica but stabilizes clays.

One U.S. gallon of Turflo will dissolve 0.92 lb of $CaCO_3$, whereas 15% HCl will dissolve 1.84 lb of $CaCO_3$. Turflo converts $CaCO_3$ to a phosphate. A crystal modifier is included to minimize phosphate scale deposition. Field stimulation results have been mixed. Reasons for favorable results have been difficult to ascertain.

Use of High Strength HCl Acid

High strength HCl is any concentration of HCl from about 20% to 35%. Dowell markets 28% HCl as "Super-X" Acid. Halliburton markets 20% to 35% HCl as concentrated acid. BJ Hughes and Western market high strength acid as 20%–30% HCl.

Advantages of High Strength Acid—Dolomite and some very dense limestones require high strength acid for dissolution. Twenty-eight percent to 33% HCl is usually employed for these types of formation.

In fracture acidizing, higher strength acid provides longer spending time, resulting in longer etched fractures.

Improved etching of fractures and greater fracture conductivity is obtained in some dense limestone and dolomite formations with nonuniform solubility.

More CO_2 is released per gallon of acid and less CO_2 is dissolved in spent high strength acid, thus providing more CO_2 gas to assist in fracture clean-out after the acid spends.

Disadvantages of High Strength HCl Acid—In fracture acidizing of very soluble limestone, over-etching may result in low flow capacity fractures. If reservoir pressure declines months or years later, overetched fractures may close, resulting in little or no permanent stimulation.

Corrosion control is difficult and expensive at temperatures above 150°F. High strength steels and high stressed steels are more subject to cracking from embrittlement.

Laboratory tests indicate problems with precipitation of significant amounts of insoluble tachyhydrite ($CaMg_2Cl_6 \cdot 12H_2O$) when treating dolomite with HCl concentrations in excess of 20%. Field results indicate severe damage in some wells treated with high strength acid.

Formation damage in dolomite from tachyhydrite precipitate can be appreciably reduced by diluting the spent acid with weak acid or water. Current field practice is to precede and follow each 10,000 gallons of high strength acid with 2,000 to 3,000 gallons of weak acid (5% to 7½% HCl), fresh water, or 1% KCl water.

In relatively dirty carbonate formations where appreciable fines are released by acid treatment, high strength acid will release more fines per volume of acid. This increases the probability of serious plugging of the formation matrix or etched fracture. The use of suspending agents, usually surfactants or polymers, for fines is important in these situations.

Sludge and emulsion plugging is considerably more severe with high strength acid. Many damage problems with both high and low strength acid fracturing are overcome by overflushing with a volume of water equal to the volume of acid used in the treatment. However, in foamed acid treatments, only enough afterflush is used to displace the foamed acid into the formation.

Summary of Use of High Strength Acid

1. Select the strength of acid required for each job. There is no particular significance in the use of 28% HCl as compared with the use of higher or lower strength acid.

2. Consider the use of high strength acid for acidizing limestone or dolomite along with other available acids from the standpoint of cost, performance, corrosion inhibition, and formation damage.

3. Special care should be exercised in selecting surfactants to prevent the formation of emulsions and sludge.

4. A suspending agent such as Halliburton's HC-2 or Dowell's F-78 should be used to reduce fracture plugging from fines in carbonate formations containing more than about 2% acid insoluble material.

5. Whenever possible, fracture etching tests at overburden pressure should be made on formation cores by oil company or service company labs prior to recommending high strength HCl for any carbonate well. These tests should include an analysis of any undissolved and reprecipitated materials in addition to determination of the etching pattern.

SANDSTONE ACIDIZING

Purpose and Acid Reactions—The primary reason to acidize sandstone wells is to increase permeability by dissolving clays near the wellbore. Clays may be naturally occurring formation clays or those introduced from drilling, completion, or workover fluids.

Hydrofluoric acid (HF) can dissolve calcium carbonate, sand, clay, shale, and feldspars. However, the only reason to use HF acid is to remove clay damage. Treating an undamaged well with HF acid will provide a maximum increase in productivity of about 30%. If the depth of clay damage is only a few inches, HF acid stimulation of a sandstone well can give production increases equal to or greater than the damage ratio.

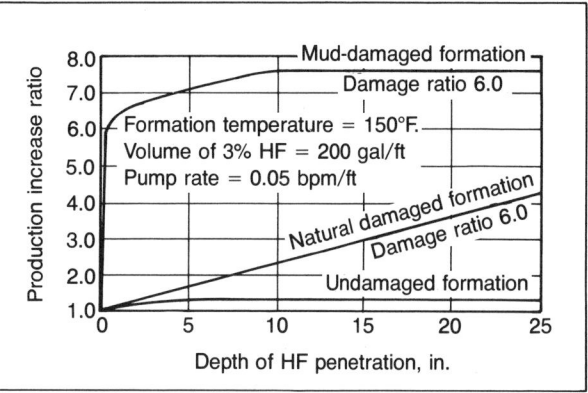

FIG. 7-11—*Effect of depth of penetration of 3% HF on production increase. Courtesy of Halliburton Services.*

The basic hydrofluoric acid treatment for sandstone wells is usually 3% HF acid + 12% HCl. This acid is sold by Dowell and Western as Mud Acid, by Halliburton as HF Acid, and by BJ Hughes as Mud Sol.

Figure 7-11 shows the effect of depth of penetration of 3% HF acid (3% HF + 12% HCl) on production increases in undamaged formation in wells with damaged formation, clays, and in wells damaged by mud invasion.

Reaction of HF Acid on Sand and Clay

$$SiO_2 + 6\,HF \rightarrow H_2SiF_6 + 2H_2O$$
(Sand) (Fluosilicic Acid)

$$Al_2Si_4O_{10}(OH)_2 + 36\,HF \rightarrow$$
(Clay)

$$4H_2SiF_6 + 12H_2O + 2H_3AlF_6$$
(Fluosilicic Acid) (Fluoaluminic Acid)

Reaction rates on sand and clay are dependent on the ratio of the surface area of rock to volume of acid in the sandstone pores. Since clay has a surface area more than 200 times an equal weight of sand, reaction of HF acid on clay is almost instantaneous.

The acids produced by the reaction HF acid on sand and clay will react with sodium, potassium, or calcium ions in NaCl, KCl, or $CaCl_2$ in the wellbore or in the sand around the wellbore and produce insoluble precipitates.

Fluosilicic Acid plus Na^+, K^+, Ca^{++}

$$H_2SiF_6 + 2Na^+ \rightarrow Na_2SiF_6 \downarrow + 2H^+$$
$$H_2SiF_6 + 2K^+ \rightarrow K_2SiF_6 \downarrow + 2H^+$$
$$H_2SiF_6 + Ca^{++} \rightarrow CaSiF_6 \downarrow + 2H^+$$

Fluoaluminic Acid plus Na^+, K^+, Ca^{++}

$$H_3AlF_6 + 3Na^+ \rightarrow Na_3AlF_6 \downarrow + 3H^+$$
$$H_3AlF_6 + 3K^+ \rightarrow K_3AlF_6 \downarrow + 3H^+$$
$$2H_3AlF_6 + 3Ca^{++} \rightarrow Ca_3(AlF_6)_2 \downarrow + 6H^+$$

As may be noted from these reactions, the insoluble precipitates formed are: Na_2SiF_6, K_2SiF_6, $CaSiF_6$, Na_3AlF_6, K_3AlF_6, and $Ca_3(AlF_6)_2$. These fluoride precipitates are gelatinous type materials and occupy a large volume of pore space. They also adhere strongly to rock surfaces and reduce well productivity.

HCl dissolves very little clay and does not dissolve sand. However, HCl can dissolve carbonates present in sandstone formations. HF acid reacts with limestone and precipitates calcium fluoride, an insoluble fine white powder. The total reaction is:

$$CaCO_3 + 2HF \rightarrow CaF_2 \downarrow + H_2O + CO_2$$

To avoid CaF_2 precipitation in sandstone acidizing, a preflush of HCl is used to dissolve the limestone and prevent calcium ions from contacting HF acid. Sandstone formations having more than 20% solubility in HCl should normally be treated with HCl only.

The ammonium ion does not form insoluble compounds with HF acid reaction products. Therefore, ammonium chloride solutions may be used as a preflush or afterflush in HF acid treatments.

Planning HF Acid Stimulation

The objective of most HF acidizing treatments is to eliminate damage around the wellbore due to:

1. Clay invasion of pores from drilling mud and well circulating or workover fluids containing a small quantity of clay.
2. Swelling, dispersion, movement, or flocculation of formation clays and mud solids.

Emulsion blocking around the wellbore may also be removed along with clay blocking.

A matrix-type treatment with injection below fracture pressure should be used. Fracture-type acidizing is not applicable in sandstone wells. The primary result of fracturing sandstone with acid is to damage the cement-to-formation bond, as illustrated in Figure 7-12, and allow channeling of unwanted fluids into the producing zone. If breakdown of perforations is nec-

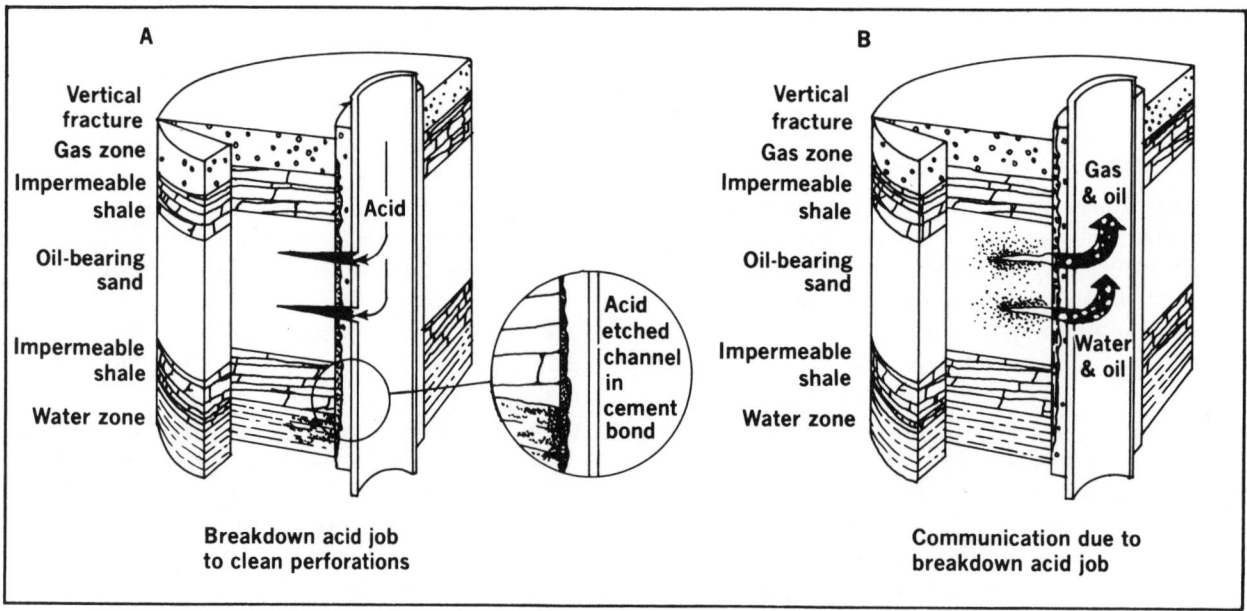

FIG. 7-12—*Effect of fracturing sandstone formations while cleaning perforations.*

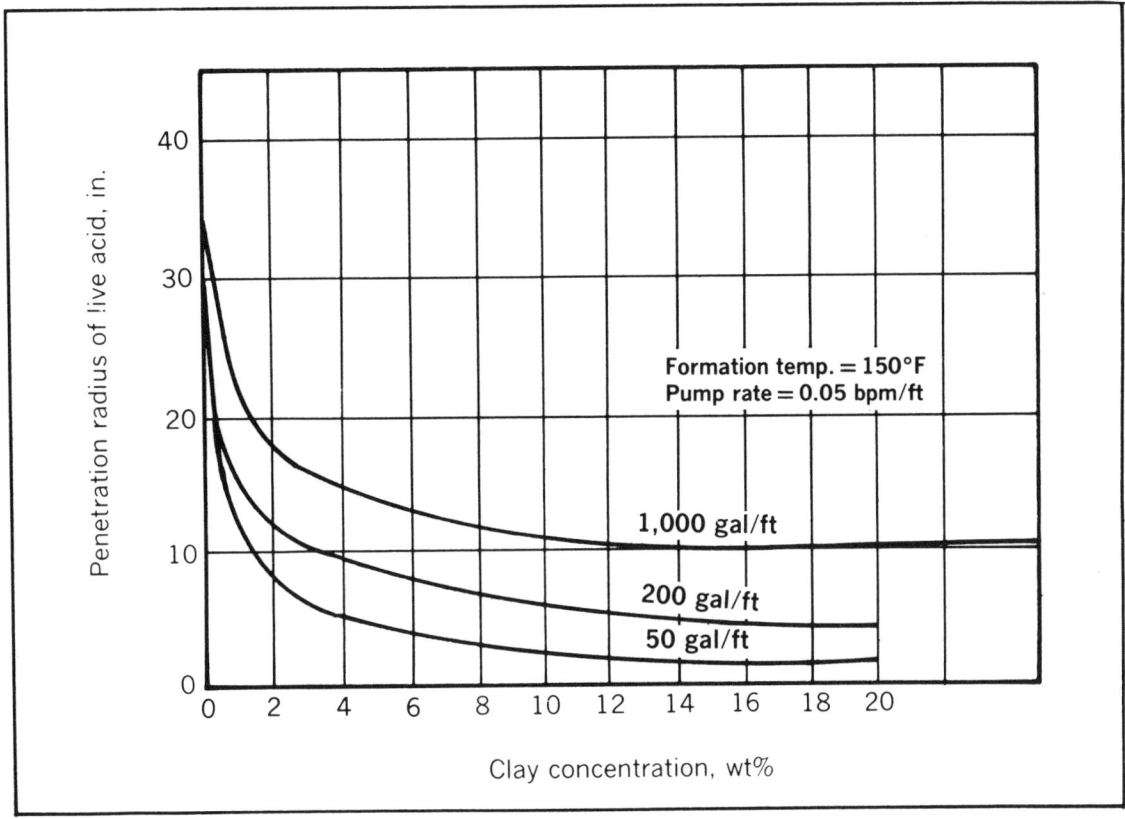

FIG. 7-13—*Penetration radius of live HF acid vs. clay concentration. Courtesy of Halliburton Services.*

essary, use clean water or clean oil, then allow fractures to heal for about thirty minutes. After fractures have healed, HCl or HF acidizing can be performed with injection pressure maintained below fracture pressure.

In planning an HF acid treatment to remove clay damage, the primary factors to be considered are (1) the depth of damage and (2) the weight percent of clay naturally occurring in the formation plus the weight of clay that has been forced into the formation pores near the wellbore from drilling or workover fluids. As clay concentration is increased in the sand around the wellbore, the depth of penetration of live acid is decreased. It may be assumed that reaction of live HF acid on clay is essentially instantaneous. The effect of clay concentration on the penetration depth of live acid, and the volume of acid required per foot of clay-damaged sandstone formation, is shown in Figure 7-13.

To achieve stimulation, the treatment should be designed to meet these requirements:

—Dissolve clays and mud solids near the wellbore.

—Prevent the precipitation of insolubles in the formation.

—Prevent the spent acid from emulsifying.

—Leave the sand and remaining fines in a water-wet condition. Moveable water-wet fines are much smaller than oil-wet particles and, therefore, can be more readily flowed from the formation after acidizing.

—Spent acid should have low surface tension and low interfacial tension to allow easy return to wellbore.

Additives for Sandstone Acidizing

Surfactants should be employed throughout the treatment to prevent emulsions, to lower surface and interfacial tension, and to water-wet sand and clay.

Corrosion inhibitiors should be selected for compatibility with surfactants employed.

Mutual solvents, materials at least partially soluble in both water and oil, have a special role in sandstone acidizing. The mutual solvent initially employed in sandstone acidizing was ethylene glycol monobutyl ether (EGMBE) and is sold under various trade names such as: Musol by Halliburton; U-66 by Dowell and J-40 by BJ Hughes. Halliburton now offers a new mutual solvent, a modified glycol ether, Musol A, which is both a mutual solvent and a good water-wetting surfactant.

Diverting Agents—Most of the diverting agents used in acidizing limestone are applicable. However, rock salt (NaCl) cannot be used with HF acid.

Clay Stabilization

As noted previously, an HF acid treatment is normally preceded by a volume of HCl to dissolve carbonates. However, the action of HCl in the formation can dislodge clays and other fines. If all of these released fines are not dissolved by HF acid, they may migrate toward the wellbore when the well is put back on production. Bridging at flow restrictions in the pores of the sand can occur and cause production delines in a relatively short time. The addition of certain cationic organic polymers to the acid and afterflush can minimize fines migration and subsequent pore plugging. Clay stabilizing polymers are available from Halliburton as Cla-Sta, from Dowell as L-53, and from Western as Clay Master.

Well Preparation Prior to Acidizing Sandstone Formations

1. Clean out debris from wellbore.
2. Remove any paraffin or asphalt in tubing or wellbore, and in formation around wellbore.
3. Remove any acid-soluble and acid-insoluble scales from tubing, wellbore, and perforations.
4. Reperforate if necessary to insure entry of acid into desired intervals.
5. Use a perforation washer or surge tool to clean out perforations if there are indications of perforations being plugged with silt, mud, or other debris. Perforations may also be broken down with water or oil using ball sealers as a diverting system.

The Role of Mutual Solvents in Acidizing Sandstone

Gidley[10] in 1970 reported many successful sandstone acidizing jobs using a mutual solvent, ethylene glycol monobutyl ether, (EGMBE). However, the exact function of mutual solvents in sandstone acidizing remained unclear. As a result, EGMBE has not been used on all jobs, and job techniques have varied between companies and areas.

In 1975 Hall[21] reported on research designed to determine the action of mutual solvents in sandstone acidizing. Experiments were made on packed silica columns using (1) $7\frac{1}{2}\%$ HCl, 0.5% surfactant, 0.3 corrosion inhibitor, with and without mutual solvents, and (2) spent HF-HCl, corrosion inhibitor, cationic surfactants, and mutual solvents.

Emulsion breaking tests, corrosivity tests, and surface tension measurements were made on fluids before and after being pumped through the silica-packed column as a means of evaluating the role of mutual solvents in sandstone acidizing.

Figure 7-14 shows changes in surface tension of a solution of $7\frac{1}{2}\%$ HCl, 0.3% corrosion inhibitor, and either cationic or anionic emulsion breakers before and after flowing the solution through a packed silica column.

It may be noted that the effluent from the tests with cationic emulsion breakers showed an appreciable increase in surface tension after being pumped through the sand and clay column, apparently because the cationics were adsorbed on the silica. Adsorption on sandstone and clays is a general characteristic of most positively charged cationic surfactants since sand, clays, and feldspars are negatively charged.

Since the cationic emulsion breakers and corrosion inhibitors were adsorbed by the sand and clay, these

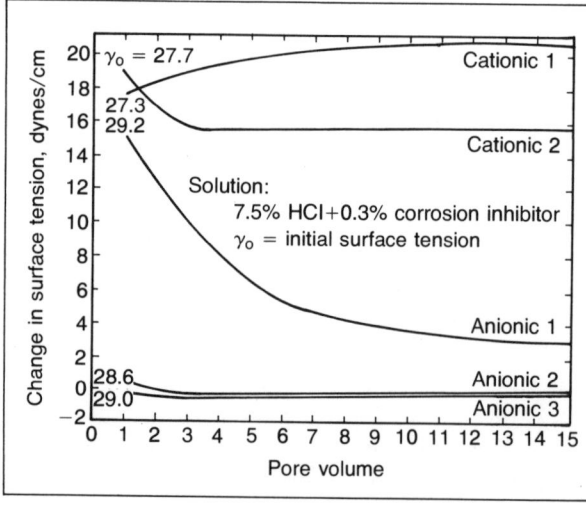

FIG. 7-14—*Change in surface tension after flowing an acid solution through a silicate column with 4.7 × 10^7 sq cm area.*[21] *Permission to publish by The Society of Petroleum Engineers.*

surfactants were not available to maintain low surface tension, prevent emulsions, or prevent corrosion. Two of the three anionic emulsion breakers tested did *not* show appreciable adsorption on the sandstone. The third anionic emulsion breaker showed initial adsorption; however, adsorption decreased with increased volumes pumped through the packed silica column.

The second phase of this experiment, illustrated in Figure 7-15, involved treating the silica-packed column with the original solution containing Cationic 1 plus various mutual solvents. BTG is butoxy triglycol; DEGMBE is diethylene glycol monobutyl ether; EGMBE is ethylene glycol monobutyl ether; and MGE is modified glycol ether.

Results in Figure 7-15 show that addition of 10% by volume of various mutual solvents to the solution of HCl, including Cationic 1, greatly reduced the surface tension of the effluent from the sand-packed column. This experiment indicated that mutual solvents preferentially adsorb on silica, thereby allowing the surfactant to be available for lowering surface tension. There was very little increase in surface tension when EGMBE was used. MGE actually lowered surface tension of the effluent.

To further support these findings, emulsion tests and corrosivity tests were run on both the original solutions and effluent used in surface tension tests, as shown in Figure 7-15. Data and results from these tests are shown in Table 7-2.

All of these tests indicate that preferential adsorption of mutual solvents on the sand and clay prevented the cationic emulsion breaker and cationic corrosion

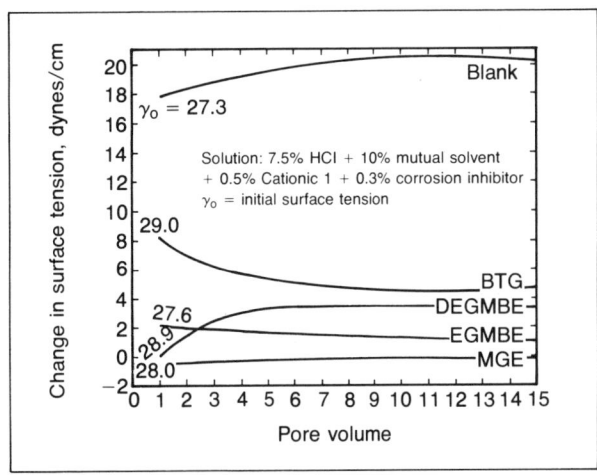

FIG. 7-15—*Change in surface tension after flowing through a silicate column with 4.7×10^7 sq cm surface area.*[21] *Permission to publish by The Society of Petroleum Engineers.*

inhibitor from being adsorbed on the clay. The cationic surfactants were then available to maintain low surface tension, prevent emulsions, and provide corrosion inhibition.

To verify the adsorption characteristics of cationics on sand and the benefits of mutual solvents in sandstone acidizing, Hall ran tests on solutions consisting of spent HF-HCl, Cationic 1 emulsion breaker, a corrosion inhibitor, and various mutual solvents. Results of the tests are shown in Table 7-3.

These data show that when a spent HF-HCl solution and cationics were pumped through the silica

TABLE 7-2

Solution: 7.5% HCl, 0.5% Cationic I, 0.3% Corrosion inhibitor, 10% Mutual solvent

Column: Length: 6 in.
Inside Diameter: 25 mm
Packing: 45 g 70–170 mesh sand +2.5 g silica flour +2.5 g bentonite
Surface area: 4.7×10^7 cm²

Mutual solvent	Initial solution			Column effluent		
	Surface tension	Emulsion break	Corrosion rate	Surface tension	Emulsion break	Corrosion rate
Blank*	27.3	90	.005	47.8	0	.055
EGMBE	27.6	90	.015	28.9	82	.020
DEGMBE	28.9	94	.009	32.1	84	.017
BTG	29.0	92	.006	34.2	74	.011
MGE	28.0	92	.011	27.6	90	.018

*No mutual solvent used on this test.

packed column, cationics were preferentially adsorbed, preventing the cationic emulsion breaker from being effective in lower surface tension and preventing emulsion.

Because the surface area of silica in the packed column of sand used in tests reported in Table 7-3 was only 1.3×10^4 cm^2 in comparison with 4.7×10^7 cm^2 in the packed column of sand, silica flour, and bentonite used on tests reported in Table 7-2, adsorption of the cationics in the sand column resulted only in raising the effluent by 1.8 dynes/cm as reported in Table 7-3. It should be noted that spent HF-HCl acid solution was used in this test, whereas, unspent HCl was used in tests reported in Table 7-2. However, when glycol type of mutual solvents were added to the spent acid, the surface tension of the effluent was lowered about 1 dyne/cm.

These findings of Hall,[21] plus other available information, suggest that for sandstone acidizing:

1. Mutual solvents can effectively prevent the adsorption of emulsion breakers and cationic corrosion inhibitors on sand and clay.
2. Mutual solvents should be used with either HCl or HF-HCl treatments when cationic emulsion breakers are used.
3. Some anionic emulsion breakers do not adsorb appreciably on silica. Therefore, mutual solvents may not be required when specific anionic emulsion breakers are employed in either HCl or HF-HCl acidizing of sandstone. However, mutual solvents should be used on acidizing of sandstone along with suitable surfactants, unless the specific surfactant does not appreciably adsorb on sandstone or clay. Mutual solvents can solubilize surfactants in acid thus increasing the effectiveness of surfactants.
4. Good nonionic surfactants are also available that have proved to be a very effective additive to all types of acids.
5. If mutual solvents are used in HCl or HF-HCl sandstone acidizing, the mutual solvent should always be in the *preflush* along with a suitable surfactant. It may also be advisable to add a mutual solvent to the regular HF-HCl phase, which is usually 3% HF and 12% HCl.
6. In most cases, the use of cationic surfactants should be avoided in sandstone acidizing because of oil-wetting of sand and clay that can cause a decrease in well productivity.

Stimulation of Oil Wells in Sandstone Formations

Some cleanup operations are carried out in sandstone wells with HCl, primarily because HCl can shrink hydrated clays. However, sandstone acidizing is usually carried out with HF acid which can dissolve clay and sand. HF acid treatments consist of three successive steps: preflush, HF-HCl, and an afterflush, as illustrated in Figure 7-16.

Preflush for Sandstone Acidizing of Oil Wells—A preflush of HCl should normally be employed ahead of HF acid to insure that carbonates are removed from the wellbore and from the formation rock adjacent to the wellbore. The preflush is also designed to flush ahead of the HF acid any CaCl$_2$, NaCl, or KCl in the tubing, wellbore, or adjacent reservoir rock, to water-wet clays and sand, and to reduce the tendency to emulsify spent HF acid.

TABLE 7-3

Solution: Spent HF-HCl acid, 0.5% Cationic Surfactant 1, 0.3% Corrosion inhibitor, 10% Mutual Solvent

Column: Length: 9 in.
Inside diameter: 25 mm
Packing: 65 gm 70–170 mesh sand
Surface area: 1.3×10^4 cm^2

Mutual solvent	Initial solution		Column effluent	
	Surface tension	Emulsion break, %	Surface tension	Emulsion break, %
Blank*	28.4	94	30.2	0
EGMBE	29.2	92	28.4	86
DEGMBE	30.0	94	28.9	92
BTG	29.8	92	28.7	90

*No mutual solvent used on this test.

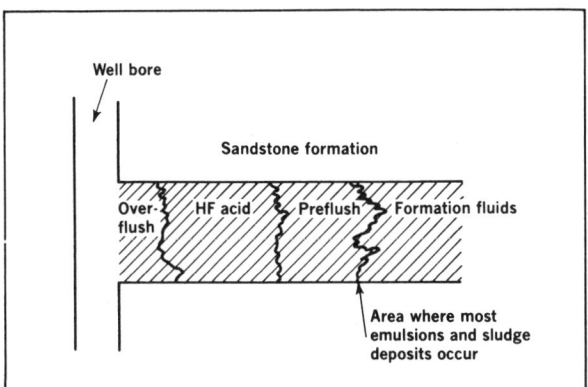

FIG. 7-16—*HF acidizing of sandstone formations. Courtesy of Halliburton Services.*

If it has been conclusively proven that there are *no* carbonate scales in the wellbore or carbonates in the formation, diesel oil, kerosene, crude oil, or ammonium chloride (NH_4Cl) can be used as a preflush. Because most sandstones contain some carbonates as cementing material or as discrete particles, it appears advisable to use an HCl preflush on essentially all jobs, with the percent and volume of HCl depending on quantity of carbonates to be dissolved.

Normal Preflush—Chemicals used: 5% to 15% HCl, surfactant (usually anionic-nonionic), corrosion inhibitor, and 5% to 10% mutual solvent (optional). Inject 50–100 gal of preflush/ft of formation at below frac pressure. If acid is being pumped down tubing in laminar flow, preflush volume should be at least one-half the tubing volume to prevent contact of HF acid with wellbore and formation water. If acid is pumped down tubing in turbulent flow, a smaller preflush volume may be used.

If mutual solvent is used, this will adsorb on sand and clay, allowing surfactants and corrosion inhibitors to be more effective. A mutual solvent should probably be used unless it has been proven that specific surfactants being employed do not appreciably adsorb on sand and clay. Use 10% mutual solvent, except for Halliburton's Musol "A," where 5% is recommended.

HF-HCl Acid Treatment for Oil Wells—Chemicals used: 3% HF + 12% HCl, surfactant, corrosion inhibitor, and 5 to 10% mutual solvent (optional). Prepare acid with fresh water only. From 1½% to 9% HF has been used; however, 3% HF acid is normal. The 2½% HF–7½% HCl acid treatment has been used extensively in California in very unconsolidated Miocene sand. Higher strength than 3% HF should usually be avoided to minimize formation collapse. Inject at below frac pressure 200 gal of 3% HF + 12% HCl/ft of sandstone formation as standard procedure.

In very permeable sand, very shaley sand, or sand with extensive clay damage, greater treatment volumes may be required. If experience proves that 200 gal/ft is not required, volume may be reduced in steps to 125–150 gal/ft of sand. Determination of clay concentration in cores by X-ray analysis can aid appreciably in determining volume of acid required. Flow tests on cores should also be part of the lab investigation.

Effect of HF Acid on Sandstone Formation—HF acid dissolves feldspar, naturally occurring formation clays, and clays deposited from drilling and completion fluids. It also reacts slowly on formation sand. HF acid reacts rapidly on carbonates to produce an insoluble precipitate, calcium fluoride.

The sandstone acid treatment may release some undissolved clays and other fines, 2 microns or smaller, which are excellent emulsion stabilizers. However, oil-wetting of clays and sand is usually necessary for emulsion stabilization in the formation. Radial flow calculations show that a 2,000 cp emulsion to a depth of one foot around the wellbore will reduce productivity to about 0.001 of its original value.[10]

Complete dissolution of cementing materials, normally silica, carbonates, or clays, between sand grains will result in disintegration of the formation matrix.

Afterflush for Oil Well Treatments—Inject about 25 gal afterflush per foot of sand. Afterflush may be 5% to 10% HCl, 2% Ammonium Chloride solution, clean filtered kerosene, diesel oil, or crude oil. All fluid should contain about 0.1% water-wetting non-emulsifying surfactant. The purpose of the afterflush is to act as a buffer between the HF acid and the pump down fluid. Sufficient afterflush should be used to thoroughly displace the HF acid into the formation.

HF acid spends very rapidly. Within one hour after afterflush is injected, swabbing, pumping, or gas lifting of the spent acid should be initiated. This will reduce the possibility of formation damage due to emulsion and insoluble precipitates.

Stimulation of Gas Wells, Gas Injection Wells, and Water Injection Wells

Treatment should follow the oil well stimulation procedure with these modifications:

1. HCl should normally be used in the preflush or afterflush. Oil should not be employed in the preflush or afterflush.

2. Gas wells should be swabbed or flowed within one hour after HF acid treatment.

3. It is not always necessary to swab water and gas injection wells following an HF acid job. Within one hour after treatment, regular injection into water or gas injection wells may be resumed.

Procedures Contributing to Successful Sandstone Acidizing

1. Select wells with positive skin damage as measured by pressure buildup and drawdown tests for sandstone stimulation.

2. Use 5% to 15% HCl preflush as standard procedure.

3. Employ 200 gal 3% HF-12% HCl per foot of sand as standard procedure. The 50 gal per foot treatment, recommended for many years, appears to be a major reason for many previous failures.

4. Use water-wetting non-emulsifying surfactants, usually anionics and nonionics, throughout treatment.

5. Use mutual solvents where indicated by API RP-42 (1977 revised) tests.

6. Start removing the spent HF acid from around the wellbore within about one hour after treatment is completed.

Emulsion Upsets in Surface Facilities Following HF/HCl Sandstone Acidizing

Although there may be little or no emulsion present in spent acid at the wellhead following an HF/HCl sandstone acid job, commingling of this spent acid with salt water from other wells has produced serious emulsion upsets in surface treating facilities.[19] This problem appears to be primarily due to mixing 0–3 pH spent acid containing precipitates, usually hydroxides and fluosilicates, with 6–7 pH water in the surface tanks. Produced fines also contribute significantly to the problem.

Coppel[19] suggests the following to avoid emulsion upsets: (1) select and use an effective demulsifier at high concentrations, (2) avoid commingling of spent and acid salt water, and (3) minimize the concentration of fines and precipitates in the system where it is not possible to avoid commingling.

In-Situ HF Generating System (SGMA)[20]

Shell's SGMA, an in-situ acid generating system was developed to allow cleanup of deep damage due to clays in sandstone formations. In normal HF treatments, clays are dissolved to a depth of perhaps 6 to 12 in. around the wellbore, depending on clay content.

SGMA involves pumping into the formation an aqueous solution of ammonium fluoride and an organic ester such as methyl formate. With time the ester hydrolyzes to produce an organic acid such as formic acid. The organic acid reacts with NH_4F to form HF acid, which rapidly dissolves clay or siliceous fines present in the pores. The system is applicable from 130° to 200°F. HF solutions up to concentration of 3.5% can be generated.

Typical SGMA treatment—Following is a typical treatment in East Bay area, Louisiana, while pumping at about $\frac{2}{3}$ gal/min/ft of perforations:

Step 1—Spearhead: Xylene, 10% HCl, and $1\frac{1}{2}$% HF–$7\frac{1}{2}$% HCl.
Step 2—Spacer: 3% ammonium chloride
Step 3—SGMA

Wells should be shut in for a long period of time following treatment, with required shut-in time decreasing with increased temperature. For example, a 150°F well is shut in for about twelve hours and a 170°F well for about four hours.

Wells are brought back into production very slowly by gradually increasing choke size over a period of several weeks.

Results reported on more than 100 wells in the Gulf Coast area of the U.S. were encouraging. The primary application for this type of treatment appears to be in wells with deep clay damage.

Sequential HF Process (SHF)[27]

Halliburton's SHF, an in-situ HF generating system, has the same objective as Shell's SGMA—to cleanup deep damage to clays in sandstone formations. The SHF process utilizes the ion-exchange properties of clay minerals to generate in-situ hydrofluoric acid on the clay particle.

In practice, a solution of HCl containing no fluoride ion is pumped into the formation. This acid solution will contact the clays in the rock and exchange protons (H^+) for the cations natural to the clay minerals, thus converting the clay to an acidic clay. Next, a neutral or slightly basic solution of fluoride ion is pumped into the formation. This solution will contact acidic clay particles and combine with protons previously absorbed to generate hydrofluoric acid on the clay minerals. In addition, some anion exchange will occur whereby fluoride ions (F^-) will be substituted for the anions natural to the clays. The in-situ generated HF acid rapidly reacts with and dissolves a portion of the clay.

Typical Treatment—A typical treatment consists of pumping the following:

1. Preflush 100 gal/ft of 5% HCl
2. 50 gal/ft of 3% HF 12% HCl
3. 25 gal/ft of 2.8% NH_4F (pH 7-8)
4. 25 gal/ft of 5% HCl
5. Afterflush of HCl, NH_4Cl, diesel oil or kerosene

Steps 3 and 4 are the in-situ acid generating steps and are referred to as sequences. After the cleanup procedures in steps 1 and 2, then 3 to 6 sequences of alternating steps of NH_4F followed by HCl are used.

Field Results—Over 200 wells have been treated with the SHF process, marketed by Halliburton as Claysol. For deep damage removal, Claysol appears to have the following advantages over conventional HF acid treatments: improved stimulation, more rapid cleanup, less corrosive, and more compatible with sand-consolidated formations.

Clay Acid

Clay acid (fluoboric acid) was developed by Dowell as a retarded acid system for use on sandstone formations. It is designed to reduce deep damage attributable to migration of clays and fines. The fluoboric acid not only dissolves clays, but also immobilizes clays and fines that are contacted but not dissolved.

Fluoboric acid hydrolyzes to generate hydrofluoric acid (HF and hydroxy Fluoboric acid) according to the following equilibrium equation:

$$HBF_4 + HOH \rightleftharpoons HBF_3OH + HF$$

Although the hydrolysis proceeds rapidly, the equilibrium allows only about 5% of the available HF to exist at any one time. This equilibrium causes the reaction on the formation to proceed slowly for the same reason that the reaction of acetic acid is slower than that of HCl. The slower reaction rate allows the fluoboric acid to penetrate a greater distance into the formation before spending. After the available HF has spent, the remaining hydroxy fluoboric acid (HBF_3OH) slowly reacts with the clays "fusing" (or bonding) them and other fines to one another and to the sand grains. As a result, fines are stabilized against dispersion by incompatible fluids and mechanical dislodgement.

The "fusion" reaction is not well-defined, but cation exchange capacity tests, flow tests with various fluids, and scanning electron microscope studies all demonstrate that the clays and fines contacted by clay acid are changed. This change is attributed to the slow secondary reaction between hydroxyfluoboric acid and the clays. Best results are achieved if wells are shut-in for up to two days following clay acid treatment, depending on the temperature. This is in contrast to the accepted practice of returning wells rapidly to production after conventional mud acid treatment.

Field results with fluoboric acid were reported in August, 1981,[37] where multiple-rate flow tests were used to determine well treatment results in a damaged formation after both mud acid and clay acid treatments. Wells treated with clay acid showed little, if any, decline even after six to eight months; wells after mud acid treatment had declined severely in less than six months. Clay acid has been of major importance in gravel-packed offshore gas wells; however, success has also been obtained both onshore and offshore in oil wells.

POTENTIAL SAFETY HAZARDS IN ACIDIZING

Hydrogen sulfide, a poison gas, may be produced from the reaction of acid on sulfide scale. Hydrogen sulfide smells like rotten eggs at low concentrations. High concentrations can paralyze the olfactory nerves and prevent detection by smell of dangerous gas concentrations. High concentrations can also paralyze other nerves in the respiratory system.

Arsenic inhibitor is poisonous if swallowed. Contact of arsenic with aluminum or magnesium may produce arsine gas in dangerous concentrations. Arsine gas is an inhalation hazard and is very deadly. Arsenic inhibitors should generally be avoided because of their toxicity and the environmental protection problems. Arsenic is very persistent as it does not decompose to harmless residue.

Acetic anhydride, used in formulating acetic acid, produces vapors which are very irritating, and direct contact will cause severe burns. If water or dilute acid is added to acetic anhydride, an explosion may occur. Always add acetic anhydride to water or dilute acid.

Dust from ammonium bifluoride used in making HF acid is very irritating. This dust will cause severe burns if combined with moisture. Contact with the dust or active HF acid should be avoided.

Most additives used in acid are toxic to varying degrees. Chemicals contacting the skin should be removed immediately by washing with soap and water. Clothing contaminated by chemicals should not be worn until laundered.

REFERENCES

1. Stone, J. B. and Hefley, D. G.: "Basic Principles in Acid Treating Limes and Dolomites," *Oil Weekly*, Nov. 11, 1940.

2. Hendrickson, A. R., Hurst, R. E., and Wieland, D. R.: "Engineering Guide for Planning Acidizing Treatments Based on Specific Reservoir Characteristics," *J. Pet. Tech.*, Feb. 1960, p. 16.

3. Knox, J. A., Lasater, R. M., and Dill, W. R.: "A New Concept in Acidizing Utilizing Chemical Retardation," 39th Annual SPE Meeting, Houston, October, 1964.

4. King, C. V. and Liu, C. L.: "The Rate of Solution of Marble in Dilute Acids," *Journal of ACS*, Vol. 55, May, 1933.

5. Pollard, P.: "Evaluation of Acid Treatments from Pressure Build-Up Analysis," *AIME Petr. Transactions*, Vol. 216, 1959.

6. Smith, C. F. and Hendrickson, A. R.: "Hydrofluoric Acid Stimulation of Sandstone Reservoirs," *J. Pet. Tech.*, Feb., 1965, pp. 215–222.

7. Nougara, J. and Labbe, C.: "Etude des Lois de L'Acidification Dans Le Cas D'un Calcaire Vacuolaire," *Revere of L'Institut Francaise du Petrole*, Vol. 10, Part 1, 354 (Jan.–June, 1959).

8. "Concentrated Acid for Improved Limestone and Dolomite Stimulation," Halliburton Technical Bulletin 6/67.

9. Gatewood, J. R., Hall, B. E., Roberts, L. D., Lasater, R. M.: "Predicting Results of Sandstone Acidizing," *J. Pet. Tech.*, June, 1970.

10. Gidley, John L.: "Stimulation of Sandstone Formations with the Acid-Mutual Solvent Method," *J. Pet. Tech.*, May 1971, p. 551.

11. Williams, B. B., and Whiteley, M. E.: "Hydrofluoric Acid Reaction with a Porous Sandstone," *Trans. SPE of AIME* (1971) 251, II-306.

12. Hayden, Brian R.: "Shell's Experience on Cedar Creek Anticline, Montana, with Blended HCl Acid—Unabeads Mixture to Continuously Divert and Stimulate Carbonate Formations," *SPE 3342*, June, 1971.

13. Williams, B. B., and Nierode, D. E.: "Design of Acid Fracturing Treatments," *Trans. SPE of AIME* (1972) 253, I-849.

14. Nierode, D. E., Williams, B. B., and Bombardieri, C. C.: "Prediction of Stimulation from Acid Fracturing Treatments," *J. Can. Pet. Tech.*, (Oct.–Dec. 1972) 31.

15. Roberts, L. D., and Guin, J. A.: "A New Method for Predicting Acid Penetration Distance," *SPE 5155* (1974).

16. Williams, B. B.: "Hydrofluoric Acid Reaction with Sandstone Formations," Paper No. 74-Pet. I, Pet. Mech. Eng. Conference, Dallas, Texas (Sept. 15–18, 1974).

17. McCune, C. C., Fogler, H. S., Lund, K., Cunningham, J. R., and Ault, J. W.: "A New Model of Physical and Chemical Changes in Sandstone During Acidizing," *SPEJ*, Oct. 1975, 361.

18. Lund, K., Fogler, H. S., and McCune, C. C.: "On Predicting the Flow and Reaction of HCl/HF Acid Mixtures in Porous Sandstone Cores," *SPEJ*, Oct. 1975, p. 248.

19. Coppel, C. P.: "Factors Causing Emulsion Upsets in Surface Facilities Following Acid Stimulation," *J. Pet. Tech.*, Sept. 1975, p. 1,060.

20. Templeton, G. C., Richardson, E. A., Karnes, G. T., and Lybarger, J. H.: "Self-Generating Mud-Acid (SGMA)," *J. Pet. Tech.*, Oct. 1975, p. 1,199.

21. Hall, B. E.: "The Effect of Mutual Solvents on Adsorption in Sandstone Acidizing," *J. Pet. Tech.*, Dec. 1975, p. 1,439.

22. Fredrickson, Sherman E., Broaddus, Gene C.: "Selective Placement of Fluids in a Fracture by Controlling Density and Viscosity," *J. Pet. Tech.*, May 1966, p. 597.

23. McCune, C. C., Ault, J. W., Dunlap, R. G.: "Reservoir Properties Affecting Matrix Acid Stimulation of Sandstones," 1975 *SPE of AIME Transactions*, Volume 259, pp. 633–640.

24. API RP 42, Revised 1977: "Recommended Practice for Laboratory Testing of Surface Active Agents for Well Stimulation."

25. Reed, M. G.: "Formation Permeability Damage of Mica Alteration and Carbonate Dissolution," *SPE 6009* (1976).

26. Ellenberger, C. W., Aseltine, R. J.: "Selective Acid Stimulation to Improve Vertical Efficiency in Injection Wells—A Case History," *J. Pet. Tech.*, Jan. 1977, p. 25.

27. Hall, B. E., and Anderson, B. W.: "Field Results for a New Retarded Sandstone Acidizing System," SPE-6871, Oct. 1977.

28. Thomas, R. L., and Crowe, C. W.: "Matrix Treatment Employs New Acid System For Stimulation and Control of Fines Migration in Sandstone Formations," SPE 7566, Oct. 1978.

29. Dill, R. W., and Keeney, B. R.: "Optimizing HCl-Formic Acid Mixtures For High Temperature Stimulation, SPE 7567, Oct. 1978.

30. Scherubel, G. A., and Crowe, C. W.: "Foamed Acid, A New Concept in Fracture Acidizing," SPE 7568, Oct. 1978.

31. Schriefer, F. E., and Shaw, M. S.: "Use of Fine Salt As A Fluid Loss Material in Acid Fracturing Stimulation Treatments," SPE 7570, Oct. 1978.

32. Brown, B. O., and Dill, F. E.: "A Self-Decentralizing Hydra-Jet Tool," Southwestern Petroleum Short Course, 1975.

33. Williams, B. B., Gidley, J. L., and Schechter, R. S.: *Acidizing Fundamentals,* Henry L. Doherty Monograph Series, SPE (1979) Vol. 6.

34. McBride, J. R., Rathbone, M. J., and Thomas, R. L.: "Evaluation of Fluoboric Acid Treatment in Grand Isle Offshore Area Using Multiple Rate Flow Test," SPE Paper No. 8399, Sept. 1979.

35. Crowe, C. W., Martin, R. C., and Michaelis, A. M.: "Evaluation of Acid Gelling Agents for Use in Well Stimulation," SPE Paper No. 9384, Sept. 1980.

36. Ghauri, W. K.: "Production Technology Experience in Large Carbonate Waterflood, Denver Unit, Wasson, San Andres Field, West Texas," SPE 8406, Sept. 1979.

37. Thomas, R. L. and Crowe, C. W.: "Matrix Treatment Employs New Acid System for Stimulation and Control of Fines Migration in Sandstone Formations," J. Pet. Tech., Aug. 1981, p. 1,491.

38. Ford, W. G. F., Burkleca, Lee F., Squire, K. A.: "Foamed Acid Stimulation: Success in the Illinois and Michigan Basins," SPE 9386, Sept., 1980.

39. Ford, W. G. F.: "Foamed Acid, An Effective Stimulation Fluid," SPE 9385, Sept., 1982.

40. Schaughnessy, C. M. and Kline, W. D.: "EDTA Removes Formation Damage at Prudoe Bay," SPE 11188, Sept., 1982.

41. Watkins, D. R. and Roberts, G. E.: "On-Site Acidizing Fluid Analysis Shows HCl and HF Contents Often Varied Substantially From Specified Amounts," *J. Pet. Tech.*, May, 1983, p. 865–871.

42. McLeod, Jr., H. O., Ledlow, L. B., Till, M. V.: "The Planning, Execution, and Evaluation of Acid Treatments in Sandstone Formations," SPE 11931, Oct., 1983.

43. Schaughnessy, C. M. and Kunze, K. R.: "Understanding Sandstone Acidizing Leads to Improved Field Practices," SPE 9388, Sept. 1980.

Chapter 8 Hydraulic Fracturing

Fracturing concepts and misconceptions
Fracture mechanics
Production increases from fracturing
Propping the fracture
Frac fluids
Frac job design
Frac job performance
Evaluation techniques

INTRODUCTION

Since hydraulic fracture well stimulation was introduced in the early 1950's, technology has increased tremendously. Frac job costs in certain situations may range upward to perhaps 100% of well drilling cost. Numerous factors must be considered to optimize a particular treatment.

Fracturing for Well Stimulation

Objective—The objective of hydraulic fracturing for well stimulation is to increase well productivity by creating a highly conductive path (compared to reservoir permeability) some distance away from the wellbore into the formation. Usually the conductivity is maintained by propping with sand to hold the fracture faces apart. Acid fracturing involves most of the same considerations as hydraulic fracturing except that conductivity is generated by removing portions of the fracture face with acid, leaving etched channels after the fracture closes.

Fracture Initiation—A hydraulic fracture treatment is accomplished by pumping a suitable fluid into the formation at a rate faster than the fluid can leak off into the rock. Fluid pressure (or stress) is built up sufficient to overcome the earth compressive stress holding the rock material together. The rock then parts or fractures along a plane perpendicular to the minimum compressive stress in the formation matrix.

Fracture Extension—As injection of frac fluid continues, the fracture tends to grow in width as fluid pressure in the fracture, exerted on the fracture face, works against the elasticity of the rock material. After sufficient frac fluid "pad" has been injected to open the fracture wide enough to accept proppant, sand is added to the frac fluid and is carried into the fracture to hold it open after the job.

A vertical fracture grows in length upward, downward, and outward. The growth upward or downward may be stopped by a barrier formation; downward growth may also be stopped by fallout of sand to the bottom of the fracture. The growth outward away from the wellbore, (as well as upward or downward) will be stopped when the rate of frac fluid leakoff through the face of the fracture into the formation equals the rate of fluid injection into the fracture at the wellbore.

When sufficient sand has been injected, the pumps are shut down, the pressure in the fracture drops, and the earth compressive stress closes the fracture on the proppant.

Sand-Out May Stop Treatment Prematurely—The width of the fracture is related to the "net fracture pressure" (pressure in excess of fracture closure pressure) working against the elasticity of the formation. As sand enters the fracture and is deposited, more fluid pressure is required to create greater stress against the fracture face to increase the frac width. If the required fluid pressure cannot be applied due to equipment or casing limitations, fluid injection rate slows, sand drops out of the fluid at a more rapid rate, and a sand-out in the fracture occurs.

Providing sufficient fracture length and width has been generated, a sand-out within the fracture is desirable from the standpoint of well productivity.

A sand-out in the casing can occur due to the fact that insufficient fracture width has been generated to accept the size sand carried in the fluid—or due to drop out of sand inside the casing closing off the perforated section. Sand-out in the casing usually occurs early in the treatment and is obviously undesirable.

Fracturing Other Than for Stimulation

Hydraulic fracturing is related to many well procedures other than stimulation, i.e., squeeze cementing, gravel packing, and many cases of lost circulation in drilling and casing operations. Commonly used terms, such as "breakdown" and "fracturing," need to be explained in terms of fracture mechanics in order to have a clear understanding of their significance.

Breakdown—What Does It Mean?—The occurrence of "breakdown" is often seen at the surface as a pressure peak. Once the pressure peak is surmounted, fluid can be injected into the formation at lower pressures. Breakdown effect is caused by the concentration of compressive stress in the formation close to the borehole. This stress concentration results when a portion of the rock is removed (by drilling the hole) while the regional rock matrix load is unchanged; thus the rock at the borehole accepts greater compressive stress.

To initiate a fracture sufficient hydraulic pressure must be applied to overcome this increased stress level at the wellbore. Use of a "penetrating" fluid, wherein fluid pressure tends to support some of the regional rock matrix load, reduces the required fluid pressure to initiate breakdown.

A pressure peak similar to "breakdown" may be a result of mud-plugged or debris-plugged perforations which do not permit pressure inside the casing to be transmitted effectively to the formation rock. Thus, the plugging material must be bypassed by the frac fluid in some manner, or must itself become the "spearhead" of the frac fluid.

Sedimentary Rock—Not a Piece of Glass—As applied to a sedimentary formation, the word fracture is sometimes thought to be an "irreparable occurrence" somewhat the same as breaking a piece of glass. This is not true. In creating a fracture, the formation matrix stress is temporarily overcome using fluid pressure.

As soon as the fluid pressure is relaxed, the fracture closes back with little if any increase in conductivity along the fracture, unless propped open by sand, (or in the case of a high pressure squeeze job, by cement filter cake).

Low Viscosity, Non-Plugging Fluid Cannot Create Much Frac Area—A low viscosity fluid injected at low rates tends to leak off so fast into the formation that it is impossible to build up sufficient pressure to create a fracture. Many times leakoff is restricted because the rock pores at the wellbore or the perforations are plugged.

In this case, frac pressure can be generated, but communication tends to occur over a small interval, (perhaps one or two perforations, or one foot in openhole). All flow occurs through this small communication path to the porous rock beyond the plugged zone. Here fluid pressure is dissipated quickly into the porous zone and very little fracture area is generated.

—Injecting a clean fluid without propping agent at above fracture pressure usually will not break through zonal barriers because the areal dimensions of the fracture are restricted.

—On the other hand, if the objective of injecting the fluid is to contact the plugged zone with a treating chemical, then obviously this purpose is not accomplished by creating the fracture.

Good Cement Sheath Does Not Fracture Preferentially—Where cement completely fills the annular space between the casing and the formation (all mud except filter cake mud on the permeable zones having been displaced) the compressive stress on the rock at the borehole should be effectively transmitted to the cement.

If this is true, the cement sheath is subject to a higher compressive stress than the formation rock several feet back from the borehole. Thus, when hydraulic pressure is built up to the point of overcoming the concentrated compressive stress in the cement sheath and in the rock very close to the face of the borehole, "breakdown" occurs and the fracture "grows" as a vertical fracture out away from the wellbore into the formation.

It moves preferentially away from the wellbore to regions of lower matrix stress rather than up or downward along the cement sheath where a higher compressive stress level exists, as shown in Figure 8–1.

Where the cement has not completely displaced the annular mud, (thus gelled mud channels exist), it may not be possible for the formation to effectively transmit compressive stress to the cement sheath. Thus, it may fracture at a lower pressure, or more likely, pressured fluid may move gelled mud through the channel. If hydraulic pressure transmitted through the

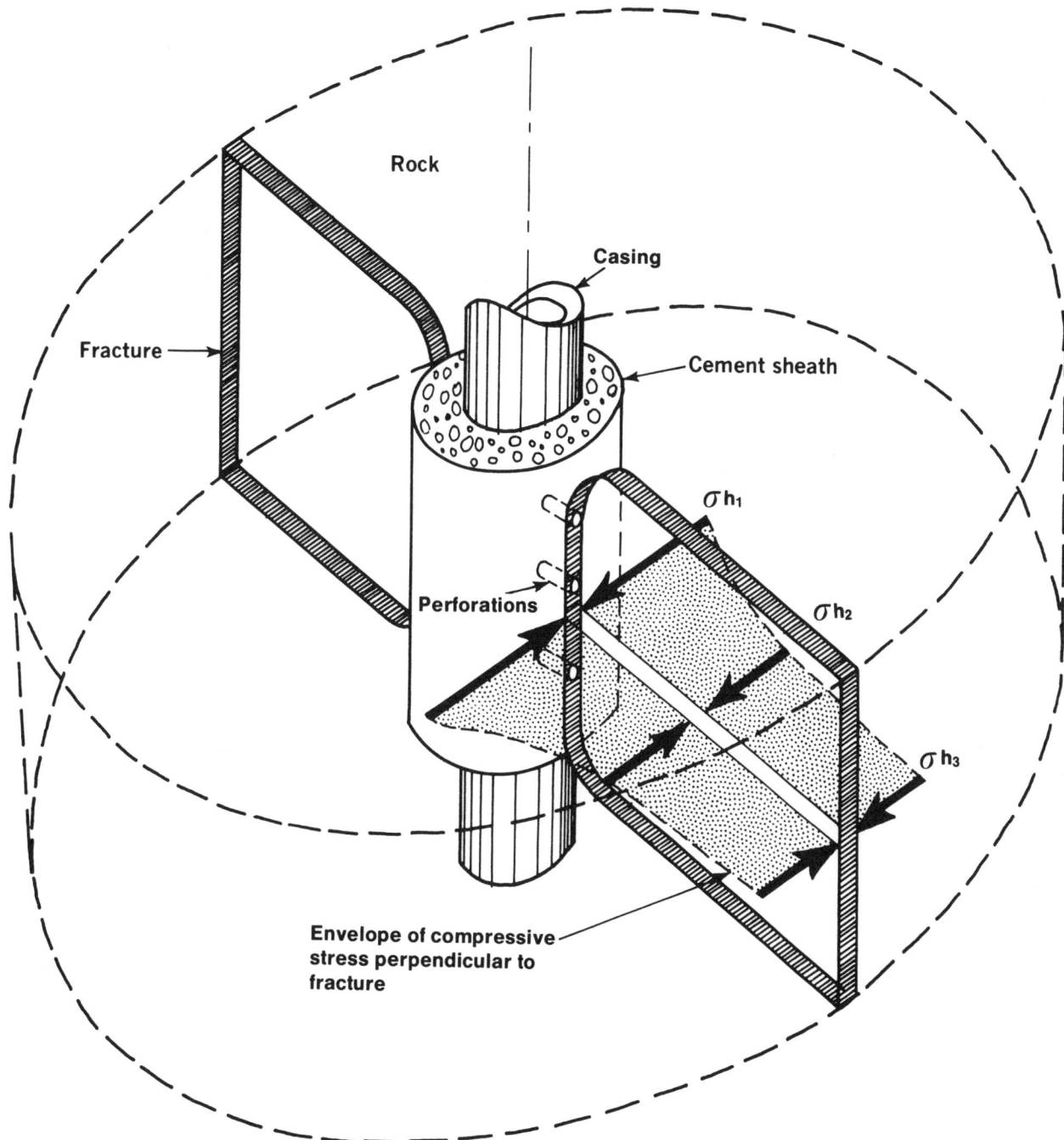

FIG. 8-1—*Fracture propagates outward away from high-compressive-stress concentration at wellbore.*

mud channel effectively contacts a nearby formation that is stressed to a lesser degree, that formation may fracture in preference to the perforated zone.

The obvious solution to the problem is a good primary cement job such that gelled mud channels are not depended on to prevent fluid communication.

MECHANICS OF FRACTURING

The mechanics of fracture initiation and extension, and the resulting fracture geometry are related to the stress condition near the borehole and in the surrounding rock, the properties of the rock, the characteristics of the frac fluid, and the manner in which the fluid is injected.

Hubbert and Willis presented a simplified fracture mechanics theory which seems to explain many of the events observed in field operations during squeeze cementing, gravel packing, hydraulic fracturing, and some instances of lost circulation during drilling. Refinements of the basic theory are being developed, but a knowledge of the Hubbert and Willis theory provides a basis for understanding formation fracturing. The work of Howard and Fast, Perkins and Kern, Nordgren, and Kristanovich form the basis for analysis of fracture extension and fracture geometry. Appendix B defines some of the basic rock mechanics parameters and briefly discusses how they can be determined.

Regional Rock Stresses

Strength Primarily Due to Stress Condition—Subsurface rocks are normally in a state of compressive stress due to the weight of the overburden. (See Figure 8–2). This overburden weight creates stresses in both the vertical and horizontal directions. Sedimentary rocks have little inherent tensile strength, rather are "held together" by compressive stresses. A fracture is extended when sufficient differential hydraulic pressure is applied to overcome these compressive stresses.

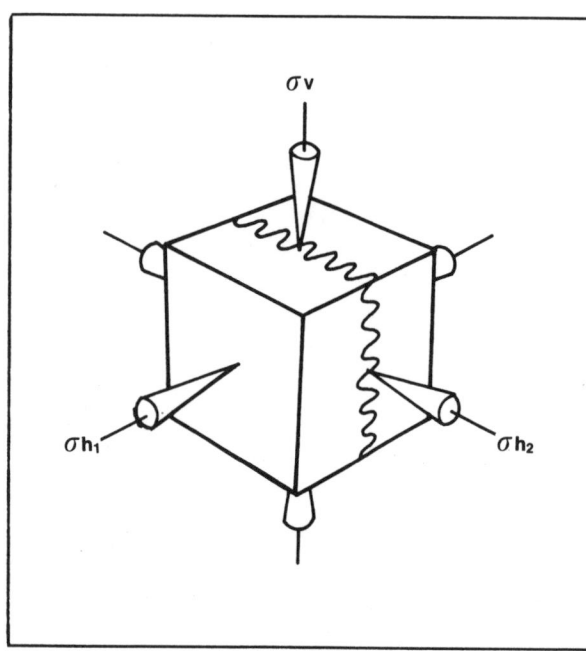

FIG. 8-2—*Tri-axial loading of rocks.*

Stress Calculation—Assuming the earth's crust is an *elastic* material in a relaxed *tectonic* condition, vertical and horizontal compressive stresses can be calculated as follows:

Total vertical compressive stress can be calculated by:

$$\sigma_v = 0.007\rho D \qquad (1)$$

where

σ_v = total vertical stress, psi
ρ = average rock density, lb/ft^3
D = depth, ft

Rock densities vary from 125 to 200 lb/ft^3; 144 lb/ft^3 is a reasonable average, and is the basis for the rule of thumb that the total vertical stress due to the overburden is 1.0 psi/ft. In the Cotton Valley area of East Texas, density log measurements indicated an average rock density to a depth of about 10,000 feet of 155 lb/ft^3. In deep water offshore, the lower density of seawater may significantly affect stress at a particular well depth measured from the ocean surface.

Effective matrix compressive stress is reduced where the formation has porosity and contains fluid. Part of the overburden load is supported by the pressured fluid.

—Effective vertical compressive stress $\tilde{\sigma}_v$, in the rock matrix can be calculated by:

$$\tilde{\sigma}_v = 0.007\rho D - P_r \qquad (2)$$

P_r = formation pore pressure, psi

As this equation indicates, vertical matrix stress is influenced by pore pressure. Matrix stress is increased by declining reservoir pressures. Abnormal pressures reduce matrix stress—thus measurement of shale density (or something related there to) is a useful indicator of abnormal formation pressure zones in drilling operations because the higher pore pressure reduces shale compaction.

—Effective horizontal compressive stress, $\tilde{\sigma}_h$ in the rock matrix can be calculated by:

$$\tilde{\sigma}_h = \frac{\nu}{1 - \nu}(\sigma_v - P_r) \qquad (3)$$

ν = Poisson's ratio

Laboratory measurements of Poisson's ratio range from about .15 to .40. Based on fracture gradients a

Midcontinent limestone may have a value of 0.27 compared with 0.33 for a softer Gulf Coast sandstone. Sonic log measurements offer possibilities for determining Poisson's ratio in-situ.

Horizontal Matrix Stress Depends on Rock Properties and Pore Pressure—In a tectonically inactive area and where rocks act as elastic materials, horizontal matrix stress is about one-third to one-half the vertical stress depending on the Poisson's ratio of the particular rock. With a value of Poisson's ratio for adjacent zones it should be possible to predict which zone would fracture preferentially.

In soft shales or unconsolidated sands horizontal matrix stress should be relatively higher. Rigid materials, such as dolomite or limestone, should fracture at lower pressures. In salt zones where Poisson's ratio may be 0.5, horizontal matrix stress may be equal to vertical matrix stress, thus high fracture pressures—and perhaps horizontal fractures should result. Casing collapse in uncemented salt zones may also be related to the high horizontal stress.

Over geologic time, relaxation or creep may affect horizontal rock stress, causing it to be greater than would be calculated from elastic theory. This is particularly true in the case of shales.

Pore pressure also affects horizontal matrix stress, since, as was previously shown, pore pressure affects vertical matrix stress, and horizontal matrix stress is a function of vertical matrix stress.

Stress Distortion Caused by the Borehole

The presence of a wellbore distorts the pre-existing stress field in the rock for a short distance away from the wellbore, and can critically affect the wellbore pressure required to initiate a fracture.

Stress Level Depends on Ratio of Horizontal Stresses—Hubbert and Willis have shown that the compressive stress at the wall of the borehole depends on the ratio of the two principal horizontal regional stresses, as well as the magnitude of the regional stresses. This is shown graphically in Figure 8-3,

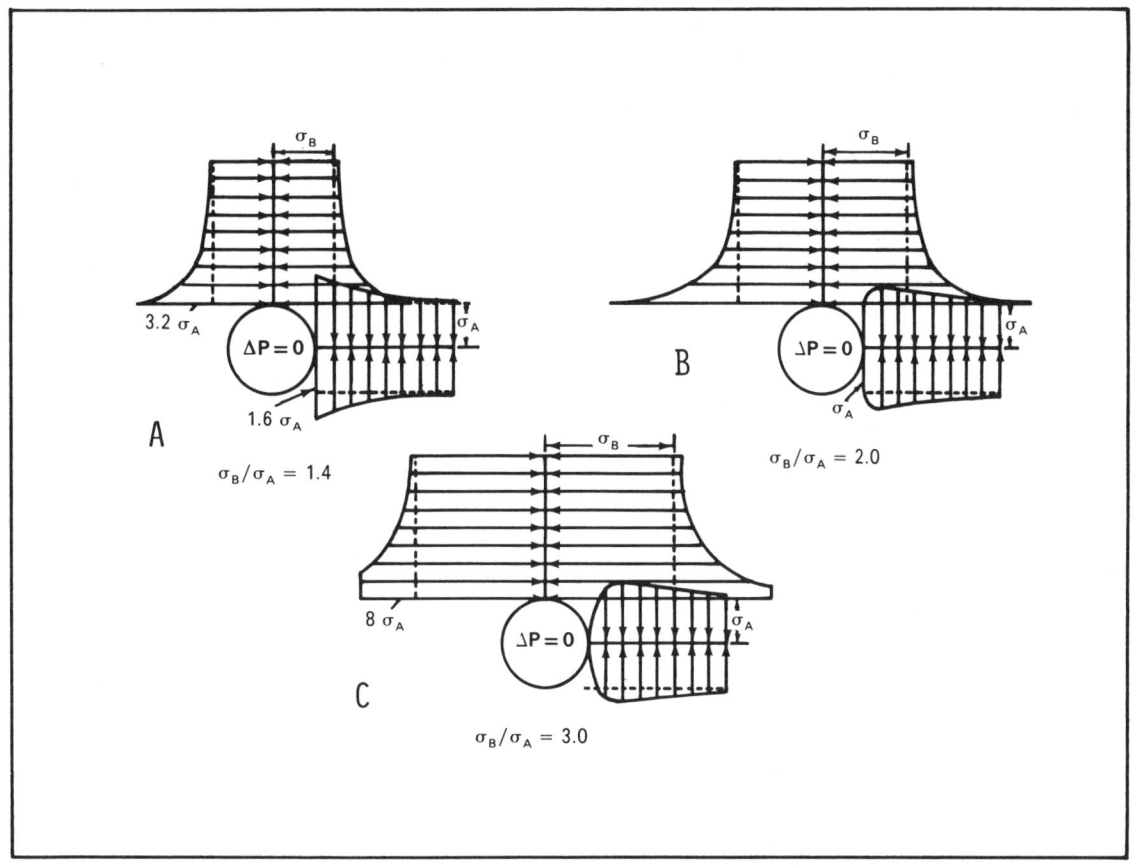

FIG. 8-3—*Stress concentration at wall of borehole.*[1] Permission to publish by The Society of Petroleum Engineers.

where σ_A and σ_B are regional horizontal stresses.

Depending on the ratio of the principal stresses, the minimum stress at the borehole can vary from twice the principal stress, when the two principal stresses are equal; to zero, when one principal stress is 3.0 times the other.

Circumferential Stress Variation Should Produce Elliptical Holes—Where the magnitude of the principal regional horizontal stresses is not equal, as in Figure 8-3, the rock at various points around the circumference of the wellbore is not stressed to the same degree. In this case, an elliptical borehole should result rather than a round borehole. Four-arm caliper surveys indicate that elliptical holes are not unusual, and further, that in some areas, the major axis of the ellipse follows a directional pattern over a rather wide area. In highly deviated directional drilling, it would seem that most holes would be elliptical. The component of vertical stress (overburden) perpendicular to the axis of the inclined hole would usually be greater than the horizontal stress, thus a stress variation would exist around the circumference of the hole. The occurrence of elliptical bore holes in highly deviated drilling has also been confirmed by four-arm caliper surveys.

Fracture Initiation

In the case of a non-penetrating fluid in open hole Hubbert & Willis presented the following equations to calculate frac initiation pressure:

Horizontal Fracture—Assuming vertical components of force are exerted against the formation, the condition necessary for horizontal fracture initiation is that the wellbore pressure must exceed the vertical stress plus the vertical tensile strength of the rock, or:

$$(P_i)_h = \bar{\sigma}_v + S_v + P_r \qquad (4)$$

where:

$(P_i)_h$ = borehole pressure required to initiate horizontal fracture
S_v = vertical tensile strength of rock

Vertical Fracture—Conditions for vertical fracture initiation depend on the relative strength of the two principal horizontal compressive stresses.

To cause formation breakdown, the pressure in the borehole must be somewhat greater than the minimum stress at the borehole, and must also overcome the tensile strength of the rock. This can be expressed as follows:

$$(P_i)_v = 3\bar{\sigma}_{h_2} - \bar{\sigma}_{h_1} + S_h + P_r \qquad (5)$$

where:

$(P_i)_v$ = borehole pressure required to initiate vertical fracture
$\bar{\sigma}_{h_1}$ = maximum principal horizontal matrix stress
$\bar{\sigma}_{h_2}$ = minimum principal horizontal matrix stress
S_h = horizontal tensile strength of rock
P_r = formation pore pressure

Penetrating Fluid Reduces Breakdown Pressure—A penetrating fluid increases the area over which pressurized fluid contacts the formation and can reduce the pressure necessary to initiate fracturing.

Laboratory and theoretical work by Fairhurst and Haimson[13] provides a basis for estimation of the magnitude of reduction in openhole. Generally reduction may be on the order of 25 to 40% in openhole.

Perforation Density and Orientation Affect Breakdown Pressure—Laboratory work simulating cased hole shows that breakdown or frac initiation pressure is affected by the number and arrangement of perforations.[22]

The existence of casing and the arrangement of perforations have little effect on created fracture orientation, but breakdown pressure is reduced by increased number of perforations. The practice of perforating with all shots in a vertical line on one side of the casing, Figure 8-4, significantly increases breakdown pressure if the perforations happen to be oriented 90° to the azimuth of the vertical fracture plane. Laboratory work supports the idea that fracture

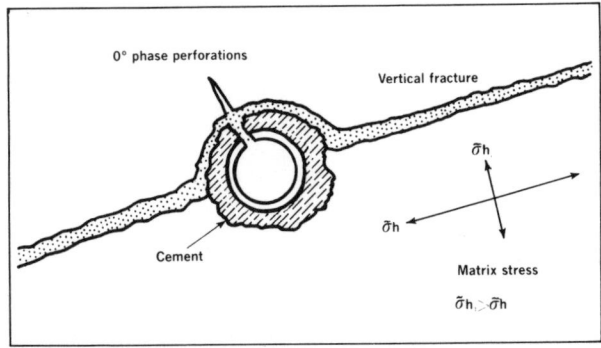

FIG. 8-4—*Orientation of perforation vs. least horizontal matrix stress. Condition resulting in highest breakdown pressure.*

initiation pressure should be minimized by a helical orientation of the perforations.

Fracture Propagation

Once the fracture has been initiated and invaded by pressured fluid, the stress concentration near the wellbore disappears.

Fracture Orientation—The fracture then propagates in a plane perpendicular to the minimum effective matrix stress. Usually the minimum stress is horizontal, and a vertical fracture results. Where horizontal matrix stresses are unequal, there will be a preferred direction for the vertical fracture. In areas of thrust faulting or where erosion has occurred after deposition, rocks may be under greater horizontal compressive stress than vertical overburden stress. If the minimum effective matrix stress is vertical, then a horizontal fracture will result. Figure 8-5 pictorially relates pressures, stresses and rock properties involved in vertical fracture propagation.

Fracture Closure Pressure—To hold the fracture open after initiation (or to just keep it from closing), the pressure in the fracture must exceed the pore pressure by an amount equal to the minimum effective rock matrix stress. This pressure is usually called the *fracture closure pressure*. The *fracture gradient* is the fracture closure pressure divided by the depth.

Fracture Propagation Pressure—As the fracture is extended, the pressure in the fracture at the wellbore (*fracture propagation pressure*) increases as a result of fluid friction required to push the frac fluid through

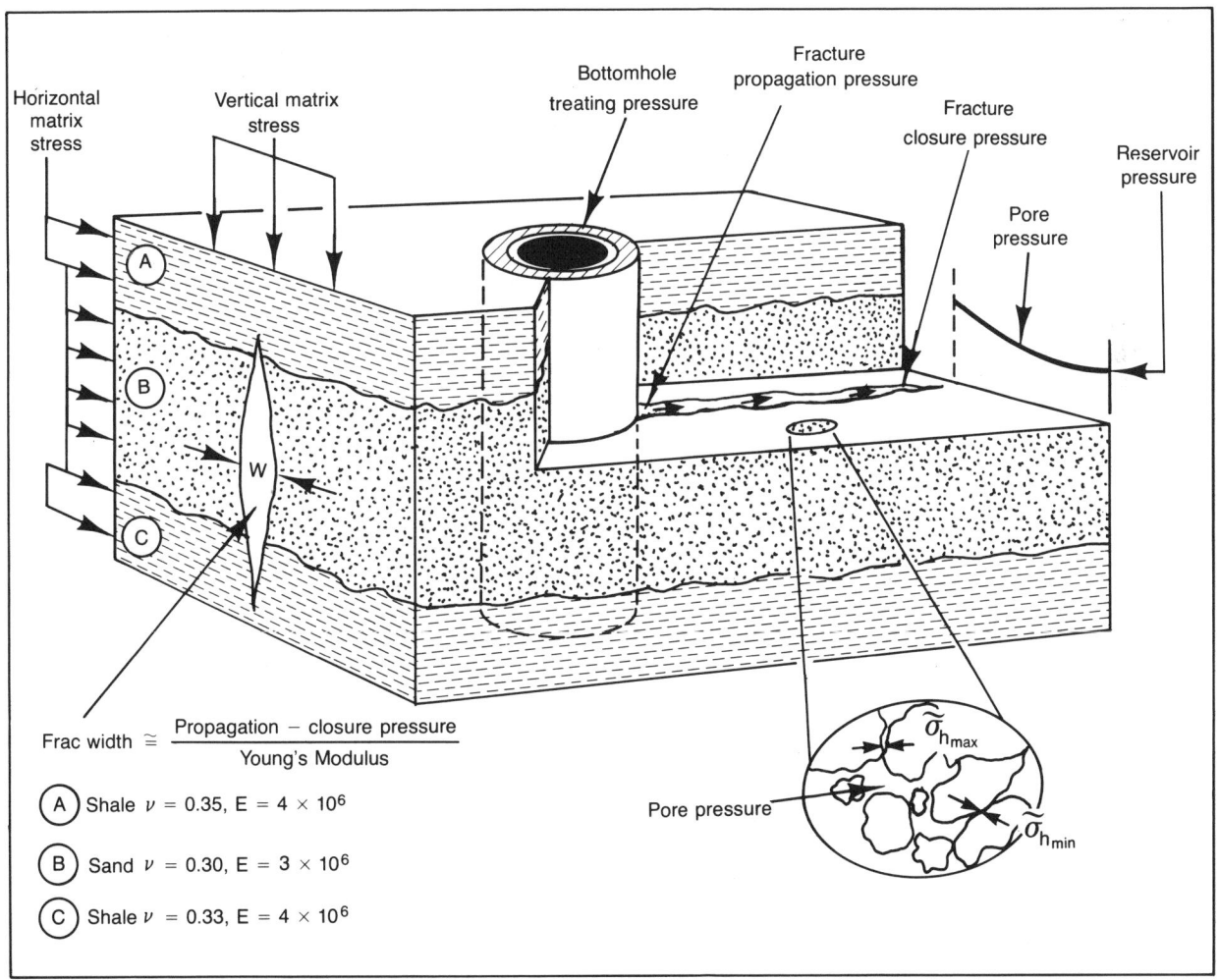

FIG. 8-5—*Pressures, stresses, and rock properties involved in vertical fracture propagation.*

an increasing distance toward the tip of the fracture.

Net Fracture Pressure—Pressure in the fracture in excess of the fracture closure pressure is the *net fracture pressure*. Net fracture pressure acts against the elasticity, or Young's modulus, of the rock to open the fracture wider. During the fracture job, the net fracture pressure can be used as an indicator of fracture extension as shown in Figure 8-6. If net fracture pressure is increasing normally, the fracture is propagating away from the wellbore with little vertical extension (vertical extension is limited by confining barriers above and below). If net fracture pressure increases abnormally, the fracture has stopped extending and is being blown up like a "balloon"; i.e., a sandout is occurring. If net fracture pressure drops, the fracture has broken (or is breaking) through a barrier zone above or below. A constant propagation pressure indicates an abnormal condition that will shortly result in a sandout or a barrier breakthrough.

Measuring Fracture Closure Pressure and Minimum Rock Matrix Stress

With the assumption of (1) a relaxed geologic area and (2) rock acting as an elastic material, fracture closure pressure can be calculated from rock density, depth, Poisson's ratio, and pore pressure. However, in many areas these assumptions may not be quite valid, and techniques of actually measuring fracture closure pressure and minimum rock matrix stress during the fracturing process are needed. Several methods can be used to provide reasonable measurements.

Instantaneous Shut-in Method—This technique involves recording the wellhead pressure immediately after the pumps are shut down following injection into the fracture (Fig. 8-7). The wellhead instantaneous shut-in pressure, plus the hydrostatic pressure of the wellbore fluid column, is the fracture propagation pressure. Fracture propagation pressure declines to the fracture closure pressure as fluid leaks out of the fracture into the formation. After closure, pressure bleeds off more rapidly, thus an anomaly occurs at the point of closure. The time between pump shutdown and fracture closure may be a few minutes—or with a tight formation may be many hours.

Pump-in Flow-back Method—A technique of obtaining the fracture closure pressure, where fracture closure time might otherwise be very long, is to bleed off the wellhead pressure using a choke to obtain a constant flow-back rate. A recording of surface pressure versus time should show a change in slope at the time the fracture closes. Figure 8-8 illustrates use of this method after a mini-frac test in the Mesaverde formation of southern Wyoming.[37]

Step-Rate Injection Method—With this technique, injection rate into the formation rock is gradually increased, and the resulting surface pressure is recorded. A break in the slope of the pressure versus rate plot indicates the surface pressure at which the fracture occurs (Fig. 8-9).[37] This surface pressure, plus the hydrostatic fluid column pressure, is the fracture closure pressure. The step-rate injection technique works best where measurements can be made in a static fluid column: for example, if injection is down tubing, pressures should be measured in the annulus to eliminate circulating pressure drop effects.

In each one of these procedures it should be remembered that the fracture closure pressure is affected by the pore pressure. Thus, if the measurement process increases the pore pressure significantly, the fracture closure pressure will also be increased. Figure 8-10 shows the effect of increasing pore pressure during sequential closure pressure measurements in a shallow partially depleted formation.[38]

The minimum rock matrix stress is the fracture closure pressure minus the pore pressure. The pore pressure is generally assumed to be the static reservoir pressure. However, the fracture closure pressure test

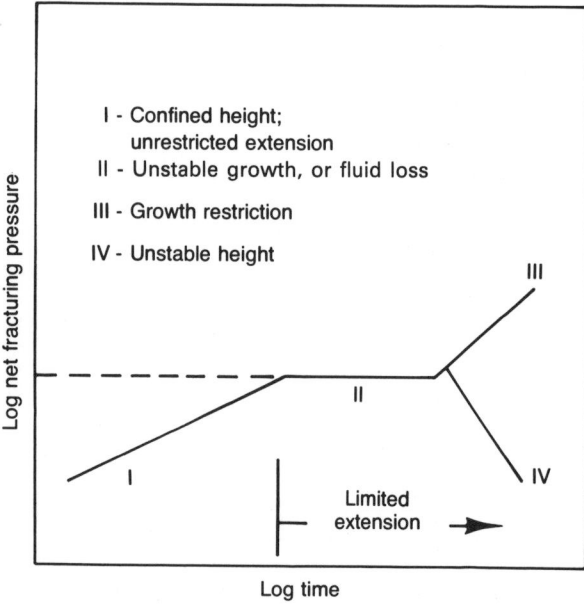

FIG. 8-6—*Net fracture pressure indicates progress of fracture extension.*

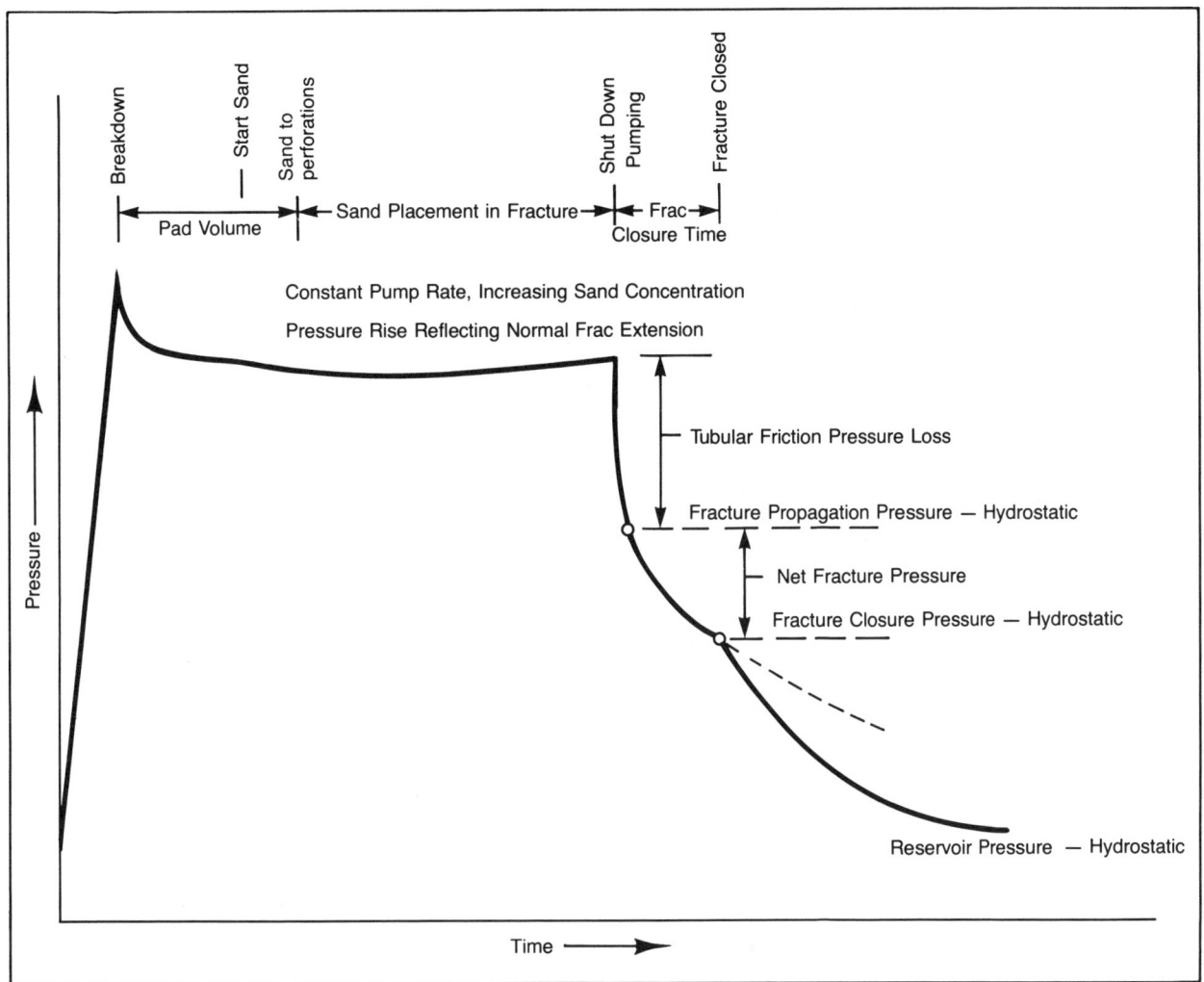

FIG. 8-7—*Idealized surface pressure during frac job.*

must be run before the pore pressure near the fracture is increased significantly by injected fluid.

Proppant Stress

The minimum rock matrix stress is of interest because it is the stress that propping agents must withstand in order to hold the fracture open. This stress increases as pore pressure is reduced. Therefore, proppant in the critical area near the wellbore is subjected to more stress than that further away, because the pore pressure near the wellbore is reduced due to drawdown in the producing process. (Fig. 8-11). This effect may be significant at high drawdown pressures. Usually proppant strength is specified on the basis of zero pore pressure.

Vertical Containment of Fracture Growth

Knowledge of the minimum horizontal matrix stress in adjacent zones is also important because it is a major factor in the extension of vertical fractures into zones above and below the pay zone.

Higher Horizontal Matrix Stress Creates a Barrier Zone—In massive fracturing (MHF), as in other fracturing situations, the existence of a barrier zone to limit upward or downward extension of a vertical fracture is important to the length of fracture obtained and to the success of the job. With Poisson's ratio values of the pay zone, and the adjacent ''barrier'' zones, it should be possible to estimate the horizontal matrix stress in each zone. The effectiveness of the barrier zone in containing the fracture can then be

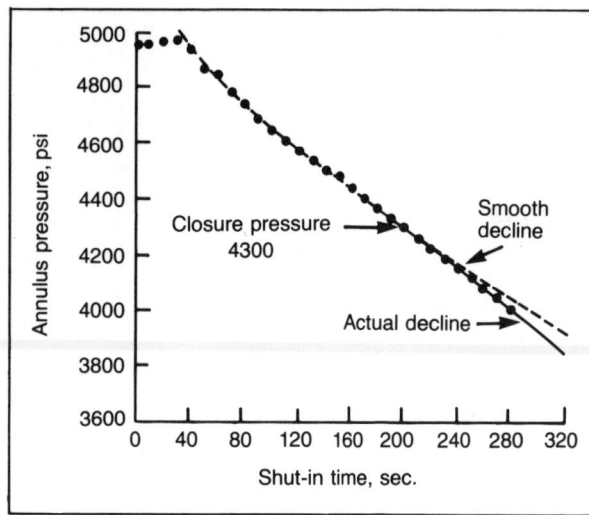

FIG. 8-8—*Post-frac pressure decline to determine closure stress.*[37] *Copyright 1981, SPE-AIME.*

evaluated. With a higher Poisson's ratio the horizontal matrix stress should also be higher. Even if tectonic forces are active, the differences in horizontal matrix stresses between zones should be meaningful.

Net Fracture Pressure Controls Vertical Extension—It should also be possible to set an upper limit on net fracture pressure to minimize vertical extension

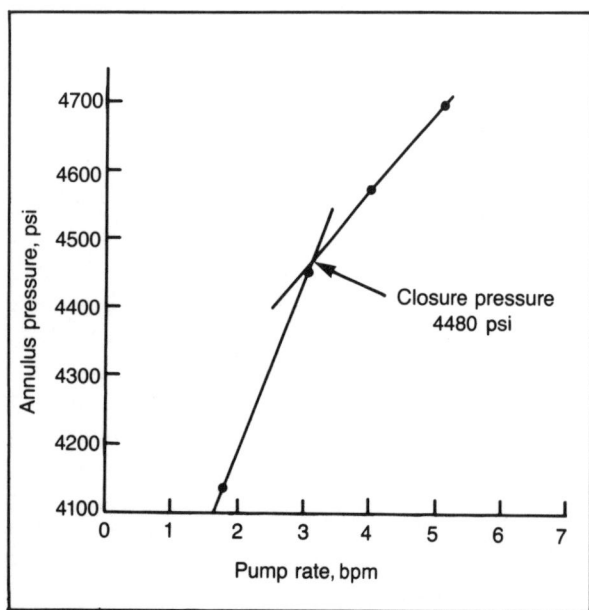

FIG. 8-9—*Step rate pump-in test to estimate closure pressure.*[37] *Copyright 1981, SPE-AIME.*

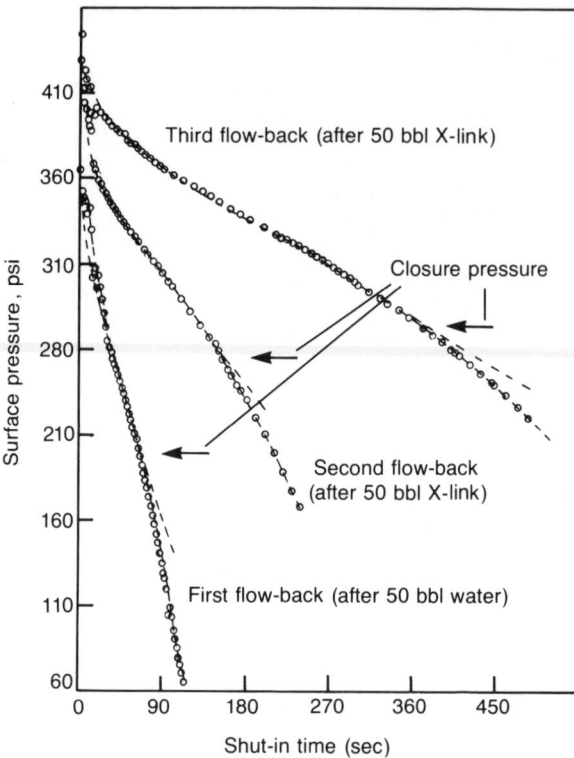

FIG. 8-10—*Three closure stress measurements showing sequential increase in closure stress.*[38] *Copyright 1981, SPE-AIME.*

of the fracture. Figures 8-12 and 8-13 show results of such an attempt in a Cotton Valley well in East Texas.[34] Predicted fracture height versus net fracture pressure, based on rock parameters determined from density log and accoustic log measurements and cor-

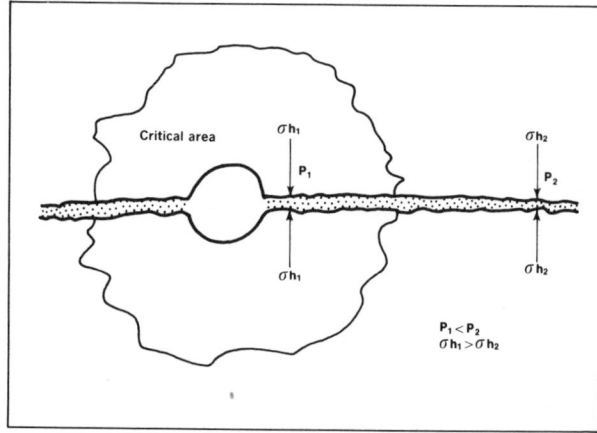

FIG. 8-11—*Proppant in critical area near wellbore is subjected to most stress.*

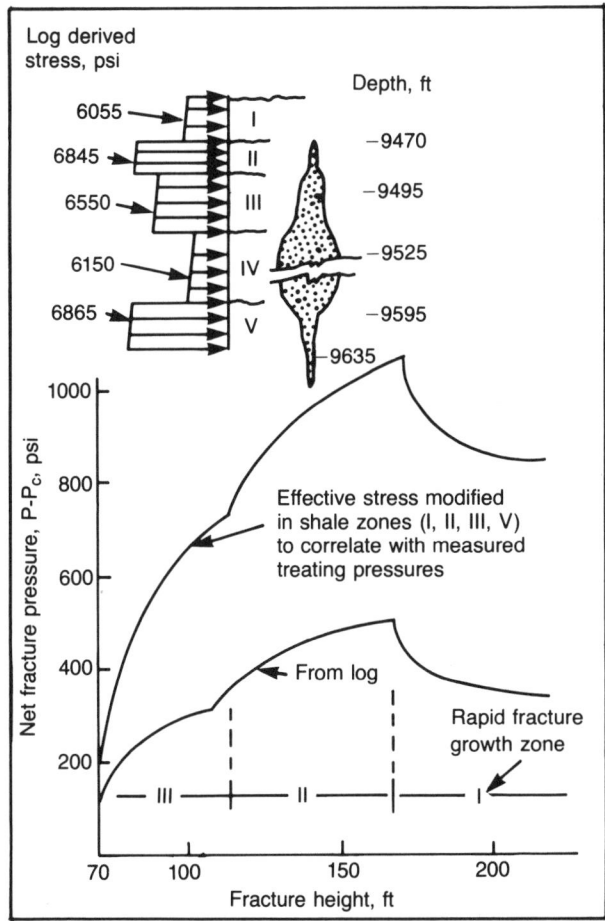

FIG. 8-12—*Predicted fracture height vs. net fracture pressure.*[34] Copyright 1980, SPE-AIME.

rected by observed treating pressure, are shown in Figure 8-12. Control of the net fracture pressure apparently was effective in limiting vertical breakthrough (Fig. 8-13), since a sharp pressure decline indicating barrier penetration did not occur.

Reduced Pore Pressure Should Contain Fracture—In a zone where the pore pressure has been partially depleted, the minimum rock matrix stress will be higher than in normal pressure zones above and below. The fracture closure pressure, however, will be lower in the partially depleted zone; thus, a fracture should be constrained within the partially depleted zone.

One additional factor, horizontal matrix stress calculated on the basis of linear elastic behavior of rocks, represents minimum values. Relaxation or creep of the material over geologic time may increase this horizontal stress. Shales would be expected to relax to a greater degree than sands; therefore shales may serve as a more effective barrier than simple linear elastic stress calculations indicate.

Azimuth Prediction

Where long fractures are needed for economic depletion of tight gas zones, the azimuth of the vertical fractures becomes important in planning well locations and resulting reservoir drainage patterns.

If the regional stress pattern can be determined from geologic features or from various measurement techniques, then the azimuth of a vertical fracture can be predicted. Techniques include: oriented core analysis, 4-arm calipers, borehole televiewers, downhole seismic monitoring, surface electric potentials, tiltmeters, surface and downhole seismic monitoring, surface lineament mapping, and impression packers. Currently tiltmeter measurements, 4-arm calipers, and to a lesser degree, downhole seismic measurements represent the most rewarding approach to azimuth determination.

In the Midcontinent United States, hydraulic fractures generally trend North 70° East perpendicular to thrust faulting in that area. Along the Gulf Coast, fractures tend to parallel normal faulting trends that run generally parallel with the coast line. In the Denver Basin fractures trend northwest-southeast generally parallel to the mountains.

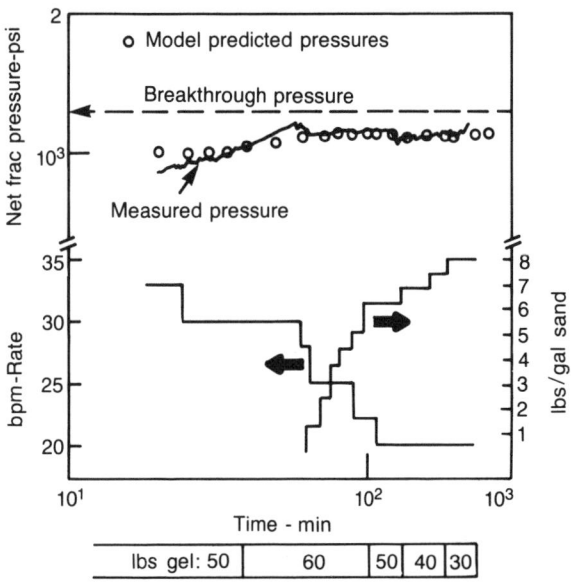

FIG. 8-13—*Net fracture pressure predicted and measured.*[34] Copyright 1980, SPE-AIME.

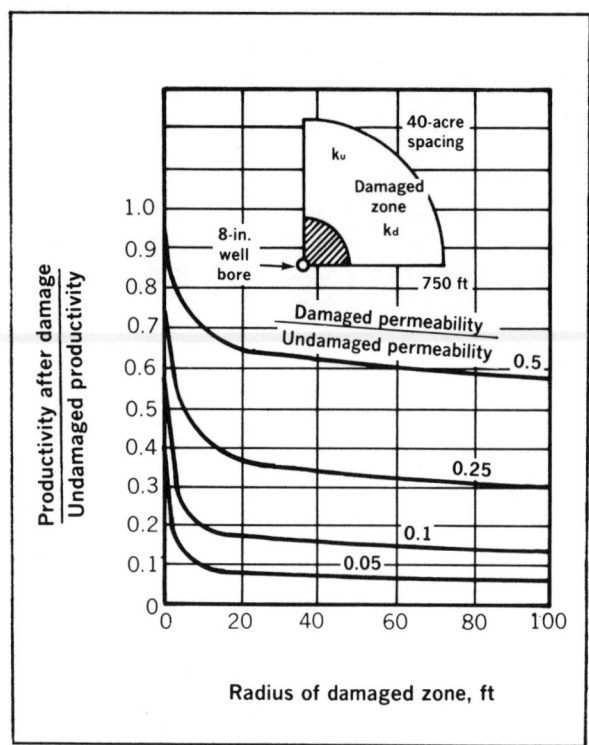

FIG. 8-14—*Production increase from bypassing reduced-permeability zone.*[3] *Permission to publish by The Society of Petroleum Engineers.*

PRODUCTION INCREASE FROM FRACTURING

Reasons for production increases from fracturing are: new zones exposed, reduced permeability bypassed, and flow pattern in reservoir changed from "radial" to "linear."

New Zones Exposed

Production increase depends on a combination of geologic and formation stress factors which many times are unknown.

In a carbonate formation where productivity depends on porosity—or in a fractured zone where primary flow capacity is related to the fracture system—or in a deltaic sand formation where permeability is related to regional depositional geometry, the possibility of increasing well productivity by fracturing into a new zone may be significant. In some situations, however, the "new zone" might be water or gas.

Bypassed Damage

Production increase from bypassing reduced permeability zone is a function of the depth of the damaged zone and the ratio of damaged to undamaged permeability. Increase can be estimated from Figure 8-14; or perhaps more effectively from transient pressure tests, or flow profiling.

It should be noted that only a short fracture is needed to bypass most damage zones, but that it is very important to prop the fracture in the area near the wellbore to provide a highly conductive path through the damaged zone.

Radial Flow Pattern Changed to Linear Pattern

Production increase from changing the flow pattern results from creation of a high conductivity fracture (relative to the formation permeability), extending a long distance from the wellbore.

Vertical Fracture Must be a Super Highway—For vertical fractures the productivity increase depends primarily upon the conductivity of the fracture relative to the formation permeability. Unless highly conductive fractures can be generated, fracture length has little significance as shown in Figure 8-15. But if sufficient permeability contrast can be developed (it can in very tight zones) then fracture length becomes quite important.

Figure 8-15 is based on well spacing of 40 acres and wellbore radius of 3 in. To convert to other geometry, use the following scaling factors:

Scaling Factors For Figure 8-15

Well spacing	Drainage Radii	(wk_f/k)	(J/J_o)
20 acres	467 ft	1.42	1.05
40 acres	660 ft	1.00	1.00
80 acres	933 ft	0.71	0.95
160 acres	1320 ft	0.50	0.91
320 acres	1867 ft	0.35	0.87
640 acres	2640 ft	0.25	0.84

For other spacing (A) and radii

$$wk_f/k = \sqrt{40/A}$$

$$J/J_o = \frac{3.095}{\log(.472)(r_e/r_w)}$$

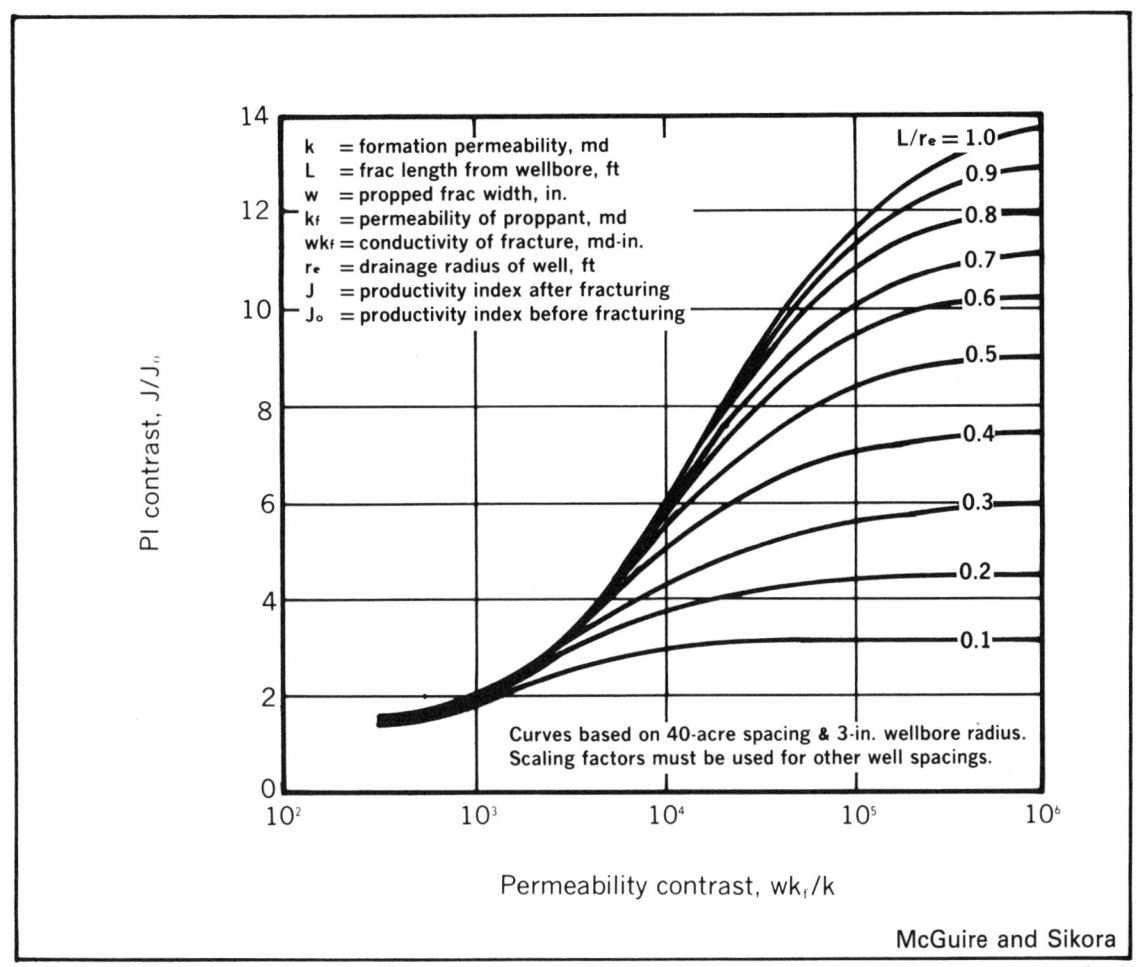

FIG. 8-15—*Effect of fracture conductivity and length.[6] Permission to publish by The Society of Petroleum Engineers.*

Referring to Figure 8-15 a permeability contrast greater than about 10,000 md in/md is difficult to attain unless formation permeabilities are quite low. For example, with a 50 md formation, fracture conductivity must be 500,000 md in. to achieve this contrast. If the fracture is packed with 8/12 mesh sand having a permeability of 1,000 darcies, closed fracture width would have to be 0.5 in. to obtain the desired conductivity. Under these conditions extending a packed fracture beyond 40 to 50% of the drainage radius would not improve well productivity significantly.

In a 0.1 md permeability gas zone, a 0.2 inch fracture packed with 20/40 mesh sand producing a conductivity of 10,000 md in, would provide a permeability contrast of 100,000 md in/md. With 320 acre spacing a fracture length of 200 ft would provide a 3.7 fold productivity increase; a 1,000-ft fracture would provide an 8.9 fold increase, or an 1,800-ft fracture would provide a 10.9 fold increase. Thus, where significant permeability contrast can be obtained the incentive to develop and pack longer fractures increases.

Near Wellbore-Conductivity is Critical—Figures 8-16 and 8-17 show the importance of propping the fracture in the critical area near the wellbore. If a small portion of the mouth of the fracture (less than about 3% of total frac length) is left unpropped, the fracture will close. Most of the stimulation which could have been achieved by the treatment is thereby lost.

Figures 8-16 and 8-17 relate stimulation (J/J_o) to the length of section closed at the fracture mouth (r_c). Conductivity of a closed fracture is difficult to estimate but permeability contrast in Figure 8-16 is taken as 12 md in./md and in Figure 8-17, 120 md in./md.

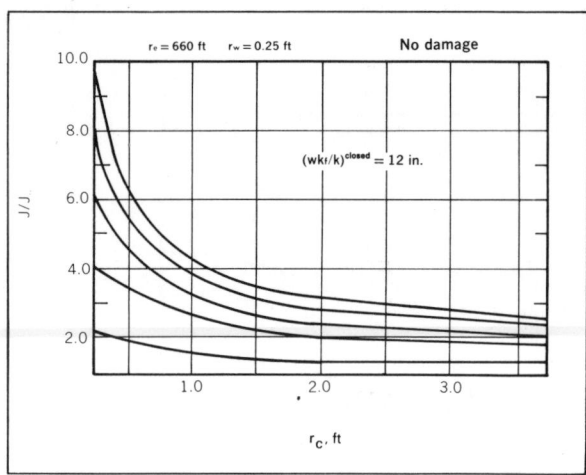

FIG. 8-16—*Effect of fracture closure near wellbore.*[18] *Permission to publish by The Society of Petroleum Engineers.*

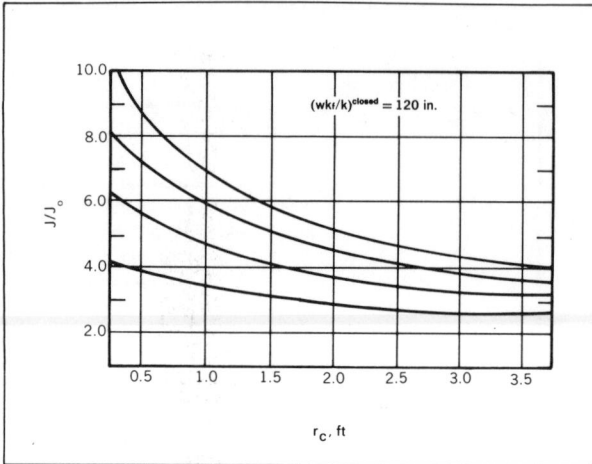

FIG. 8-17—*Effect of fracture closure near wellbore.*[18] *Permission to publish by The Society of Petroleum Engineers.*

Importance of Fracture Height—Figure 8-15 assumes that the fracture is propped throughout the full thickness of the producing formation, and from the wellbore to the extreme tip. Figures 8-18, 8-19, and 8-20, relate productivity increase (J/J_o) to relative capacity (permeability contrast) for fractures propped from the wellbore to the tip in each direction, but not through the full thickness of the formation. These figures assume a homogeneous formation. In a layered formation, fillup of the fracture is much more important than these figures indicate.

Frac Conductivity is the Problem With Permeability > 5-10 md—The fracture conductivity obtained with typical low to medium viscosity frac fluids and typical proppant sizes is shown by the shaded area of Figure 8-21.

To produce significant results in formations having permeabilities much greater than 5 to 10 md, every effort must be made to maximize fracture flow capacity.

Non-Steady State Flow Conditions Require Reservoir Simulation—The McGuire and Sikora curves

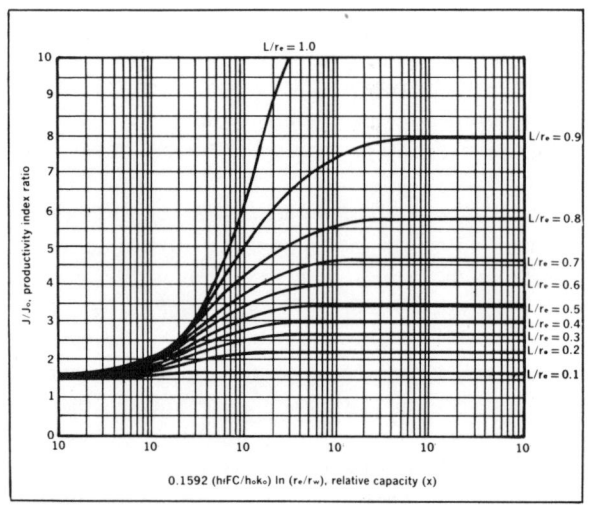

FIG. 8-18—*Effect of propped fracture height, 25% propped.*[12] *Permission to publish by The Society of Petroleum Engineers.*

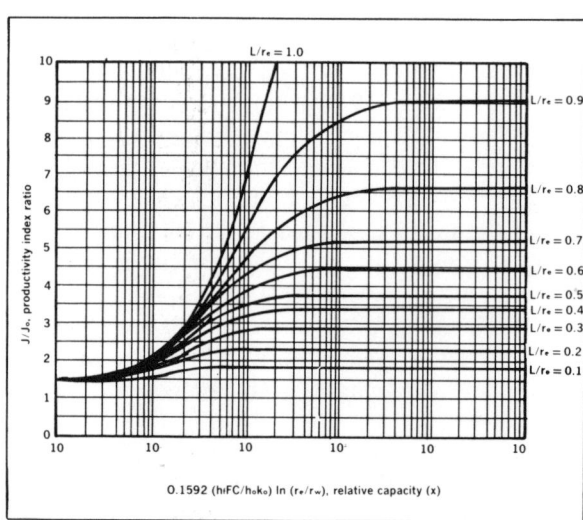

FIG. 8-19—*Effect of propped fracture height, 50% propped.*[12] *Permission to publish by The Society of Petroleum Engineers.*

Hydraulic Fracturing

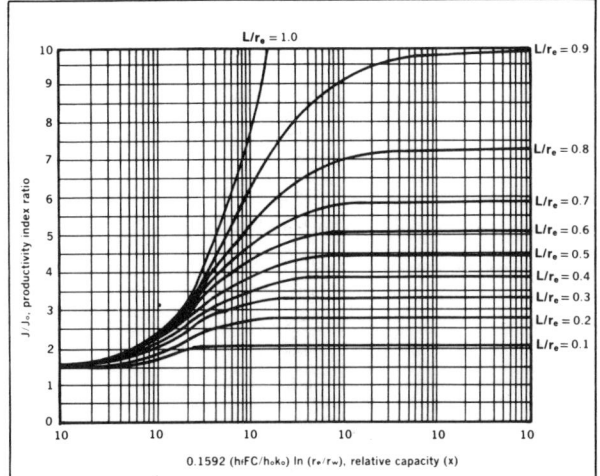

FIG. 8-20—*Effect of propped fracture height, 100% propped.*[12] *Permission to publish by The Society of Petroleum Engineers.*

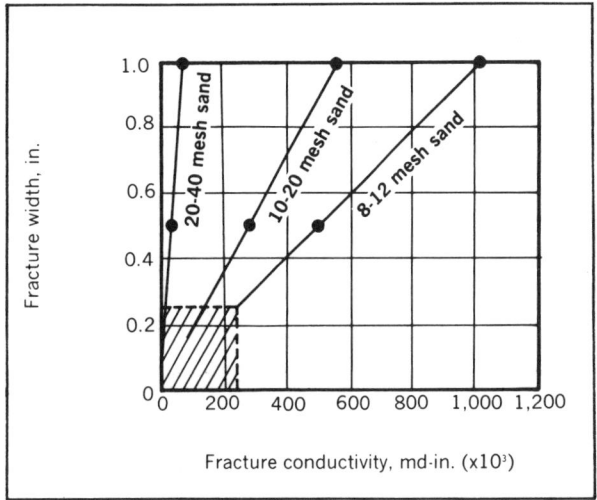

FIG. 8-21—*Fracture conductivity vs. frac width and prop permeability.*[15] *Permission to publish by The Society of Petroleum Engineers.*

(Fig. 8-15) apply to steady state flow conditions that are approximated in most reservoirs. However, in very tight gas zones having permeabilities in the range of 1–100 microdarcies, flow deviates significantly from steady state conditions. Here, reservoir simulation using numerical techniques must be used to predict the effect of fracture length, height, and conductivity on productivity improvement and on recovery.

PROPPING THE FRACTURE

The objective of propping is to maintain desired fracture conductivity economically. Fracture conductivity depends upon a number of interrelated factors: type, size, and uniformity of the proppant; degree of embedment, crushing, and/or deformation; and amount of proppant and the manner of placement.

Commonly used propant types and size ranges are:

Sand

Mesh	Range, in.
70/140	.0083 × .0041
40/70*	.0165 × .0083
30/50	.0234 × .0117
20/40*	.0331 × .0165
16/30	.0469 × .0234
12/20*	.0661 × .0331
8/16	.0937 × .0469
6/12	.132 × .0661

*Primary sizes

Sintered Bauxite

Mesh	Range, in.
40/60	.0165 × .0078
20/40	.0331 × .0165
12/20	.0661 × .0331

Desirable Properties For Propping Agents

Size and Uniformity—Decreasing size range increases the load that can be supported, and also the permeability of a packed fracture.

Significant quantity of fines can seriously reduce fracture permeability. For example, 20% material finer than 40 mesh will reduce the permeability of 20/40 sand by a factor of 5. Sand graded to 10/16 has a permeability about 50% greater than 10/20 sand.

Typical size specifications for 12/20 frac sand are:

12/20 Sand Specifications

U.S. sieve no.	% passing (by wt.)
10	100
12	98 Minimum
20	1 Maximum
30	0

The following table shows *no-load* permeability of popular graded sand size ranges. Permeability values must be used with care since differences of grade distribution within a particular size range may change permeability significantly.

Permeability of Good Quality Frac Sand
(Low Confining Load)

Size—U.S. mesh	Permeability—darcies
4/6 —(Angular)	3,000
8/12—(Angular)	1,740
10/20—(Angular)	880
10/20—(Round)	325
10/30—(Round)	190
20/40—(Round)	120
40/60—(Round)	45

Strength—Significant reductions in permeability of proppants occur as the closure stress or horizontal matrix stress increases. It is less of a problem in a pack than in a partial monolayer. Strength of sand grains varies according to the source of the sand and the size range. The following table shows comparative strengths of single grains based on L/D^2 ratio.

Sand size and source	L/D^2 strength, psi
20/40 Ottawa	9,200–15,500
20/40 San Saba	7,400–19,000
10/20 San Saba	4,800–9,300
8/12 San Saba	3,400–6,000

Chemical and Temperature Stability—Metallic pellets may corrode. Plastic beads have a temperature limit of about 180°F.

Density—Low specific gravity reduces falling rate and promotes placement of uniform partial monolayers.

Cost—Sand is by far the lowest cost proppant. Sintered bauxite costs about 10 to 15 times as much.

Fracture Conductivity vs. Proppant Concentration and Load

The ideal concept of the "Optimum Partial Monolayer" is illustrated in Figure 8-22. With a confining load of 2,000 psi, optimum partial monolayer concentration of 10/20 Ottawa sand would approximate 60 lb per 1,000 sq ft. This would provide a fracture conductivity of about 16,000 md ft. Thus, the optimum partial monolayer provides very high conductivity, and does not require a very wide fracture or much proppant. With low to medium viscosity frac fluids, sand fall-out characteristics are such that a dispersed monolayer packing is probably not achieved. With the advent of very high viscosity fluids it may be a possibility.

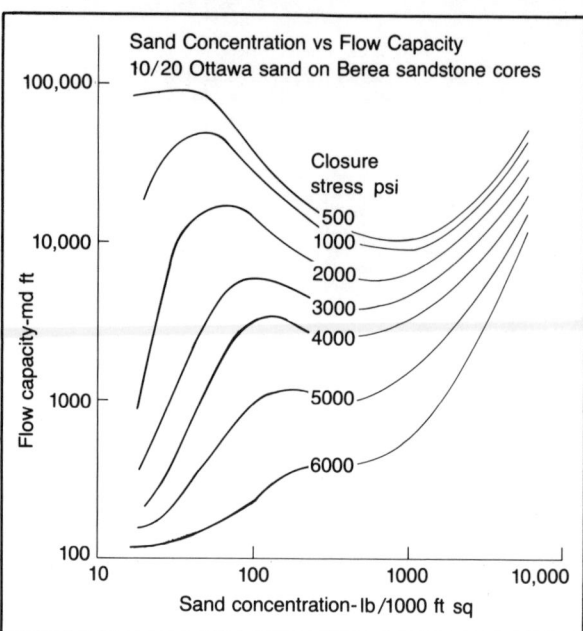

FIG. 8-22—*Typical fracture conductivity vs. prop concentration.*

Another approach to the partial monolayer concept is the "pillar frac" technique where an attempt is made to place "clumps" of small sand in the fracture, leaving open spaces between for flow.

With proppant concentrations above 400 to 600 lb per 1,000 sq ft a multilayer packing results. Conductivity then increases with fracture width.

Measurements of Fracture Conductivity—Flow measurements under formation conditions are necessary in selecting the type and concentration of proppant from the standpoint of (1) ability to prop the fracture without crushing or embedding, and (2) providing necessary conductivity to achieve desired stimulation.

In a vertical fracture a proppant is subjected to a

TABLE 8-1
Flow Capacity (Darcy ft)

Confining load	2000 psi		6000 psi	
Proppant concentration lb/1000 ft²	10/20 Sand	12/20 Sintered Bauxite	10/20 Sand	12/20 Sintered Bauxite
500	5.8	4.0	1.1	2.0
1000	6.5	6.8	1.5	5.3
2000	8.8	9.5	3.0	8.5
4000	20.0	19.0	8.0	17.0

TABLE 8-2
Flow Capacity—(Darcy-ft)
(2000 psi Confining Load—Berea Sand Core)

Proppant concentration lb/1,000 ft²	Sand		
	20/40 round	10/20 round	8/12 angular
500	2.1	6.1	11.0
1000	2.4	7.0	12.0
2000	3.2	9.0	17.0
4000	5.5	15.0	27.0

load equal to the fracture closure pressure less the formation pore pressure. It should be noted that the pore pressure near the wellbore during production may be significantly less than the shut-in reservoir pressure.

The effect of proppant concentration and proppant load on flow capacity with Berea sandstone is shown in the following tables. Table 8.1 shows that under low confining load 12/20 sintered bauxite provides about the same flow capacity as 8/12 sand with the same concentration. At 6,000 psi confining load, 12/20 sintered bauxite provides about twice the flow capacity as 8/12 sand. At higher loads sintered bauxite would look even more favorable in comparison.

Sintered bauxite—or similar higher-strength proppants—while expensive, is justified where high fracture conductivity is needed, and where confining loads are high.

The effect of proppant size is shown in Table 8-2. With smaller sand much higher concentrations per unit frac area, and much wider fractures are required to provide the same flow capacity as larger sand sizes.

It should be noted that Table 8-2 applies to low confining load (2,000 psi). At higher proppant loads (5,000 psi), 20/40 sand may provide more flow capacity than 8/12 sand. Tables 8-1 and 8-2 are intended to show general relations only. It is emphasized that conductivity vs. load tests should be run on available proppants to aid in selection.

Ideal Situation (Disregarding Placement Problems)—For soft formations, embedment dictates the use of multilayer packing. With a multilayer packing proppant concentrations on the order of 2000 to 4000 lb per 1000 sq ft may be desirable. In soft sands five layers are considered desirable to maintain fracture conductivity despite embedment. If a fracture wider than five layers of a particular proppant size can be generated, then a larger diameter proppant size should be considered.

For hard formations where embedment is not the problem, and under low confining loads where crushing is not a problem, high conductivity is provided by the optimum partial monolayer of the largest particle that can be placed in the fracture.

Placement of Proppant

Proppant Transport—Placement of proppant in a vertical fracture in any pattern other than a packed condition is difficult to achieve. With low viscosity fluids, model studies show that the first portions of sand entering the fracture drop to the bottom of the fracture, near the wellbore. As more sand enters the fracture, the pack height increases to some equilibrium point. Additional sand is then carried over the pack and deposited further out.

Where stimulation ratio depends on propping a long fracture, or in areas where fractures may extend below the productive zone, predicting proppant transport becomes quite significant. This is even more important when the fracture does not close quickly after the pumps are shut down, and the proppant continues to settle until the fracture does close, or until it is trapped by concentration.

Attempts to Model Sand Transport—As pointed out by Novotny,[25] prediction of proppant transport is complicated. As the fluid moves through the fracture, the fluid and proppant are heated, and the rock is cooled. Proppant concentration is increased by fluid lost to the rock. Fluid velocity decreases, and apparent viscosity decreases with increased shear rate, temperature and time, but increases with sand concentra-

TABLE 8-3[25]
Well Data for Treatment Comparison

Formation Properties:	
Reservoir depth	10,000 ft
Fracture gradient	0.70 psi/ft
Gross fracture height	60 ft
Reservoir sand thickness	20 ft
Porosity	0.10
Permeability	1.0 md
Reservoir temperature	200°F
Young's modulus	10^7 psi
Well spacing	80 acre
Reservoir fluid properties:	
Viscosity	2.0 cp
Density	50.00 lb/ft³
Compressibility	0.0002 psi^{-1}
Pressure	5000 psi

tion. Even in the laboratory, it has proved difficult to measure proppant settling rates as a function of fluid properties and fracture flow conditions. Three computer-simulated example treatments from Novotny show the importance of proppant transport. Tables 8-3 and 8-4 give well data and treatment conditions, and Figure 8-23 shows a pictorial representation of the proppant in the fracture as the treatments progress. In all treatments, injection rate was 10 bbl/min. A polyemulsion fluid (1/3 brine, 2/3 oil) was used with increasing amounts of polymer in treatments A, B, and C, respectively. Injected volumes consisted of 100 bbl of pad, then 150 bbl fluid with 2 lb/gal 20/40 sand and finally 150 bbl fluid with 2 lb/gal 10/20 sand.

In treatment A, the 20/40 sand settled only slightly as it moved to its final position at the end of pumping. When the fracture closed 65 minutes later, the 20/40 sand still extended well up into the productive zone. The 10/20 sand, which began to enter the fracture 25 minutes after the job started, settled quickly and much of it had fallen out below the productive zone even before the end of pumping. A void space, containing no sand, developed between the 10/20 sand closer to the wellbore and the 20/40 sand, which had been pushed further into the fracture. If the fracture had closed immediately at the end of pumping, enough of the 10/20 sand would have been trapped in the productive zone to give a stimulation ratio of 3.7. However, during the 65 minutes required for closure, the 10/20 sand settled below the productive zone, and no stimulation resulted. Evidently, the 20/40 sand, as used, contributed nothing to the job success. However, if 20/40 sand had been used for the entire job, or even if the 10/20 stage had been eliminated, significant stimulation would have resulted. In treatment C, with a high viscosity fluid having better sand transport characteristics, the 20/40 sand again moved to its final position with almost no settling. The 10/20 sand settled somewhat to the end of pumping, and somewhat more in the next 69 minutes until the fracture finally closed; but sufficient height of 10/20 sand remained through the productive zone to provide a stimulation ratio of 4.1. In treatment B, most of the 10/20 settled below the productive zone, but enough remained in the area near the wellbore to

TABLE 8-4[25]
Treatment Data and Results

Treatment	A	B	C
Fluid—Polyemulsion—33% brine, 67% oil			
Lbs of gelling agent/bbl of brine	0.5	1.0	1.5
Fluid properties at 200°F			
$\quad$ K'	0.81	2.65	8.53
$\quad$ n'	0.81	0.71	0.60
Proppant settling velocities (cm/sec)			
$\quad$ 10/20 sand	4.26	1.30	0.28
$\quad$ 20/40 sand	0.69	0.18	0.032
Dynamic fracture length (ft)	906	852	797
Dynamic fracture width @ wellbore (in.)	0.192	0.211	0.234
$\quad$ Fluid efficiency (%)	60	62	64
Stimulation ratio			
$\quad$ Predicted this study	1.01	2.61	4.10
$\quad$ Predicted if closure time = 0	3.68	4.00	4.10
Data common to all treatments:			
$\quad$ Injection rate	10 bpm		
$\quad$ Injection temperature	100°F		
$\quad$ Pad volume	100 bbl		
$\quad$ First stage	150 bbl w/2 lb/gal sand		
$\quad$ Second stage	150 bbl w/2 lb/gal sand		
$\quad$ Permeability of 20/40 sand	76 darcies @ 5000 psi stress		
$\quad$ Permeability of 10/20 sand	120 darcies @ 5000 psi stress		

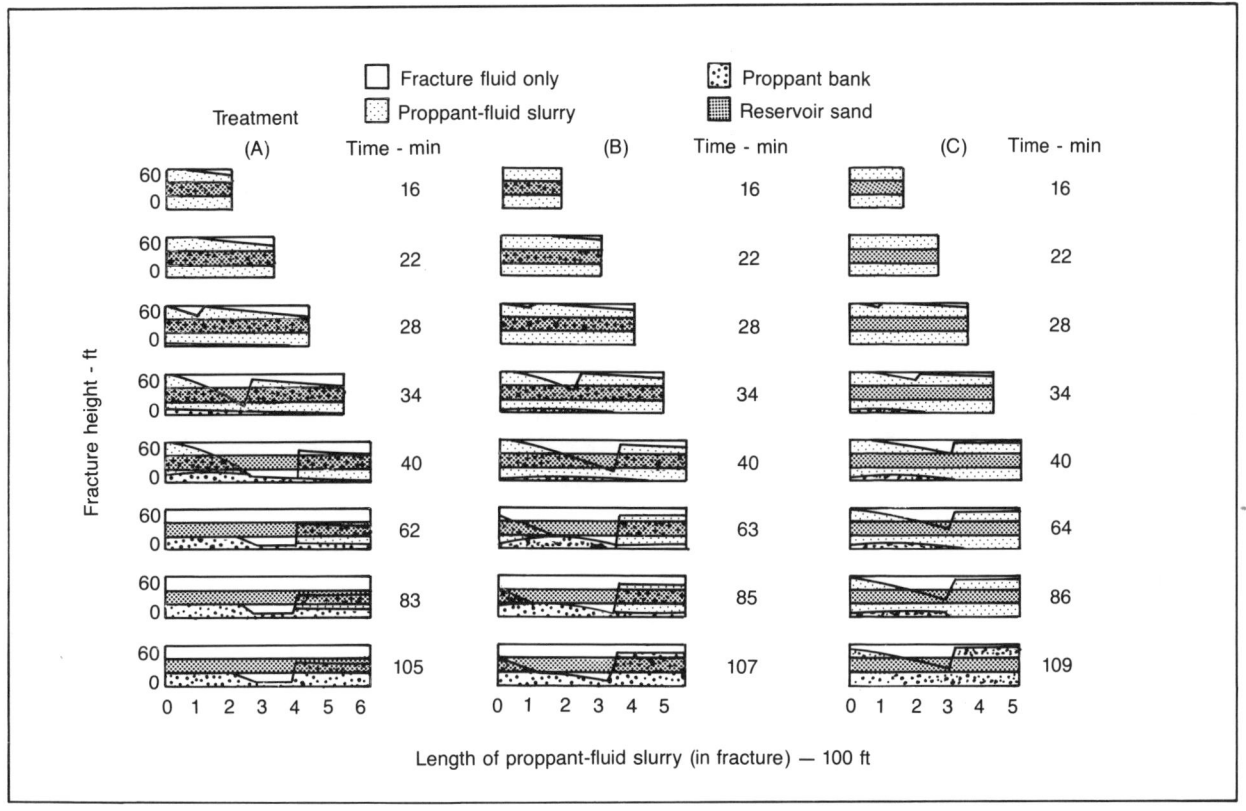

FIG. 8-23—*Proppant slurry behavior during and after hydraulic fracturing treatment.*[25] Copyright 1977, SPE-AIME.

provide a stimulation ratio of 2.6.

Scheduling Sand Size and Concentration for Packed Fracture—Sand scheduling requires a knowledge of the dynamic fracture width and the loss of fluid to the formation at various intervals along the fracture. Proppant size and concentration can then be selected to maintain as much of the fracture width and height as possible after the pumps are shut down.

Critical Area Near Wellbore—The area adjacent to the wellbore is most important as far as proppant placement is concerned. This is particularly true where formation damage is a major reason for stimulation. Obviously "overflushing" with sand-free fluid is undesirable because it pushes sand back away from this critical area.

Ideally this near wellbore region should be packed with sand by achieving a sand-out in the fracture. Techniques for creating a sand-out in the fracture involve: increasing sand concentration, reducing frac fluid viscosity, reducing injection rate and increasing fluid loss.

It should be noted that reducing fluid viscosity and/or injection rate or increasing fluid loss results in decreasing dynamic frac width, thus may not be a desirable solution. Usually, increasing sand concentration is the best solution.

FRAC FLUIDS

Basically oil or water fluids are used to create, extend, and place proppant in the fracture. Our ability to tailor the properties of fluids to achieve desired results has improved tremendously with recent advances aimed at providing much higher fluid viscosities, better high temperature stability, and minimizing formation damage effects. Usual modifications include: fluid loss control, gelling or thickening, crosslinking of gelling agents, emulsification, and foaming.

Generally these comparative statements can be made:

1. Crude oil fluids are cheap and have inherent viscosity which makes them advantageous for relatively low injection rate, shallow to medium depth fracturing. Pressure loss down the casing and safety

consideration are often limiting factors.

2. Gelled water fluids (linear aqueous gels) have special advantages due to their higher density and lower friction loss in deeper wells, and where higher injection rates are needed. Where high temperatures are involved reasonable viscosity can be maintained up to 250°F.

3. Crosslinked aqueous gels have very high viscosity to create fracture width, and to provide the proppant carrying capacity for producing highly conductive fractures needed in stimulating higher permeability zones or in propping the long fractures needed in stimulating low permeability zones.

Compared to linear aqueous gels, crosslinked gels can provide similar viscosity with lower polymer concentration, thereby reducing cost and formation damage. With high temperature crosslinkers and stabilizers, they can maintain relatively high viscosities for extended pumping time at high temperatures.

4. Emulsion fluids provide good viscosity and proppant carrying capacity and very good fluid loss and cleanup at reasonable cost.

5. Gelled-oil fluids have primary application in water-sensitive sand zones.

6. Foamed fluids have primary application in low permeability gas zones. Usually, sufficient nitrogen is added to produce 65 to 75% quality foam. A primary advantage of foam is excellent cleanup because of the small amount of liquid and the large quantity of energy represented by the high pressure nitrogen. Use of foamed fluids is relatively new; however, application is increasing. Initial problems of low sand concentration and short foam half-life are being overcome by centrifugal sand concentrators and by use of polymers for foam stabilization.

7. Alcohol fluids have primary application in low permeability dry gas zones where, due to relative permeability, oil should not be used.

8. Weak acid gels have application in dirty sands where clay stabilization may be important.

Fluid Properties and Modifications

Selection and modification of fluid properties to fit a specific well situation is an important part of frac job design. A basic consideration is "void creating" ability, or fracture width and length. This is a function of fluid efficiency (or fluid loss control) and "fracture viscosity." High viscosity fluids with attendant high pressure required to force them through the fracture, tend to promote very wide fractures (Fig.

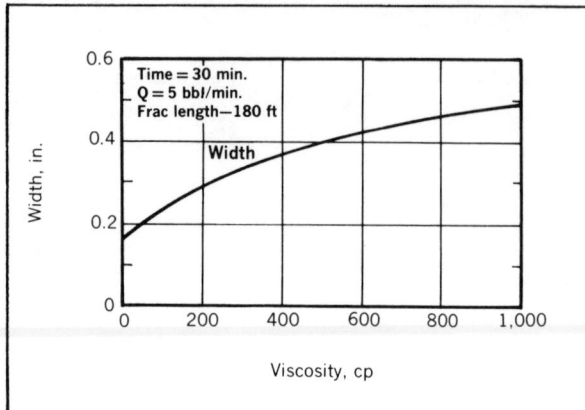

FIG. 8-24—*How viscosity affects fracture width.*[16] *Permission to publish by API Production Department.*

8-24). In a confined vertical fracture, fluid loss control tends to promote fracture length.

Other frac fluid considerations are friction-loss-down-the-pipe; proppant carrying ability, formation damage, fracture cleanup, temperature stability, mixing and storage problems and cost.

Fluid Loss Control—During fracturing some of the injected fluid leaks off into the formation matrix, and is not available to extend the fracture. Rate of fluid leakoff is influenced by three factors: compressibility of the reservoir fluids, viscosity of the frac fluid, and bridging materials in the frac fluid.

The combined effect of the controllable factors (viscosity and bridging) for a particular frac fluid is measured in the laboratory against actual or synthetic cores to determine the efficiency of that fluid. Results are reported in terms of spurt loss and fluid loss coefficient.

Figure 8-25 shows an idealized lab measurement. In massive frac work where long fractures are required, fluid loss becomes very important. Techniques have been developed for estimating actual fluid loss or fluid efficiency during a preliminary calibration frac. Several thousand gallons of frac fluid (with no proppant) are injected above fracture pressure. Analysis of the pressure bleed-off curve after pumping is stopped can provide a reasonable value of actual fluid loss, as well as other parameters needed in frac job design.[30]

Fluid-loss additives are generally finely divided solids that form a filter cake on the fracture face. Effective control requires a size range of small inert particles for bridging plus a plastering agent, usually a polymer, to plug the voids in the bridge.

Hydraulic Fracturing

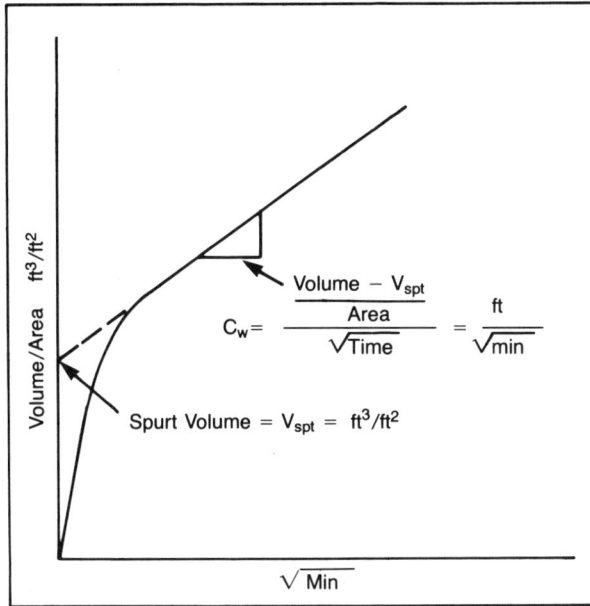

FIG. 8-25—*Typical fluid loss test result*

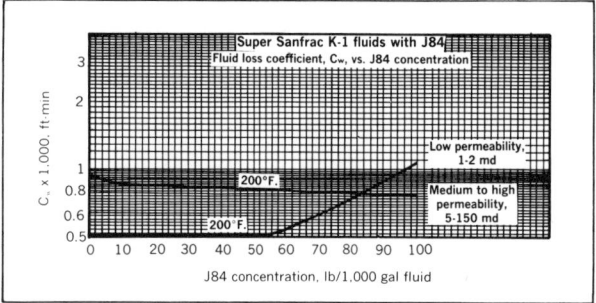

FIG. 8-27—*Effect of silica flour on fluid loss, polymer-emulsion fluid. Courtesy of Dowell.*

For crude or refined oil Adomite Mark II, calcium carbonate coated with an oil soluble surfactant, is the most commonly used additive.

For water base fluids silica flour is commonly used. Particle size for frac fluids is very fine, typically 92% passes a 325 mesh screen, (43 microns).

Dispersed hydrocarbon such as diesel or condensate can be used in concentrations of 3 to 5% to provide fluid loss control without the need for solids. Surfactant-oil mixtures (Halliburton WAC-12L) also provide fluid loss control at concentrations as low as 0.5%.

Emulsified or high viscosity fluids inherently have low fluid loss, and usually do not require use of additives.

Figures 8-26 and 8-27 show fluid loss coefficients for a low viscosity gelled water fluid and a polymer emulsion fluid with varying concentrations of silica flour.

Viscosity Control—A variety of fluid thickening polymers and mechanisms are used to provide needed viscosity control. Briefly, for gelled aqueous fluids, the natural polymers, guar, hydroxypropyl guar (HP guar), or hydroxyethylcellulose (HEC) are currently popular, with HP guar having perhaps the widest range of application. Crosslinking provides additional benefits. Emulsion fluids have inherent viscosity, but usually higher viscosity is provided by gelling the water phase. Soap-type gelling agents are used to thicken kerosene, diesel, or light crude oil. With acid fluids, biopolymer or synthetic polymers build vis-

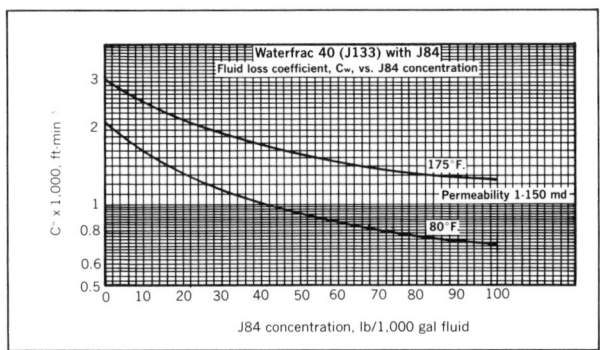

FIG. 8-26—*Effect of silica flour on fluid loss, low-viscosity gelled-water fluid. Courtesy of Dowell.*

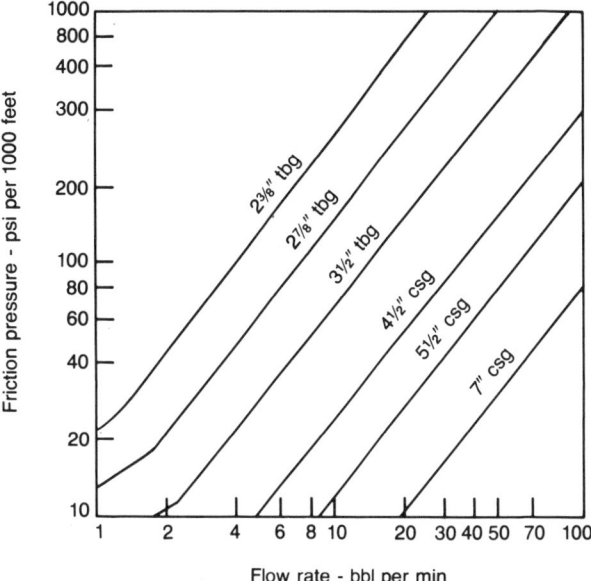

FIG. 8-28—*Pressure loss relations for 40 lb/1000 gal guar system. Courtesy of Dowell.*

TABLE 8-5
Comparison of Aqueous Fluids

Fluid System Dowell Trade Name and (Chemical Code)	Polymer	Polymer Concentration lbs/1000 gal	Suitable Base fluid			Mixing Procedure		Breaker System		Bottom-hole Temperature Range °F
			Fresh Water	Brine	Produced Brine	Batch	Continuous	Enzyme	Chemical	
Linear Gels										
Waterfrac (J111)	Guar	10–100	x	x	x	x			x	125–225
Waterfrac (J133)	Guar	10–110	x	x	x		x	x		100–225
Waterfrac (J160)	HEC	20–80	x	x			x	x		60–200
Widefrac YF "HC" (J301F+J303D)	HEC	60–120	x	x			x		x	125–450
Waterfrac (J347)	HP Guar	20–80	x	x	x	x			x	60–200
Crosslinked Gels										
Widefrac YF "G" (J111–J133)	Guar	40–80	x	x		x			x	100–275
Widefrac YF "G" (J331)	Guar	40	x	x			x		x	100–225
Widefrac PSD (J346)	HP Guar	40–60	x	x	x		x		x	100–275
Widefrac PSD (J347)	HP Guar	20–60	x	x	x	x			x	100–275
Crosslinked Gel + Hi-temp Stabilizer										
Widefrac YF "400" (J347+J353)	HP Guar	40–10	x	x		x			x	200–350

cosity. With foamed water, polymers can provide greater half-life, or foam stability.

Table 8-5 compares typical aqueous gel fluids as to polymer type, base fluid, mixing procedure, breaker system, and temperature range. Appendix C presents a detailed discussion of water soluble polymers.

Linear Aqueous Gels are usually prepared by mixing guar, HP guar, or HEC with fresh water containing 1–2% KCl, or with brine. Polymer concentrations of 20–50 lb/1000 gal are common, 100–150 lb/1000 gal are maximum. Gelling reduces friction loss, increases apparent viscosity, and provides some fluid loss control. Figure 8-28 shows friction loss versus flow rate relationships for a typical 40-lb guar fluid.

Table 8-6 shows rheological properties and break times for various guar concentrations and formation temperatures. The effect of temperature on apparent viscosity is evident, as is the need for a viscosity breaker at lower temperatures.

Crosslinked Aqueous Gels are usually prepared by crosslinking guar or HP guar with borate, or, where temperatures exceed about 210°F, with an organo-metallic crosslink. By crosslinking, much higher viscosity can be obtained for a given polymer concentration. Table 8-7 shows typical rheological properties for a borate crosslinked HP guar. Break times are not shown; however, these fluids should not be used without an effective breaker selected for the specific formation temperature conditions. Low temperature breaker systems are available to provide break times as short as 6 hours at temperatures as low as 50°F.

Figure 8-29 shows viscosity versus time for a 50-lb/1000 gal HP guar, with an organo-metallic crosslink and a high temperature stabilizer. Reasonable viscosity is maintained for 6 to 8 hours even with formation temperatures up to 300°F.

Polyemulsions usually consist of two-thirds oil and one-third water, emulsified so that water is the ex-

TABLE 8-6
Rheology of Typical Linear Aqueous Gel Fluids[a]

Polymer Concentration lb/1000 gal	Temp °F	Power Law Fluid Parameters		Apparent Viscosity cp		Break Time[b] hr
		n'	K'	160 sec^{-1}	479 sec^{-1}	
20	80	0.70	1.4×10^{-3}	13	10	15
20	100	0.93	6.5×10^{-4}	10	8	10
20	125	0.95	1.9×10^{-4}	6	6	3
20	150	1.00	7.5×10^{-5}	4	4	1
30	100	0.60	4.7×10^{-3}	30	20	24
30	125	0.63	3.5×10^{-3}	26	17	8
30	150	0.67	2.3×10^{-3}	20	14	4
30	175	0.75	1.2×10^{-3}	15	12	2
40	125	0.51	9.4×10^{-3}	37	21	30
40	150	0.56	5.7×10^{-3}	29	18	19
40	175	0.62	3.1×10^{-3}	22	14	8
40	200	0.74	1.3×10^{-3}	16	12	2
60	125	0.40	4.9×10^{-2}	109	56	50
60	150	0.42	3.5×10^{-2}	88	47	20
60	175	0.48	2.3×10^{-2}	76	43	40
60	200	0.58	8.4×10^{-2}	48	30	3

(a) Dowell J111 or J133 guar
(b) Low temperature breaker can be used for break time of six hours or less at 50°F.

TABLE 8-7
Rheology of Typical Crosslinked Gel Fluids[a]

Polymer Concentration lb/1000 gal	Temp °F	Power Law Fluid Parameters		Apparent Viscosity cp	
		n'	K'	160 sec^{-1}	479 sec^{-1}
15	80	0.945	0.0024	85	80
15	100	0.915	0.0018	85	51
15	125	0.640	0.0085	65	44
15	150	0.282	0.0730	91	41
20	100	0.757	0.0127	177	136
20	125	0.948	0.0022	82	77
20	150	0.640	0.0139	107	72
20	175	0.253	0.1420	154	68
40	100	0.48	0.185	633	358
40	125	0.62	0.054	374	246
40	150	0.37	0.241	471	236
40	175	0.25	0.401	426	188
40	200	0.39	0.027	58	30

(a) Dowell J347 HP guar

TABLE 8-8
Rheology of Typical Polyemulsion Fluids[a]

Polymer Concentration lb/bbl of Brine	Temperature °F	Power Law Fluid Parameters		Apparent Viscosity cp	
		n'	K'	160 sec^{-1}	479 sec^{-1}
1	100	0.700	0.0169	176	127
1	125	0.746	0.0105	132	104
1	150	0.779	0.0065	101	78
1	200	0.826	0.0025	50	41
2	150	0.627	0.0226	163	108
2	200	0.667	0.0113	100	69
2	250	0.694	0.0066	67	48
2	300	0.713	0.0043	48	35
2	350	0.727	0.0029	35	26

[a] 2/3 diesel, 1/3 brine (9.2 lb/gal NaCl w/J 133 guar thickener)

ternal phase. The oil may be kerosene, diesel, crude oil, condensate, or refined oil. The water may be fresh water or light brine. The emulsifier should be selected so as to leave the frac sand water-wet. The fluids should be bench tested to make sure that a stable emulsion can be formed. Guar, HP guar, or HEC may be used to thicken the water phase. Polymer concentration depends on temperature, but usually 1 lb/bbl of water is used below 180°F, and 2 lb/bbl of water above 180°F. A gel breaker may be needed depending on the type of polymer. The water external emulsion breaks due to adsorption of the emulsifer. Table 8-8 shows rheology of typical polyemulsion fluids.

Gelled oil fluids can be prepared with kerosene, diesel and many crude oils. The concentration of gelling agent must be determined for the specific oil, but usually is in the range of 6–10 gal/1000 gal oil. Breaker requirements depend on the type of oil and the formation temperature and should be determined by bench testing. Below a temperature of 180°F, apparent viscosity @ 160 sec^{-1} should be about 150–200 cp; @ 479 seconds^{-1}, apparent viscosity drops to about 70–80 cp.

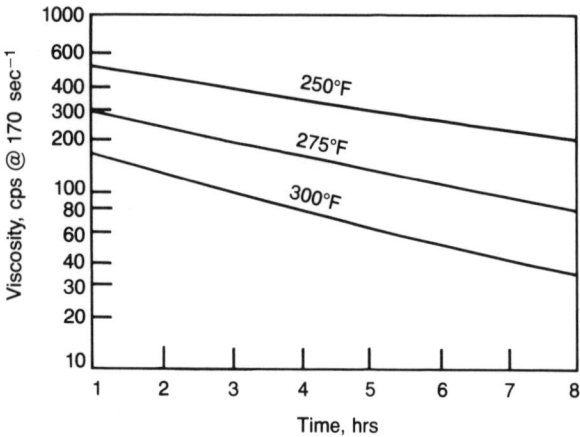

FIG. 8-29—*Viscosity vs. time relations for 50 lb HP guar system with organo-metallic crosslink and 10 lb stabilizer per 1000 gal. Courtesy of Dowell.*

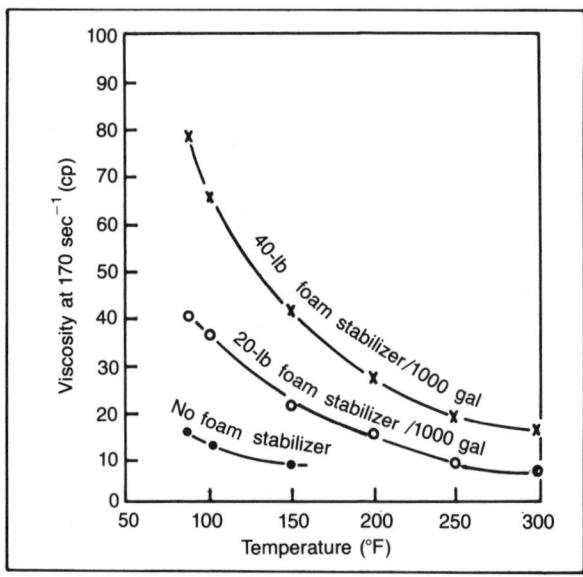

FIG. 8-30—*Viscosity of 75% quality foams at temperature and 1000 psi.*[41] Copyright 1981, SPE-AIME.

Foamed fluid characteristics are quite different than gelled fluids, in that foams have relatively low viscosity but good particle transport ability. Polymer stabilizers may be needed to increase viscosity and extend half-life. Figure 8-30 shows viscosity characteristics of 75% quality foam at 1000 psi with varying amounts of stabilizer.

Rheology

Rheological properties become an important factor with high injection rates, or high viscosity fluids. Rheology of frac fluids is complex, particularly the rheology of viscoelastic, emulsion, and foamed fluids. Basically fluids are classified as:

Newtonian—those which exhibit constant viscosity independent of flow rate, and
Non-Newtonian—those which exhibit "apparent viscosity" which changes with flow rate, generally becoming lower at higher flow rates.

Frac fluids are described mathematically by the Power Law model in which the actual shear stress—shear rate relation plotted on log-log paper is approximated by a straight line.
Flow behavior index (n') is the slope of the line. An n' value of 1.0 indicates a Newtonian fluid. More non-Newtonian, or shear thinning behavior, is indicated as n' drops below 1.0. Measured on a field model Fann V-G meter at 300 and 600 rpm it becomes:

$$n' \text{ (dimensionless)} = 3.32 \log_{10} \frac{(600 \text{ rpm reading})}{(300 \text{ rpm reading})} \quad (6)$$

Consistency index (K') is the shear stress intercept at shear rate = 1.0 sec^{-1}. It is a measure of fluid "thickness." From Fann V-G meter values it is:

$$K' \text{ (lb F sec}^{n'}/\text{ft}^2) = N \frac{(300 \text{ rpm reading})}{100 \times 511^{n'}} \times 1.066 \quad (7)$$

N = spring factor of the Fann (usually 1.0)

Apparent viscosity for non-Newtonian fluids is determined as follows:

$$\mu_a = \frac{4.788 \times 10^4 K'}{(\text{Shear rate})^{1-n'}} \quad (8)$$

μ_a = apparent viscosity, cp
K' = consistency index
n' = flow behavior index

Shear rate is related to fluid flow velocity in various portions of the wellbore, fracture and formation, and in measuring devices as follows:
Viscometer relation between shear rate and Fann rpm:

Fann rpm	Shear rate (sec^{-1})
100	170
200	341
300	511
600	1022

For circular pipe, perforation, and pore channels:

$$\text{Shear rate, sec}^{-1} = \frac{96 V}{D} = \frac{1642 Q}{D^3} \quad (9)$$

For rectangular fractures:

$$\text{Shear rate, sec}^{-1} = \frac{72 V}{w} = \frac{40.3 Q}{h w^2} \quad (10)$$

where:

V = flow velocity, ft/sec
D = conduit diameter, in.
w = fracture width, in.
h = fracture height, ft.
Q = total injection rate, bpm

Calculating shear rate in a fracture requires a knowledge of fracture width and height. Since these factors are usually unknown fracture shear rate is often assumed to be in the range of 10 to 170 sec^{-1}.
Temperature can drastically affect the viscosity of Newtonian fluids or apparent viscosity of non-Newtonian fluids.

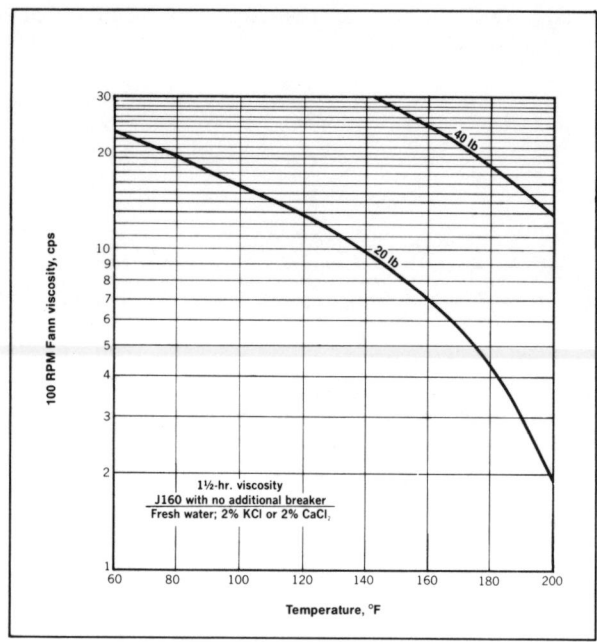

FIG. 8-31—*Temperature-viscosity relation of non-Newtonian water base, gel-thickened fluid. Courtesy of Dowell.*

Temperature-viscosity relation of a typical Newtonian Midcontinent refined oil is:

Temp. °F	Viscosity – cp
100	7000
140	800
210	75

Temperature-viscosity relation of a typical non-Newtonian water base, gel-thickened fluid is shown in Figure 8-31.

Rheology of emulsions (or water-oil dispersions), viscoelastic fluids, and foamed fluids is the subject of considerable research. With an emulsion fluid, "slippage behavior" occurs above a relatively low shear rate which drastically reduces apparent viscosity, Figure 8-32. Viscoelastic fluids apparently exhibit similar characteristics, such that friction pressure loss in smooth wall pipe flow is much less than the apparent viscosity relation would indicate. Rheological studies of foam show power law behavior represented by the following relation:[41]

$$\frac{3D\Delta P}{L} = K' \frac{2V^{n'}}{3D}$$

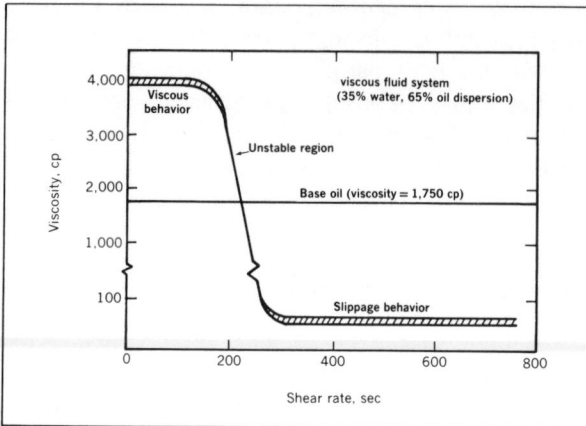

FIG. 8-32—*Slippage behavior of emulsion fluid drastically reduces apparent viscosity.*[17] *Permission to publish by The Society of Petroleum Engineers.*

D = Tubular inside diameter, in
ΔP = Pressure drop, psi
L = Tubular length, ft
V = Flow velocity, ft/sec
K' = Shear stress @ 1 sec^{-1}, lb F sec$^{n'}$/ft^2
n' = Log slope of shear stress—shear rate curve, dimensionless

Ability to Carry Proppant

Proppant-carrying ability is largely a function of viscosity, the density difference between the proppant and fluid, and the size of the proppant. Stokes law, Figure 8-33, can be used as an indicator of sand falling rate of single grains; however, it predicts rates higher than actually occur in the fracture due to interference effects between grains, and viscous effects particularly with emulsified fluids. Studies to determine more realistic correlations between sand falling rates and fluid properties are under way in several laboratories.

Sand concentrations which can be carried by the frac fluid are affected by viscosity as well as fluid type as shown in Table 8-9.

Mixing, Storage, and Handling

Ease of mixing and stability in storage are important when gelled or emulsified fluids are used. Some crudes are not suitable for emulsions and some gelling agents do not "yield" properly in certain waters. Fluids must be bench tested ahead of time. (See Job Performance.)

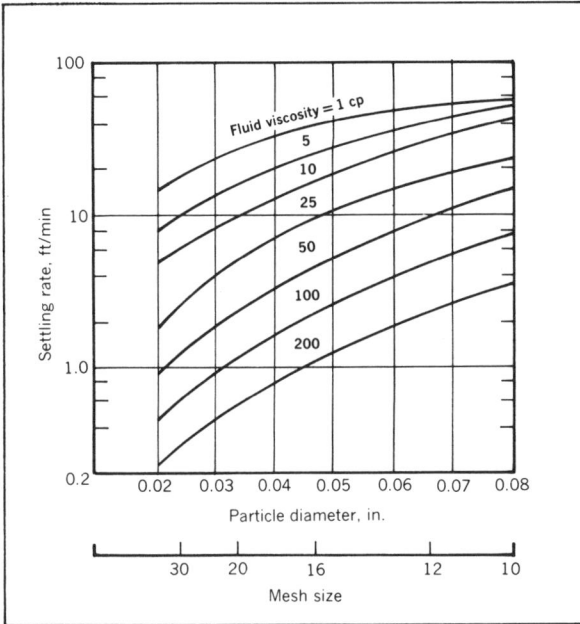

FIG. 8-33—*Sand settling rate from Stokes law.*

Where large volumes of fluids must be mixed, continuous mixing rather than batch mixing simplifies the problem of fluid disposal if the job cannot be completed as planned. However, effectiveness of viscosifiers and crosslinkers must be inferred from results of bench tests. Stability in storage can be a problem with gelled water. Bacteria can breed rapidly, and they and the attendant enzymes can destroy the gel or interfere with crosslinking. This problem can be controlled with clean tanks and bactericide.

Handling of ultra-high viscosity refined oils which must be maintained at high temperature can create difficult problems in many areas.

TABLE 8-9
Typical Sand Concentration (lb sand per gal liquid)

	Average	Maximum
Gelled water	1–1½	2
Crude oil (<100 cp)	2	3
Crude oil (100–250 cp)	2–3	4
Crude oil (250–500 cp)	3–4	5
Crude oil (>500 cp)	4–5	6–8
Crosslinked gelled water	4–5	6–8
Polymer emulsion 20–40; 10–20	4–5	6
Polymer emulsion 8–12	2–3	3–5
Foam (w/concentrator)	2–3	4–6

Cost

The cost of frac fluids varies considerably. Crude oil is normally the cheapest, while the high viscosity water gels designed for high temperature conditions are the most expensive. Emulsion fluids and gelled water fluids fall in the mid-cost ranges. Foamed fluids depending on treating pressure are also expensive. Table 8-10 attempts to compare frac fluids on the basis of polymer concentration, cost per 1,000 gal, and cost per unit of viscosity. A more detailed comparison is shown in Appendix C.

Formation Damage

An ideal frac fluid should have no tendency to plug or block the formation. Reduction in matrix permeability along the fracture wall is not as serious as it would be in a radial flow situation around the wellbore. The critical concern is plugging of the propped fracture itself with any of several possible mechanisms related to the formation, formation fluids or the frac fluid.

Incompatibility Between Fluids—Solubility, cementaceous materials, salt content, clay materials, reservoir fluid constituents, should be considered. Softening of the formation, or dissolving cementing materials may release fines. Dissolving salt may help or hinder. Any change in ionic environment may let clays move and block pores.

If cores are available, laboratory tests should be run to evaluate damage problems. Tests involve:

1. X-ray diffraction to determine clay content and scanning electron microscope to determine clays and clay positioning in the pore spaces.
2. Flowing prospective frac fluids through the core, and short-term immersion in the frac fluids with permeability determinations before and after—and observation of fines release;
3. Petrographic analysis of a thin section to show grain size, pore size and constituents between the grains and within the pores.

In many situations laboratory tests confirmed by field results show clean fresh water with 2% KCl to be an excellent frac fluid based on compatibility with formation clays and frac fluid gelling agents.

In most cases a surfactant should be used to prevent emulsions particularly with water base fluids in oil reservoirs. It should be selected on the basis of pilot tests—must be good water-wetter—must be compat-

TABLE 8-10
Cost Comparison of Frac Fluid Systems[a]

Fluid System	Polymer Concentration lb/1000 gal	Apparent Viscosity cp @ 160 sec^{-1} 150°F.	Cost $/1000 gal	$/cp @ 160 sec^{-1}; 150°F
Linear Aqueous Gel				
Guar	20	4	60	16.70
	40	29	120	4.00
HP Guar	20	14	85	6.30
	40	37	165	4.50
	80	189	330	1.75
HEC	20	6	110	18.50
	40	36	220	6.00
	80	175	440	2.50
Crosslinked Aqueous Gel				
Guar	20	97	90	0.90
	40	667	175	0.33
	60	1750	265	0.15
HP Guar	20	107	90	0.85
	40	471	195	0.45
Polyemulsion				
1/3 Brine, 2/3 Oil	8	101	85[b]	0.85
	16	163	105[b]	0.65
Gelled Oil				
	NA	165	150[b]	0.90
Foam Frac				
70% Quality	@ 1000 psi BHTP		74	—
w/Nitrogen	@ 10,000 psi BHTP		460	—

(a) Fluid property data from Dowell; costs Dowell Midcontinent U.S. April 1981 (b) assumes full recovery of oil

ible with frac fluid additives—and must not be lost due to plating out on the formation surfaces.

Fluid Retention—Fluid retention is most severe in low permeability, low porosity situations.

High viscosity oil fluids often rely on dilution by formation crude for removal. A real problem can occur if the formation oil acts to strip the light ends from the frac oil resulting in much higher viscosity rather than dilution.

Failure of gels to break can be a cause of fluid retention damage. Breaker mechanisms include: enzymes, acids, and oxidizing agents. Formation temperature is an important factor in breaker selection.

Formation rock, fines, and frac sand must be left in a water-wet condition to speed recovery of frac fluids, to minimize emulsion problems, and to provide favorable oil relative permeability for subsequent oil production. Oil or condensate should not be used in a dry gas zone due to relative permeability effects if an oil saturation is established.

Capillary pressure is the most common cause of fluid retention. Capillary pressure effect can be reduced by lowering surface tension of frac fluid, or by energizing the formation with CO_2 or N_2.

Surface tension is best lowered by using a fluid having inherent low surface tension—methanol or high gravity oil. Surfactants may lose effectiveness by adsorbing on silicate surfaces.

Energizing the frac fluid with nitrogen or carbon dioxide may be the best solution to fluid retention in a low pressure, low permeability formation.

Nitrogen is inert, thus will not react with frac fluid, formation fluids, or the formation. N_2 has lower solubility in hydrocarbons than CO_2.

Carbon dioxide can be introduced into the frac fluid stream using normal pumping equipment. It will react with the frac fluid to lower pH to 3.3 to 3.7 which is good from the standpoint of clay swelling and holding soluble Fe or Al salts in solution. Due to pH reducing effect CO_2 cannot be used with some gelling materials. CO_2 is usually cheaper than N_2.

Residue in Matrix or Fracture—Potential sources include: insoluble particles in the base fluid, fluid loss control agents, residue from the gelling agent, fines in the fracture proppant, fines released by reaction of the fluid with the formation.

1. Base fluids: water must be solids free. Hydrocarbon must not have paraffins or asphaltenes.
2. Mixing tanks, pumps, and lines must be clean.
3. Proppant fines must be limited by specifications, careful checking, and careful handling.
4. Guar gum has inherent residue which is not water soluble. HP guar has much less residue; HEC has almost no residue.
5. Anionic gelling agents can react with divalent calcium and magnesium to form precipitate—this can be checked in lab immersion tests.
6. Cold frac fluid can precipitate paraffin in the fracture or matrix.
7. Use of insoluble fluid loss agents should be minimized.

Effect of residue in the fracture can be minimized by establishing high fracture flow capacity using high concentration of large proppant—and by putting well on production at gradually increasing rates.

Frac Fluid Trade Names

Based on names applied by Halliburton Services the following types of frac fluids are in common usage:

1. *Water Base Fluids*
   Water Frac—2% KCl fluid with guar gum or polymer gelling agent to provide relatively low viscosities at moderate temperatures.
   My-T-Gel—Crosslinked guar gum system providing ultra high viscosity at normal formation temperatures. My-T-Gel III is a recent improvement providing higher temperature stability.
   Versafrac—Low residue HP guar system which, similar to My-T-Gel, can be crosslinked providing ultra high viscosity. Is compatible with CO_2 or N_2, or methanol. Has higher temperature stability than My-T-Gel I.
   Hy-gel—A two-component polymer system in 1% KCl brine, in which the second polymer is buffered to require heat before it reacts; thus reasonably high viscosity is provided at 300°F formation temperature. Hy-Gel 500 is a new system for improved temperature stability above 300°F.
   Logel—A *CMC* polymer fluid having the advantage of low residue, intended for moderate treating temperature.
   Kleer Gel—Synthetic polymer, residue free, that can be crosslinked to provide very high viscosity. Is compatible with CO_2 or N_2.

2. *Oil Base Fluids*
   Oil Frac—Selected crude oil with no other additives except Adomite Mark II for fluid loss control.
   My-T-Oil—Gelled diesel or selected crude oil crosslinked to provide ultra high viscosity fluid. My-T-Oil II permits wider selection of crude oils.

3. *Emulsion Fluids*
   Super Emulsifrac—A water external emulsion of 67% light crude oil and 33% polymer-gelled water, having reasonable viscosity at relatively low cost. Water should contain 1% NaCl or KCl. For cleanup, emulsifier plates out on formation or will hydrolyze or break. Gelled water also contains breaker.

4. *Weak Acid Base Fluids*
   Acidgel Frac—Guar or synthetic polymer used to thicken weak HCl. Acidgel Frac II uses a crosslinked synthetic polymer to give higher viscosities.

5. *Alcohol Fluids*
   Alcogel I—A medium viscosity mixture of 30% methanol and 70% water intended for gas wells where oil fluids might reduce relative permeability.
   Alcogel II—A crosslinked synthetic polymer system for moderate to high bottomhole temperature application.

6. *Aerated Fluids*
   Foam Frac—An aerated fluid (water, alcohol-water, oil, or acid), wherein high sand concentrations are mixed in the liquid to be subsequently diluted when injected into the nitrogen stream. Advantage is better cleanup in low permeability formations—particularly gas zones.

FRAC JOB DESIGN

Select the Right Well

The first and often most important step is the selection of a proper well situation where the benefits of fracturing can be maximized and the possible problems from creating undesired communication can be avoided. General comments are as follows:

The well and reservoir situation must allow maximum sustained production increase—Pick the good wells rather than poor producers—unless the poor wells are damaged.

Consider the risks involved—Conditions that increase risks are:

1. Less than 15–20 ft of shale between the frac interval and a gas or water sand.
2. Other things being equal fractures tend to move upward due to sand fallout to the bottom of the fracture—sometimes this effect can be maximized.
3. Water or gas contact nearby and located in a direction so that fracture would go toward it.
4. Well producing HGOR or WOR are poor candidates for fracing unless it is free gas or water from a zone which can be shut off.

Design for the Specific Well

Too many times a treatment technique successful in one area is applied in another area without regard to formation differences. Detailed knowledge of the specific reservoir geology, reservoir drive, characteristics of the reservoir fluids, petrography of the reservoir rock is the key to successful stimulation.

Design parameters to be considered are:

1. *Lithology and Mineralogy of the Formation:*

—Porosity and Permeability—to determine the relative importance of frac conductivity and fracture length.
—Rock properties—hardness or compressive strength for clues to proppant embedment or crushing, and subsequent generation of fines and frac width reduction. Rock strength may indicate application of a partial monolayer or pillar frac-type proppant placement.
—X-ray diffraction, scanning electron microscope, and Petrographic analysis—grain and pore sizes, clay mineral types and location within the matrix for clues to possible formation damage problems that may dictate a particular frac fluid.
—Solubility and Immersion tests—for physical evidences of the effect of a prospective fluid on the formation rock. Highly nonuniform solubility in acid may indicate application of an acid frac treatment to obtain etched conductivity rather than a propped fracture conductivity.
—Fluid loss tests—for physical evidence of the "efficiency" of various frac fluid modifications in creating fracture areas or length.

2. *Fracture Geometry Parameters*

—Young's modulus—related to fracture width and the possibility of obtaining highly conductive fractures using packed fracture techniques.
—Poisson's ratio—related to horizontal matrix stress and fracture gradient or bottomhole treating pressure in the producing formation.
—Horizontal matrix stress in boundary formations—related to the possibility of the fracture extending upward or downward out of the desired zone. A zone with a low horizontal matrix stress (low Poisson's ratio) probably would not serve as an effective barrier to fracture extension. A "plastic" shale (high Poisson's ratio) on the other hand might confine the fracture to a more rigid limestone or dolomite producing zone.

3. *Reservoir Fluids and Reservoir Energy*

—Oil—viscosity, emulsifying tendencies, asphaltene content, paraffin forming characteristics must be considered in selecting and modifying a frac fluid.
—Gas—liquid content or reservoir characterization, possibility of an oil frac fluid remaining in the rock as an oil phase to reduce relative permeability to gas.
—Reservoir pressure—is it sufficient to expel the frac fluid after the treatment.

4. *Physical Well Configuration*

In a new well the well completion and hookup should be designed with subsequent stimulation problems in mind—specifically:

—Size and pressure capacity of casing and well head fittings.
—Perforation pattern to permit effective coverage.
—Tubing and packer configuration to return well to production after the treatment such as to minimize formation damage.

In an older well the frac treatment may have to be

modified in accord with limitations imposed by the well completion itself.

Optimize Design Over Several Jobs

Proper design of a frac treatment involves an optimizing process to balance anticipated productivity increase against the cost of obtaining the productivity increase. Usually the experience gained in several carefully designed and evaluated jobs is necessary to achieve an optimum design.

1. *Treatment cost* depends on:

—Type and volume of frac fluid—gelling agents, fluid loss control agents.

—Quantity and type of propping agent.

—Amount of hydraulic horsepower required.

2. *Productivity increase* depends on numerous factors, many of which are known only approximately. Major factors include:

—Degree of formation damage present.

—Coverage obtained in multizone sections.

—Fracture conductivity and perhaps fracture length—which depends on fracture geometry, as influenced by the frac fluid quality, injection rate, treatment volume, and type and arrangement of proppant.

Utilize Calculation Procedures As A Guide

Treatment design must specify the following parameters:

1. Frac fluid type
2. Fluid volume
3. Fluid viscosity and fluid loss schedule
4. Proppant size and type
5. Proppant schedule
6. Injection rate schedule

Basic design procedures are as follows:

Determine Required Fracture Length and Conductivity—Computer programs based on McGuire and Sikora, or similar curves, are available to relate productivity after fracturing to productivity before fracturing, at steady state conditions. These can be used to compare relative performance of various fracture lengths and conductivities. For low permeability formations where steady state conditions would require a long time period, productivity forecasts can be made with reservoir simulators or type-curves.

Determine Frac Fluid Characteristics and Injection Rates—Several calculation procedures are available to estimate fracture geometry based on treatment variables. Some combine the Howard and Fast leakoff and fracture area concept, with the fracture width relation of Perkins and Kern (which was modified by Nordgren), or that of Kristianovich. Other calculation procedures may combine the equations of Williams with those of Geertsma and de Klerk, or Daneshy.

Determine a Treatment Pumping and Proppant Injection Schedule—Injection rate and pad volume determined previously are used in a proppant transport program, which models the effect of proppant addition on proppant penetration and concentration along the fracture.

Service companies have computerized these programs to enable many combinations of variables to be considered with minimum effort. Table 8-11 shows a typical example program from Halliburton Services.

Long Zones Need Special Consideration

In multiple zones or long sections special techniques are needed to obtain coverage of entire interval. These techniques include:

1. *Limited entry* involves perforating a small number of holes in each zone (perhaps one or two holes per zone). The pressure drop across the perforations tends to distribute the frac fluid to all perforations, thus insuring a fracture in each zone.

2. *Multiple staging* with ball sealers involves perforating each zone with the same number of holes. Each zone is then considered as a separate, though identical, frac treatment. Each zone receives its own pad volume and graduated sand schedule; and at the end of each stage enough ball sealers are dropped to cover the perforations in one zone. The ball sealers are then followed by the pad volume (fluid with no sand) to initiate the fracture in the next zone.

3. *Zone isolation* with packers and/or sand fill involves, in effect, a series of separate frac treatments. Usually the lowest zone is fractured first through tubing, using a packer set above the lower perforations. When this treatment is completed, sand is allowed to drop out inside the casing, covering the lower set of perforations, and the tubing packer is reset above the next perforated interval.

TABLE 8-11

Job Type—Waterfrac Fracturing Service

Basic Input Information

Ajax Oil Company; Ajax No. 1; 7,000 feet
50 lb WG-12, 25 lb WAC-9, 1 gal TRI-S per 1000 gal 2% KCl water

Well & Formation Data
    Youngs Modulus   3.00E+06   psi
    Permeability   0.5000   md
    Porosity   12.0   %
    Reservoir fluid compressibility   2.50E−05   1/psi
    Reservoir fluid viscosity   1.00   cp
    BHTP   5000.   psi
    Reservoir fluid pressure   2800.   psi
    Closure pressure   3000.   psi
    Gross fracture height   50.   ft
    Net fracture height   35.   ft
    Wellbore diameter   5.50   in.
    Drainage radius   933.   ft
    Well spacing   80.   acres
    Bottom hole temperature   160.   deg F

Treatment Data
    Type of gel   WG-12
    Gel concentration   50 lb/mgal
    Injection rate   20.0   bpm
    Treatment fluid SP GR   1.020
    N   0.6000
    K(Slot)   0.008000   lbf-sec**N/sq ft
    CW—fluid loss coeff   0.00200   ft/sqrt(min)
    Spurt volume   0.0500   gal/sq ft
    CVC—spurt loss coeff   0.00147   ft/sqrt(min)
    Spurt time   2.54   min
    Damage ratio   1.0
    Apparent viscosity   62.   cp, at 0.410 in. width

Overall Design Comparisons

Design No	Volume Total (gal/1000)	Volume Pad (gal/1000)	Created Length (ft)	Width Avg (in.)	Prop Length (ft)	Prop Height (ft)	Prop Total (sx)	Dimensionless Capacity	Prod Incr	Fluid eff-%
1	28.9	7.0	716.0	0.410	696.0	43.7	425.	4.26	6.6	63.2
2	48.8	10.0	980.0	0.490	976.0	40.3	840.	5.33	7.5	61.3

TABLE 8-11 (cont.)

Bed Deposition for Design No. 1

Pumping Schedule
7000.0 gallons of pad volume
5000.0 gallons with 1.00 lb/gal of 20/40 mesh sand
5000.0 gallons with 2.00 lb/gal of 20/40 mesh sand
5000.0 gallons with 2.50 lb/gal of 20/40 mesh sand
5000.0 gallons with 3.00 lb/gal of 20/40 mesh sand
425. sacks total prop

Deposition Profiles
At the end of pumping:
Carry distance   696.0 ft
Max bed height   1.2 ft
Avg bed height   0.7 ft
% prop deposited   9.3%

Proppant Transport

Distance from Well, ft	Deposited Prop Bed Height, ft		Suspended Prop		
	End of Pumping	Final	Height ft	Concentration lb/gal	lb/sq ft
4.0	1.2	11.2	50.0	3.1	0.88
36.0	1.2	11.1	49.8	3.1	0.88
68.0	1.2	11.1	49.4	3.1	0.88
100.0	1.2	11.1	48.9	3.1	0.88
132.0	1.2	11.0	48.4	3.1	0.88
164.0	1.2	11.0	47.8	3.1	0.88
196.0	1.0	9.4	47.1	2.8	0.77
228.0	0.9	9.3	46.6	2.8	0.77
260.0	0.9	9.4	46.1	2.8	0.77
292.0	1.0	9.5	45.6	2.8	0.77
324.0	1.0	9.6	45.1	2.8	0.77
356.0	0.7	7.7	44.0	2.6	0.61
388.0	0.6	7.7	43.5	2.6	0.61
420.0	0.5	7.8	43.0	2.6	0.61
452.0	0.6	8.0	42.5	2.6	0.61
484.0	0.6	8.3	41.9	2.6	0.61
516.0	0.4	4.6	39.1	1.7	0.28
548.0	0.2	3.9	38.5	1.7	0.28
580.0	0.2	4.1	37.9	1.7	0.28
612.0	0.2	4.6	37.3	1.7	0.28
644.0	0.3	5.4	36.7	1.7	0.28
676.0	0.4	6.9	36.1	1.7	0.28

Equivalent Bed
Length = 696. ft
Height = 43.7 ft
Bed Concentration = 698. lb/1000 sq ft
Flow Capacity = 1282. md-ft

Courtesy of Halliburton Services

TABLE 8-11 (cont.)

Bed Deposition for Design No. 2

Pumping Schedule
  10000.0 gallons of pad volume
  7000.0 gallons with 1.00 lb/gal of 20/40 mesh sand
  7000.0 gallons with 2.00 lb/gal of 20/40 mesh sand
  7000.0 gallons with 2.50 lb/gal of 20/40 mesh sand
  7000.0 gallons with 3.00 lb/gal of 20/40 mesh sand
  7000.0 gallons with 3.50 lb/gal of 20/40 mesh sand
  840. sacks total prop

Deposition Profiles
  At the end of Pumping:
    Carry distance   976.0 ft
    Max bed height   2.2 ft
    Avg bed height   1.5 ft
    % prop deposited   15.4%

Proppant Transport

Distance from Well, ft	Deposited Prop Bed Height, ft		Suspended Prop		
	End of Pumping	Final	Height ft	Concentration lb/gal	Concentration lb/sq ft
4.0	2.2	13.3	50.0	3.6	1.21
52.0	2.2	13.2	49.6	3.6	1.21
100.0	2.2	13.1	49.0	3.6	1.21
148.0	2.2	13.0	48.2	3.6	1.21
196.0	2.2	12.9	47.3	3.6	1.21
244.0	2.0	11.6	46.2	3.4	1.09
292.0	1.9	11.5	45.5	3.4	1.09
340.0	1.9	11.5	44.7	3.4	1.09
388.0	2.0	11.6	43.8	3.4	1.09
436.0	1.6	9.7	42.3	3.2	0.93
484.0	1.4	9.6	41.5	3.2	0.93
532.0	1.4	9.7	40.6	3.2	0.93
580.0	1.4	9.9	39.8	3.2	0.93
628.0	1.1	7.7	37.7	3.1	0.72
676.0	1.0	7.8	36.9	3.1	0.72
724.0	0.9	8.1	36.0	3.1	0.72
772.0	1.0	8.6	35.1	3.1	0.72
820.0	0.7	5.4	29.8	3.2	0.45
868.0	0.6	5.5	28.5	3.2	0.45
916.0	0.5	6.6	27.2	3.2	0.45
964.0	1.0	12.2	25.9	3.2	0.45

Equivalent Bed
  Length = 976. ft
  Height = 40.3 ft
  Bed Concentration = 1069. lb/1000 sq ft
  Flow Capacity = 1869. md-ft

Courtesy of Halliburton Services

Frac Job Performance

Job performance is the key to success in any oil field operation—but this is particularly true in large frac jobs involving millions of dollars in equipment and materials and the attendant personnel which must be mobilized to perform an operation lasting several hours, then demobilized quickly to permit subsequent well operations to proceed.

Detailed planning, checking, and rechecking must be the order of the day. Appendix A is a check list for planning and executing a massive hydraulic frac (MHF) job, which with modifications can be adapted for most frac operations.

Problems can be grouped into two categories: *logistical* problems involving physical preparation of the wellsite, equipment, and materials prior to starting the job, and *operational* problems involving fluid rate and sand concentration monitoring, pressure variations, and equipment malfunctions during the job.

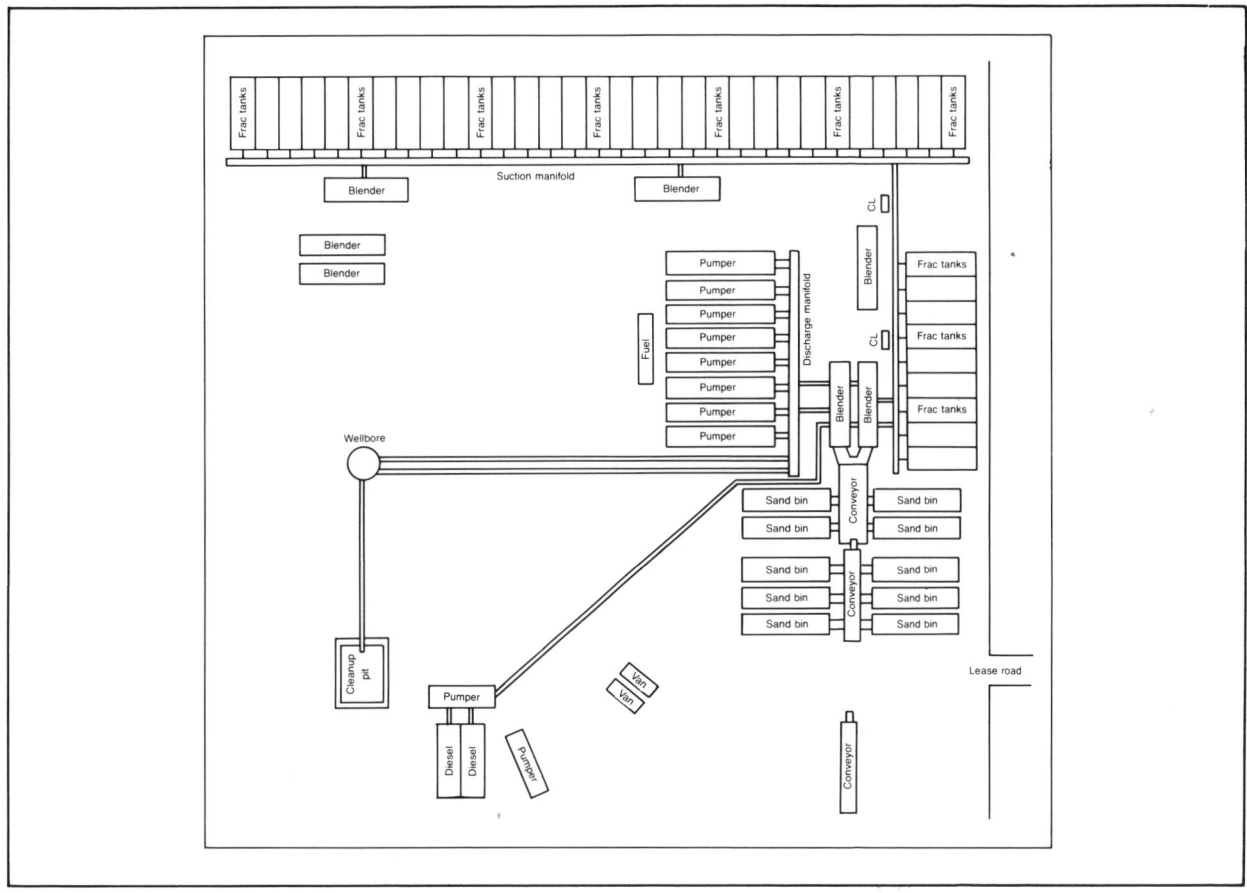

FIG. 8-34—*Typical equipment layout for massive frac operation.*[39] *Copyright 1981, SPE-AIME.*

Logistical Problems

Wellsite Preparation—The location should be large enough to easily accommodate the required frac fluid and sand tanks and stimulation equipment. Scale drawings are helpful. Figure 8-34 shows a typical location plan (250 ft × 350 ft) used one operator for East Texas MHF jobs.[39] Manned equipment should be at least 50 feet from the well head. Pads for frac tanks should be sloped toward the suction outlet. At least 5% excess fluid should be allowed for tank bottoms under ideal conditions. Space should be provided around frac tanks to permit refilling tanks or emptying contaminated tanks. Unobstructed space must be provided for emergency vehicles from service roads. Personnel escape routes must be established from each piece of manned equipment.

Standby Equipment—To insure continuous pumping, sufficient standby equipment should be hooked up and operational. Two pumping blenders should be rigged up side-by-side and positioned so that either can be replaced by a standby blender if necessary. With 40 to 50 frac tanks on an MHF job, transfer blenders are desirable to pump fluid to working tanks close to the pumping blenders. Standby sand conveyor belts are needed on MHF jobs, since sand conveyor problems are as prevalent as blender problems. Standby high pressure pumps, crosslinker pumps, and instrument vans should be hooked up and operational before starting the job. High pressure pumps should be hooked up so that they can be isolated, to permit changing valves and seats during the job, without shutting down other equipment. Experience shows that the additional cost of standby equipment is minimal compared to overall job cost.

Quality Control—Proppant should be analyzed to ensure that it meets specifications as to sieving and acid solubility. Water (and oil) should be checked to ensure compatibility with the gelling system. These checks should be run at the source—and again after delivery to the location.

Fluid Preparation—All tanks must be steam cleaned and flushed with fresh water. Bacteria can be a major problem. Sulfate reducing bacteria produce H_2S and sulfite ions, with the possibility of precipitating iron sulfide in the formation. But, perhaps more important, the bacteria produce enzymes that act as breakers and interfere with crosslinking. Bactericide should be placed in the frac tanks before filling. Enzymes are *not* eliminated by bactericides but can be removed by reducing pH to less than 2. After filling, water compatibility checks should be run again. Gelling should not begin until the well is ready for treatment. After gelling, pilot crosslinking tests should be run periodically, up to the start of the job, to make sure that bacterial problems are under control.

Erosion of Tubulars—Critical areas to watch for sand erosion are those where changes of fluid velocity or direction occur. Wellhead isolation tools can be problems. If fluid is injected through tubinghead valves, a blast joint should protect the tubing. If a special frac tree is to be installed, it should be tested to a maximum pressure above the expected frac pressure. All lines must be staked down, and lines to the wellhead should be chained to the christmas tree. To protect the casing, pressure relief valves should be installed on the casing annulus connection if frac fluid injection is to be down a tubing-packer hookup.

Operational Problems

Fluid and Sand Rate Monitoring—Most instruments used to monitor fluid and sand rate are accurate to ±5%. Verification should be made at frequent intervals during the job. Turbine flowmeters should not be the only source of fluid rate verification. Counting pump strokes provides at least a backup approximation. The most accurate verification is tank strapping to measure the volume actually removed from a frac tank in a given time period, usually 5 to 6 minutes. However, levels in other tanks, which may be hooked into the pump suction, should be observed to see that manifold valves are not leaking.

For sand rate verification, blender augers or conveyors should be calibrated before the job if they are to be relied on. Radioactive densimeters have inherent possibilities for monitoring sand concentration, but currently are a qualified success. The most reliable method is to gauge sand tanks. There should be a schedule for emptying individual compartments at predetermined stages during the treatment as an additional check on sand usage.

Pressure Deviations—One of the most critical areas to evaluate during a treatment is the fluctuation of treating pressure. Often the cause must be identified and remedial action taken quickly to prevent job failure. Bottomhole recording pressure equipment makes analysis much better because hydrostatic and tubular friction problems are eliminated.

Sources of pressure deviation include mechanical problems, changes in gel properties, variation in sand concentration, or formation response. An abnormally high pressure initially, while pumping pad, may be due to flow restriction through perforations. This can sometimes be verified by shutting down to check fracture closure pressure. Subtracting this from the bottomhole treating pressure should provide an approximation of pressure loss through perforations.

When surface pressures fluctuate, several factors could be at fault:

1. Varying sand concentration—higher sand concentration increases hydrostatic head and fluid viscosity.

2. Varying gel concentration—higher polymer loading usually increases friction pressure loss. After mixing, unhydrated polymer may rise to the top of the tank—and increase polymer loading when the liquid level falls as fluid is pulled from the tank.

3. Failure of crosslinking mechanism—this usually causes surface pressure to drop due to reduced viscosity. A sandout could result, therefore, this should be corrected quickly.

4. Pressure changes that correlate with the time to displace fluid from the surface to the bottom usually are related to changes in fluid properties.

5. Pressure changes that do not correlate with displacement time may be related to fracture geometry and dynamics. These changes can be interpreted by the technique of Nolte and Smith[36] observing the slope of the net fracturing pressure versus time plot. Briefly, a pressure drop may mean that a confining barrier has failed. A pressure rise may signal a sandout.

6. A sudden rise may mean a sandout near the wellbore usually due to mechanical or gel problems, an unusually high sand concentration, perhaps a blender failure, or crosslink failure. If a sandout is to be avoided, remedial action must be taken quickly. Decreasing sand concentration may help—but may also reduce fracture conductivity. Injection rate may be increased to provide added fracture width through

increased net fracture pressure—but this may cause penetration of a confining barrier.

With most of these problems, the alternatives need to be thought out ahead of time, so that the most favorable course of action can be taken.

FRAC JOB EVALUATION TECHNIQUES

To evaluate the success of a frac treatment and help design succeeding treatments, it is necessary to know:

1. What sustained production increase was obtained.
2. What zone or zones were actually stimulated.
3. For vertical fractures, what was the fracture height and azimuth.
4. What was the fracture length.
5. What was the fracture conductivity.

Extended Production Results

Production tests and decline curves are one of the most important methods of measuring the effectiveness of a frac job, and modifying treatment technique for subsequent jobs.

1. *Insufficient Initial Response* may be related to:

—Fracture conductivity.
—Fracture length assuming high frac conductivity compared to formation matrix permeability.
—Damage mechanisms restricting flow even though adequate fracture geometry was obtained. Damage may be related to the treatment fluids—or to damage occurring as the well is returned to production after the treatment.
—Insufficient coverage in multi or thick zones.
—Fracturing out of zone.

2. *Rapid Production Decline* may be related to:

—Limited reservoir which is quickly exhausted.
—Very tight reservoir that is approaching steady state conditions.
—Reduction of fracture conductivity with time due to movement of fines from the face of the matrix into the fracture; or gradual crushing of proppant or formation frac face as regional matrix stresses are readjusted.
—Fracture closure at the wellbore due to inadequate propping, (or overflushing) as the wellbore stress concentration is reestablished over a period of time.

Pressure Analysis

Analysis of surface pressures during the frac job can provide clues as to fracture extension, penetration of zonal barriers, sandout, or the effectiveness of ball sealers in generating multifractures. The density of the fluid column, which varies with proppant concentration, must be carefully considered to obtain bottom-hole pressure. Pressure information is more meaningful if obtained through a static fluid column or with bottom-hole instrumentation.

Pressure decline after the pumps are shut down can be analyzed using type-curve methods presented by Nolte[30] to estimate the fluid loss coefficient, the fracture length and width, fluid efficiency, and the time for the fracture to close. Transient pressure measurements can sometimes be used before a frac treatment to indicate the extent of wellbore damage and the capacity of tight zones. Similar tests after the frac treatment should indicate whether or not damage was overcome, and the extent of increased capacity developed by fracturing.

Determination of Fracture Height

Determination of the fracture height in the zone or zones that were fractured has always been an impor-

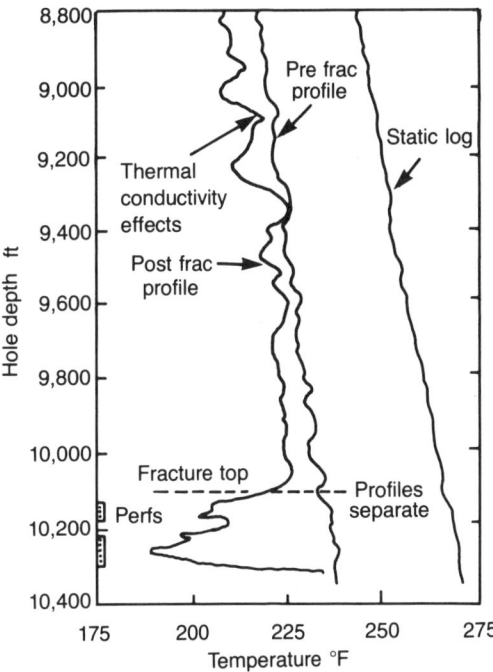

FIG. 8-35—*Pre and post frac temperature profiles show thermal conductivity effects.[29] Copyright 1979, SPE-AIME.*

tant question for fracture analysis. It is particularly important in massive fracturing, where fracture height must be constrained in order to develop fracture length. Temperature surveys after fracturing have been used for this purpose for many years, and with proper interpretation, are still the most definitive tool. In many cases interpretation, which involves detection of temperature anomalies in comparison with a static temperature log, has been hindered by differences in thermal conductivity of various rocks. Variation in thermal conductivity can also cause anomalies. Figure 8-35 shows one method of improving interpretation. It involves circulating fluid through the wellbore below frac pressure at rates similar to those that will be used on the frac job, then running a pre-frac temperature profile. Formation thermal conductivity will affect this survey just like it does the post frac temperature survey. Thus, differences between pre- and post-frac surveys must be the result of the frac job, and a more reliable fracture height determination can be made. Sometimes, comparison of the post-frac temperature survey with a lithology log will eliminate the need of the pre-frac temperature survey.

Radioactive sand (0.5–1.0 millicuries/1000 lb) and a post-frac gamma ray log can also be used to identify zones. Tagged sand must be added at the blender throughout the job to be meaningful, since the first frac sand does not necessarily enter the same zone as the last. The tagged sand should be stopped shortly before the end of the job, so that it is cleared out of the tubing before the end of pumping. Figure 8-36 shows a comparison of the post-frac gamma ray with the post-frac temperature survey. The tagged sand technique should normally be used only to supplement the temperature technique. Sometimes the temperature log cannot be run after the frac until sand is cleaned out of the wellbore—and then it may be too late to see the anomalies.

REFERENCES

1. Hubbert, King M., and Willis, D. G.: "Mechanics of Hydraulic Fracturing," AIME Trans., 1957, p. 153.

2. Howard, G. C., and Fast, C. R.: "Optimum Fluid Characteristics for Fracture Extension," *Drilling & Prod. Practices,* Vol. 261, 1957.

3. Van Poollen, H. K., Tinsley, and Saunders: "Hydraulic Fracturing—Fracture Flow Capacity vs. Well Productivity," AIME Trans., 1958.

4. Kristianovich, et al.: "Theoretical Principles of Hydraulic Fracturing of Oil Strata," Paper 23, Fifth World Petroleum Congress, 1959.

5. Kern, Perkins, and Wyant: "Mechanics of Sand Movement in Fracturing," AIME Trans. Vol. 216, 1959.

6. McGuire, W. J., and Sikora, V. J.: "The Effect of Vertical Fractures on Well Productivity," J. Pet. Tech., Oct. 1960, p. 72.

7. Murphy, W. B., and Juch, A. H.: "Pin-Point Sand Fracturing—A Method of Simultaneous Injection into Selected Sands," J. Pet. Tech., Nov. 1960.

8. Darin and Huitt: "Effect of Partial Monolayer of Propping Agent on Fracture Flow Capacity," AIME Trans. Vol. 219, 1960.

9. Perkins, T. K., and Kern, L. R.: "Widths of Hydraulic Fractures," J. Pet. Tech., Sept. 1961.

10. Dunlap, I. R.: "How Propping Agents Affect Packed Fractures," Petroleum Engineer, Nov. 1965.

11. Raymond, L. R., and Binder, G. G., Jr.: "Productivity of Wells in Vertically Fractured, Damaged Formations," J. Pet. Tech., Jan. 1967, p. 120.

12. Tinsley, J. M., Williams, J. R., Jr., Tiner, R. L., and Malone, W. T.: "Vertical Fracture Height—Its Effect on Steady State Production Increase," J. Pet. Tech., May 1969, p. 633.

13. Haimson, B., and Fairhurst, C.: "Hydraulic Fracturing Porous-Permeable Materials," J. Pet. Tech., July 1969.

14. Geertsma, J., and de Klerk, F.: "A Rapid Method of Predicting Width and Extent of Hydraulically Induced Fractures," J. Pet. Tech., Dec. 1969.

15. Kiel, Othar, M.: "A New Hydraulic Fracturing Process," J. Pet. Tech., Jan. 1970, p. 89.

16. Holtmyer, Marlin, D., and Githens, Charles, J.: "Field Performance of a New High-Viscosity Water Base Fracturing Fluid," API Paper No. 875-24-E, April 1970.

17. Sinclair, A. R.: "Rheology of Viscous Fracturing Fluids," J. Pet. Tech., June 1970, p. 711.

18. Wood, D. B., and Junkin, George II: "Stresses and Displacements Around Hydraulically-Fractured Wells," SPE

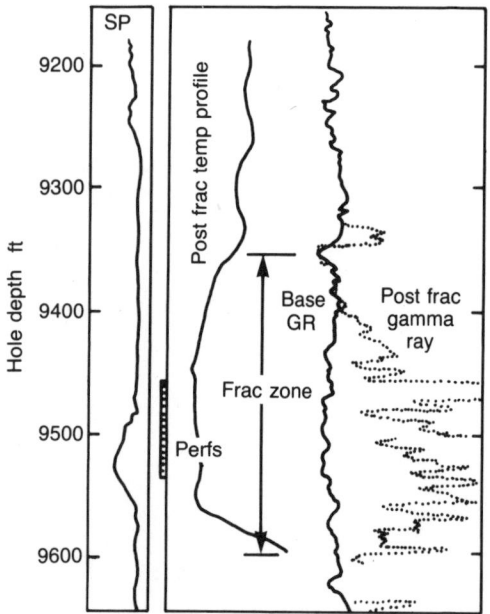

FIG. 8-36—*Post frac gamma ray log agrees with post-frac temperature profile.*[29] Copyright 1979, SPE-AIME.

Paper No. 3030, Houston, Oct. 1970.

19. Nordgren, R. P.: "Propagation of a Vertical Hydraulic Fracture," Soc. Pet. Eng. J., Sept. 1972.

20. Williams, B. B., Nieto, G., Graham, H. L., and Leibach, R. E.: "A Staged Fracturing Treatment for Multi-Sand Intervals," J. Pet. Tech., Aug. 1973, p. 897.

21. Cooke, C. E., Jr.: "Conductivity of Fracture Proppant in Multiple Layers," J. Pet. Tech., Sept. 1973, p. 1, 101.

22. Daneshy, A. A.: "Experimental Investigation of Hydraulic Fracturing Through Perforations," J. Pet. Tech., Oct. 1973, p. 1, 201.

23. Tiner, R. L., Stahl, E. J., and Malone, W. T.: "Developments in Fluids to Reduce Potential Damage from Fracture Treatments," SPE Paper No. 4790, New Orleans, Feb. 1974.

24. Fast, C. R., Holman, G. B., and Colvin, R. J.: "A Study of the Application of MHF to the Tight Muddy "J" Formation, Wattenberg Field, Adams and Weld Counties, Colorado," SPE Paper No. 5624, Dallas, Sept. 19, 1975.

25. Novotny, E. J.: "Proppant Transport," SPE Paper No. 6813, Oct. 1977.

26. Simon, D. E., Kaul, F. W., and Culbertson, J. N.: Anadarko Basin Morrow- Springer Sandstone Stimulation Study," SPE Paper No. 6757, Oct. 1977.

27. Black, H. N., Ripley, H. E., Beecroft, W. H., and Pamplin, L. O.: "Drilling and Fracturing Improvements for Low Permeability Gas Wells in Canada," SPE Paper No. 7926, May 1979.

28. "API Recommended Practice on Standard Procedure for the Evaluation of Hydraulic Fracturing Fluids," API RP-39, 1979.

29. Dobkins, T. A.: "Methods to Better Determine Hydraulic Fracture Height," SPE Paper No. 8403, Sept. 1979.

30. Nolte, K. G.: "Determination of Fracture Parameters From Fracturing Pressure Decline," SPE Paper No. 8341, Sept. 1979.

31. Smith, M. B.: "Effect of Fracture Azimuth on Production with Application to the Wattenberg Gas Field," SPE Paper No. 8298, Sept. 1979.

32. Rosepiler, M. J.: "Determination of Principle Stresses and Confinement of Hydraulic Fractures in Cotton Valley," SPE Paper No. 8405, Sept. 1979.

33. Chatterji, Jitem, and Borchardt, John K.: "Applications of Water-Soluble Polymers in the Oilfield," SPE Paper No. 9288, Sept. 1980.

34. Veatch, Ralph W., Jr. and Crowell, Ronald F.: "Joint Research-Operations Programs Accelerate Massive Hydraulic Fracturing Technology," SPE Paper No. 9337, Sept. 1980.

35. Parker, C. D.: "Logistic and Operational Considerations for Massive Hydraulic Fracturing," J. Pet. Tech., July, 1981, p. 1189.

36. Nolte, K. G. and Smith, M. B., "Interpretation of Fracturing Pressures," J. Pet. Tech., Sept. 1981.

37. Dobkins, Terry A. "Procedures, Results and Benefits of Detailed Fracture Treatment Analysis," SPE paper No. 10130, Oct. 1981.

38. Smith, M. B.: "Stimulation Design for Short, Precise Hydraulic Fractures—mhf," SPE paper No. 10313, Oct. 1981.

39. Schlottman, B. W., Miller, W. K., II, and Lueders, R. K.: "Massive Hydraulic Fracture Design for the East Texas Cotton Valley Sands," SPE paper No. 10133, Oct. 1981.

40. Vaegele, M. D., Abou-Sayed, A. S., and Jones, A. H.: "Optimization of Stimulation Design Through the Use of In Situ Stress Determination," SPE paper No. 10308, Oct. 1981.

41. Wendorff, C. L. and Ainley, B. R.: "Massive Hydraulic Fracturing of High Temperature Wells with Stable Frac Foams," SPE paper No. 10257, Oct. 1981.

42. Recommended Practices for Testing Sand Used in Hydraulic Fracturing Operations, API RP   Dec. 1981.

43. Penny, Glenn S.: "Nondamaging Fluid Loss Additives for Use in Hydraulic Fracturing of Gas Wells," SPE Paper No. 10659, Fifth SPE Symposium on Formation Damage Control, March, 1982.

44. Nolte, K. G.: "Fracture Design Considerations Based on Pressure Analysis, SPE 10911, Cotton Valley Symposium, May, 1982.

45. Nierode, D. E.: "Comparison of Hydraulic Fracture Design Methods to Observed Field Results," SPE 12059, Oct., 1983.

Appendix A

Check List for Planning and Executing Frac Treatment

PLANNING

A. Selecting the service company
  1. Obtain the basic engineering data pertinent to the well and treatment.
  2. Contact service companies, provide them basic data and objectives, and request treating plans and detailed cost estimates.
  3. Conduct in-house fracture design calculations.
  4. Select the service company.

B. Initial pre-job review with service company and operator personnel to discuss the following:
  1. Well data
     a. Wellhead and wellbore configuration
     b. Estimated pressure and rate limits
     c. Bottom-hole temperature
  2. Proppant specifications
     a. Size
     b. Sieve analysis
     c. Acid solubility
     d. Permeability versus load

3. Fluid specifications
   a. Base fluid
   b. Inhibitors
   c. Gelling additives
   d. pH control
   e. Breakers
   f. Bactericides
   g. Surfactants
   h. Effect of fluid on wettability of proppant
   i. Service company should demonstrate properties of prospective fluids with bench tests
4. Results of service company fracture design program
   a. Fracture cool-down program
   b. Pad volume and pad depletion program
   c. Fracture penetration program
   d. Tubular friction pressure loss data
   e. Expected productivity increase
5. Service company cost estimate.
6. Modify program as necessary based on results of in-house and service company fracture design programs and fluid data.
7. Select source of fluid and proppant to be used.
8. Specify necessary modifications of wellhead or well configuration

C. Final pre-job review with service company, and operator engineering and field personnel covering the following points:
1. Proppant
   a. Quality check results
   b. Permeability versus load test results.
2. Fluid
   a. Quality check results from service company bench tests.
   b. Evidence of compatibility of the proposed gel systems with the mixing water and/or oil to be used.
   c. Expected viscosities in the storage tanks and at the blender.
   d. Viscosity versus time data for the prospective fluid (mixed with the water and/or oil to be used) both with and without breakers at various temperatures.
   e. Breaker or stabilizer schedule, to determine required shut-in time at bottom hole temperature.
   f. Revise pressure, rate, and horsepower estimates, if necessary.
3. Determine liquid storage requirements.
   a. Provide separate tanks for pad and flush fluids
   b. Provide overage of 15%–20% for "tank bottoms." If tank pads can be sloped toward suction valve, 5% overage should be enough.
   c. Employ outside visual tank volume gauges where possible.
4. Determine sand storage requirements.
   a. Make sure handling system is adequate for rates required.
   b. Provide overage as required by bulk tank "bottom" configuration.
   c. Provide for water wetting sand, if required by oil wetting-type frac fluid.
5. Determine equipment needs
   a. Pumping equipment, mixing, transfer and injection pumps
   b. Blenders
   c. Manifold
   d. Sand handling equipment.
   e. Provide excess equipment for breakdown, based on previous experience.

EXECUTION

Installation of Storage Equipment and Materials

A. Storage
1. Locate according to layout plan.
2. Tanks must be cleaned by steaming and flushing with fresh water.
3. Check adequacy of tank outlets and manifolding.

B. Fluid hauling
1. Check fluid-hauling tanks to insure cleanliness.
2. Instruct hauler regarding fluids to be used.

C. Fluids
1. Recheck water quality.
2. Determine need for bactericide, and if required, type and concentration.
3. Have service company rerun bench tests of fluids on location.

D. Proppants
1. Review sieve analysis of each tank on location.
2. Check solubility in acid.

Day Before the Job

A. Fluid

1. Inhibit fluid with proper concentration of KCl or NaCl.
2. Check fluid viscosity after gelling, and periodically until job started. Check crosslinking periodically until job started.

B. Pumping and Mixing Equipment
1. Check equipment placement—pump trucks, blenders, sand trucks, and manifolds.
2. Check blender calibration for specific proppant to be used.
3. Install bleed connection, so that fluid samples be obtained during pumping.
4. Check pump packing to insure it is adequate for anticipated pumping time.
5. Check facilities for isolating each pump to replace valves during long jobs.
6. Make sure fuel been arranged for.

C. Well Hookup
1. Check installation of injection lines. Pressure test lines and wellhead. May need to set tubing plug for adequate test.
2. Make provision to reverse sand-laden fluid out of tubing in event of sandout. Check ball sealer equipment for injecting, and recovering balls.
3. Make sure that well can be flowed back conveniently according to gel breaking schedule.
4. Make provision for cleaning up well, if the well does not flow back after job.
5. Check installation of production equipment.
6. Ensure that all lines have been staked down.
7. Install pressure recorders to record pressure falloff after frac. Two recorders should be used.

Day of the Job

A. Start early enough to allow enough time to do the job as planned in daylight.

B. Pre-job meeting—operator and service company personnel
1. All service company supervisory and operating personnel must know the job schedule, staging details, displacement schedule.
2. Check adequacy of ground communications.
3. Check quantity and location of safety and fire equipment.

C. Make sure everyone understands alternate plans for job execution, what changes will be made if:
1. Rate is low and job cannot be completed as planned before dark
2. If pressure approaches maximum
3. If fluid viscosity is not adequate
4. If sand contains excessive fines
5. If shortage of fluid exists
6. If blender malfunctions
7. If sandout occurs, or appears to be imminent
8. If confining barrier appears to have failed

D. Pressure test lines and equipment

E. Recheck Materials
1. Fluid viscosity, crosslinking action
2. Sand quality, sand water-wetting, if required.

F. Finalize polymer and breaker schedule with service company

During the Job

A. Operator personnel should be in, or at control truck at all times to continually observe job and make changes as required.

B. Monitor rates and pressure during injection of pad to ensure that they are satisfactory to continue job.

C. Check sand quality during job
1. Make periodic visual checks, plus periodic sieve analysis for fines.
2. Take duplicate sample for laboratory analysis.

D. Check viscosity of fluid during all stages

E. Check fluid volumes—meter versus tank gauge.

F. Check sand volumes

G. Check rates and pressures during flush. Slow down towards end of job to prevent over displacement.

AFTER THE JOB IS COMPLETED

A. Record final tank gauges, and total volumes of fluid and sand used.

B. Check proper operation of instrumentation recording pressure falloff following pump shutdown.

C. Confirm breaker times with service company to determine when, and how, well will be flowed back.

D. Run temperature survey and/or gamma ray log for frac height determination.

E. Flow back at moderate rates, minimize drawdown.

F. Observe flow-back fluids to determine if gel is broken, and if sand is flowing back.

G. Make sure replacement chokes are on location. Larger chokes may be needed if ball sealers were used.

H. If ball sealers have been used, check ball catcher to ensure that it is operating. Check balls for evidence of effective sealing action.

Appendix B

Rock Mechanics

Rocks generally obey the same laws of mechanics as other materials such as metals, but because of discontinuities and inhomogeneities, they are sufficiently different that a relatively new branch of engineering, rock mechanics, has developed. Rock mechanics and the related rock stress condition is important in many areas of oil and gas production operations.

In drilling, penetration rate, lost circulation, abnormal pressures, hole problems, and hole eccentricity are related to rock mechanics. In primary cementing, the maximum displacement rate, cement column height, and lost circulation are related to rock mechanics. In squeeze cementing, accidental fracturing and cement placement are related to rock mechanics. In sand control, formation strength and gravel placement are related to rock mechanics.

In hydraulic fracturing, fracture initiation and propagation pressures, fracture geometry (length, width, and height), required proppant strength, and fracture conductivity are related to rock mechanics. In fracture acidizing and matrix acidizing, rock mechanics play a similar role. In reservoir engineering, even the basic concepts of porosity and permeability are related to rock mechanics.

Various rock properties and elastic constants can be measured on cores in the laboratory simulating downhole conditions. Downhole logging measurements present an approach to in situ determination of rock properties. In an ideal *relaxed geologic area*, rock stress conditions could be approximated from simple relations using these measurements. Since earth tectonics also influence in situ stress conditions, simple calculations per se are usually insufficient to fully characterize actual stress conditions. Actual rock stress conditions can be determined more directly by step-rate or flowback measurements during hydraulic fracturing or other well treatments involving pressures in excess of the fracture pressure.

The following discussion defines some rock mechanics parameters and the relations between them, and briefly describes how they can be determined.

Stress-Strain, Ultimate Strength Relations

Any material subject to tensile or compressive load will deform. Load or force per unit area is called

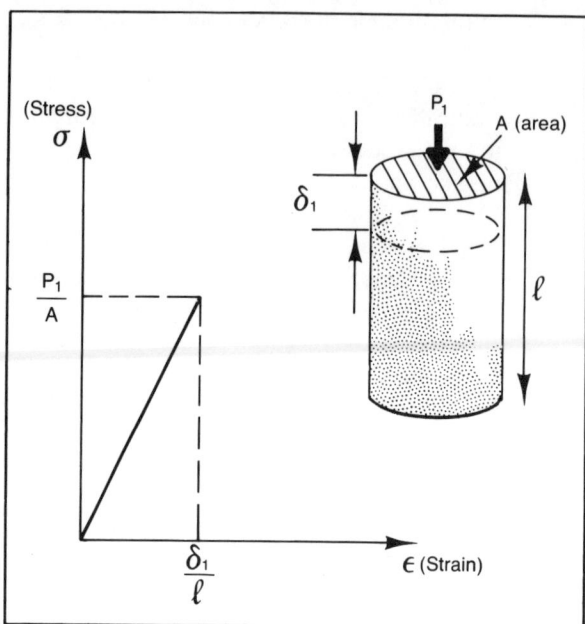

FIG. 8-37—*Stress-strain relationship for linear elastic materials.*

stress, (σ). Under stress, the deformation or change in length, (δ), compared to the original length, (l), is called strain, (ε). At low stress levels, a plot of stress versus strain is a straight line for linear elastic materials such as metals. Figure 8-37 shows this relationship.

Obviously, there is a maximum stress that a material can sustain before failure. Ductile materials, such as steel, exhibit a plastic, or yielding, behavior such that as the yield strength is exceeded, strain increases disproportionately with stress until ultimate failure results. With brittle materials such as rock, this failure occurs suddenly with little additional strain. The compressive stress required to cause failure is called the uniaxial compressive strength, (C_o) of the material.

Young's Modulus

The amount of strain caused by a given stress is a function of the stiffness of a material. Stiffness can be represented by the slope of the axial stress-strain plot and is termed the Young's modulus (E).

$$E = \frac{\sigma}{\varepsilon} = \frac{\text{stress}}{\text{strain}} \quad \left[\frac{\text{lb/in}^2}{\text{in/in}}\right] = \text{lb/in}^2$$

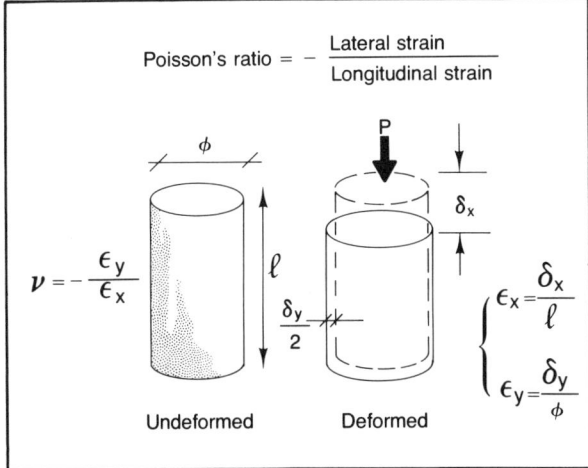

FIG. 8-38—*Measurement of Poisson's ratio.*

For mild steel, the Young's modulus or modulus of elasticity is 30×10^6 psi. For rock, E values range from 0.5 to 12×10^6 psi.

Poisson's Ratio

Compressive stress applied to a block of material along a particular axis causes it to shorten along that axis but also to expand in all directions perpendicular to that axis as illustrated in Figure 8-38. The ratio of the strain perpendicular to the applied stress, to strain along the axis of applied stress, is termed Poisson's ratio (ν).

$$\nu = \frac{\text{lateral strain}}{\text{axial strain}} \quad \left[\frac{\text{in/in}}{\text{in/in}}\right]$$

A material that under stress deforms laterally as much as it does axially would have a Poisson's ratio of 0.5. A material that does not deform laterally under axial load would have a Poisson's ratio of 0.0. Mild steel has a Poisson's ratio of about 0.3. In general, limestone, sandstone, shale, and salt exhibit Poisson's ratios of approximately 0.15, 0.25, 0.4, and 0.5, respectively.

Shear Modulus

Shear stress applied to a particular plane surface in a block of material causes that plane to move with respect to a second parallel plane some perpendicular distance away as shown in Figure 8-39. The ratio of the applied shear stress to the resulting angle of deformation is a measure of the rigidity of the material. This ratio is termed the shear modulus (G).

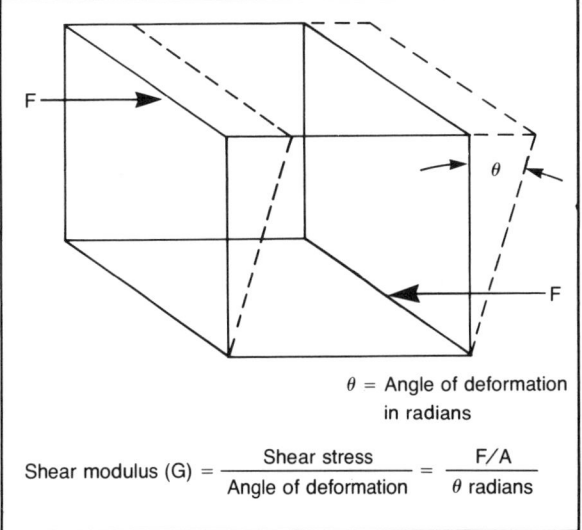

FIG. 8-39—*Definition of shear modulus.*

$$G = \frac{F/A}{\theta} = \frac{\text{shear stress}}{\text{resultant angle of deformation in radians}} \quad \left[\frac{\text{lb/in}^2}{\text{radian}}\right]$$

For a fluid, $G = 0$; for a solid, G is a finite number.

Bulk Modulus

Compressive load applied on all sides of a block of material, as occurs in a hydrostatic condition,

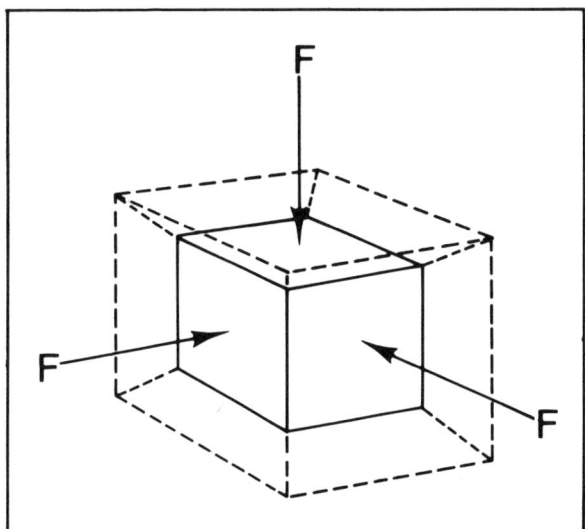

FIG. 8-40—*Definition of bulk modulus.*

causes a reduction in total bulk volume (Fig. 8-40). The ratio of the stress applied (force per unit surface area) to the change in volume per unit of original volume is termed the bulk modulus (K).

$$K = \frac{F/A}{v/v} = \frac{\text{Force/Surface Area}}{\text{Change in volume/original volume}}$$

$$\left[\frac{\text{lb/in}^2}{\text{in}^3/\text{in}^3}\right]$$

Bulk modulus is the reciprocal of compressibility.

Mechanical Properties of Rocks

Rocks are inhomogeneous composite materials containing different crystals. Discontinuities and micro cracks are randomly oriented throughout this material. Since rocks have very low tensile strength and since they are normally under compressive stresses in the earth, most measurements of rock properties are made by applying compressive loads.

As compressive load is applied to a rock, the micro cracks, particularly those perpendicular to the load, begin to close. Much of this initial strain is not related to the deformation of the crystals themselves. This is represented by Region I in the stress-strain curve of Figure 8-41. Region II of Figure 8-41 corresponds to compression of the rock grains themselves, illustrating a linear stress-strain relation. This linear region continues to about 75% of the ultimate strength. New microcracks then appear, nonreversible deformations occur, and finally, macrocracks indicate failure.

The consequence of this nonlinearity is that both Young's modulus and Poisson's ratio are not constant over the complete range of stress from zero to ultimate. Generally, these rock properties are determined at 50% of the ultimate strength of a rock.

Young's modulus (the slope of the axial stress-strain plot is usually determined from the slope of a line tangent to the stress-strain curve at 50% of ultimate strength (tangent modulus $E_{t,50}$).

Effect of Anisotropy—Most sedimentary rocks have been deposited in layers, and as such, their properties are directionally dependent. Where such anisotropy exists, Young's modulus and Poisson's ratio must be measured perpendicular and also parallel to the bedding planes in order to fully characterize the material.

Effect of Confining Stress—During a uniaxial compression test, if the sides of the rock are re-

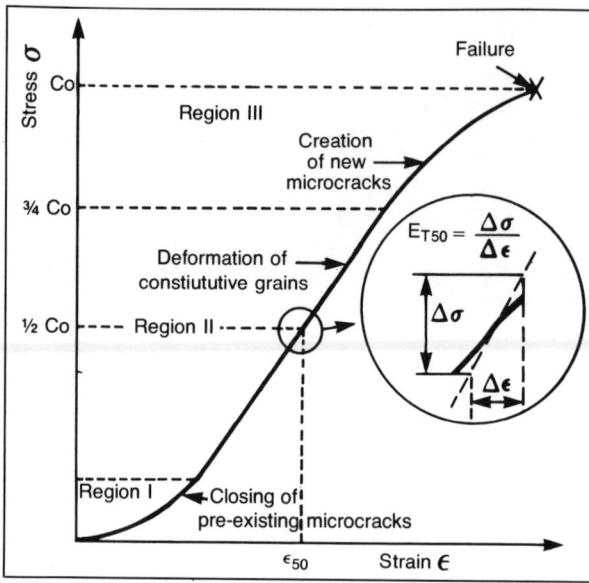

FIG. 8-41—*Stress-strain relationship for rock.*

strained, the rock will become more ductile and will exhibit higher compressive strength. Since rocks at some depth in the earth are subject to triaxial loading, laboratory tests of mechanical properties must simulate these conditions. Figure 8-42 shows the typical effect of increasing confining stress.

It is important to measure mechanical rock properties at the actual confining stress levels on the rock as they exist in the formation. This in-situ confining stress is the consequence of the overburden load, the stiffness of the rock, previous tectonic activity, and the pore pressure in the rock.

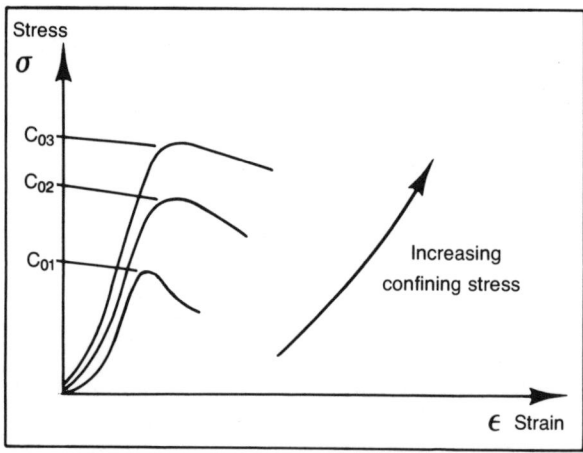

FIG. 8-42—*Effect of confining stress on stress-strain relation. Courtesy of Dowell.*

Effect of Temperature—Temperature is also an important factor in rock behavior. As temperature increases, rocks tend to creep. Thus, ultimate compressive strength is reduced, Young's modulus is normally reduced, and Poisson's ratio is normally increased.

Effect of Dynamic Conditions—Rock is usually assumed to be under static stress. The elastic constants discussed previously are related to static conditions. The process of hydraulic fracturing is usually considered to be a static condition since loads are applied relatively slowly.

Rocks subject to dynamic loading behave quite differently than those subject to static loading. Rocks will withstand much higher dynamic stress without failure. Young's modulus is usually higher for rocks under dynamic stress. Poisson's ratio may be higher or lower.

Lab Measurements of Rock Properties under Static Conditions—Techniques are available for measuring rock properties in the laboratory using cores under the simulated downhole conditions of confining stress and temperature. Figure 8-43 shows the equipment used in the Dowell rock mechanics lab to measure Poisson's ratio, Young's modulus, and ultimate compressive strength. Figure 8-44 is a schematic of the load cell. Axial and lateral strain gauges transmit deformations as axial stress is applied. Confining stress can be applied as desired. Figure 8-45 shows measurements made during one such test, and the resulting calculations of Young's modulus, Poisson's ratio, and compressive strength. Table 8-12 shows rock properties of typical North American formations as measured by Halliburton Services.

Tensile strength can be measured in the laboratory by means of a burst test. A suitable setup is shown in Figure 8-46. With suitable confining load, overburden load, and pore pressure applied to the rock, borehole pressure is increased until bursting occurs. Tensile strength will then be given by the following relation:

$$T = \frac{2\sigma_3 b^2 - p_l(a^2 + b^2)}{b^2 - a^2}$$

T = Rock tensile strength, psi
b = Diameter of rock core, in
a = Diameter of hole in rock, in
σ_3 = Confining pressure, psi
p_l = Borehole pressure to cause failure, psi

FIG. 8-43—*Equipment used in measuring rock properties. Courtesy of Dowell.*

Tensile strength of sedimentary rocks is normally quite low; however, it is a second order factor in fracture initiation. Knowledge of tensile strength is important when hydraulic fracturing is used for in situ stress determination.

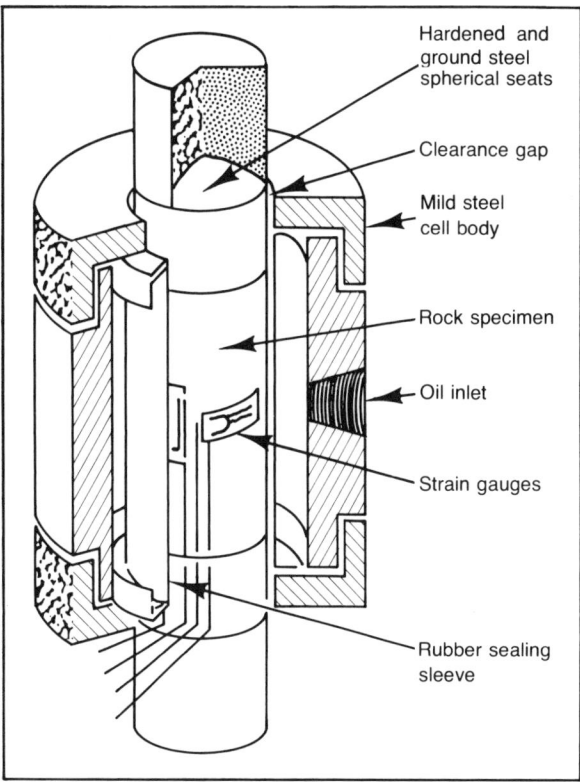

FIG. 8-44—*Load cell for stress-strain measurement under confining stress. Courtesy of Dowell.*

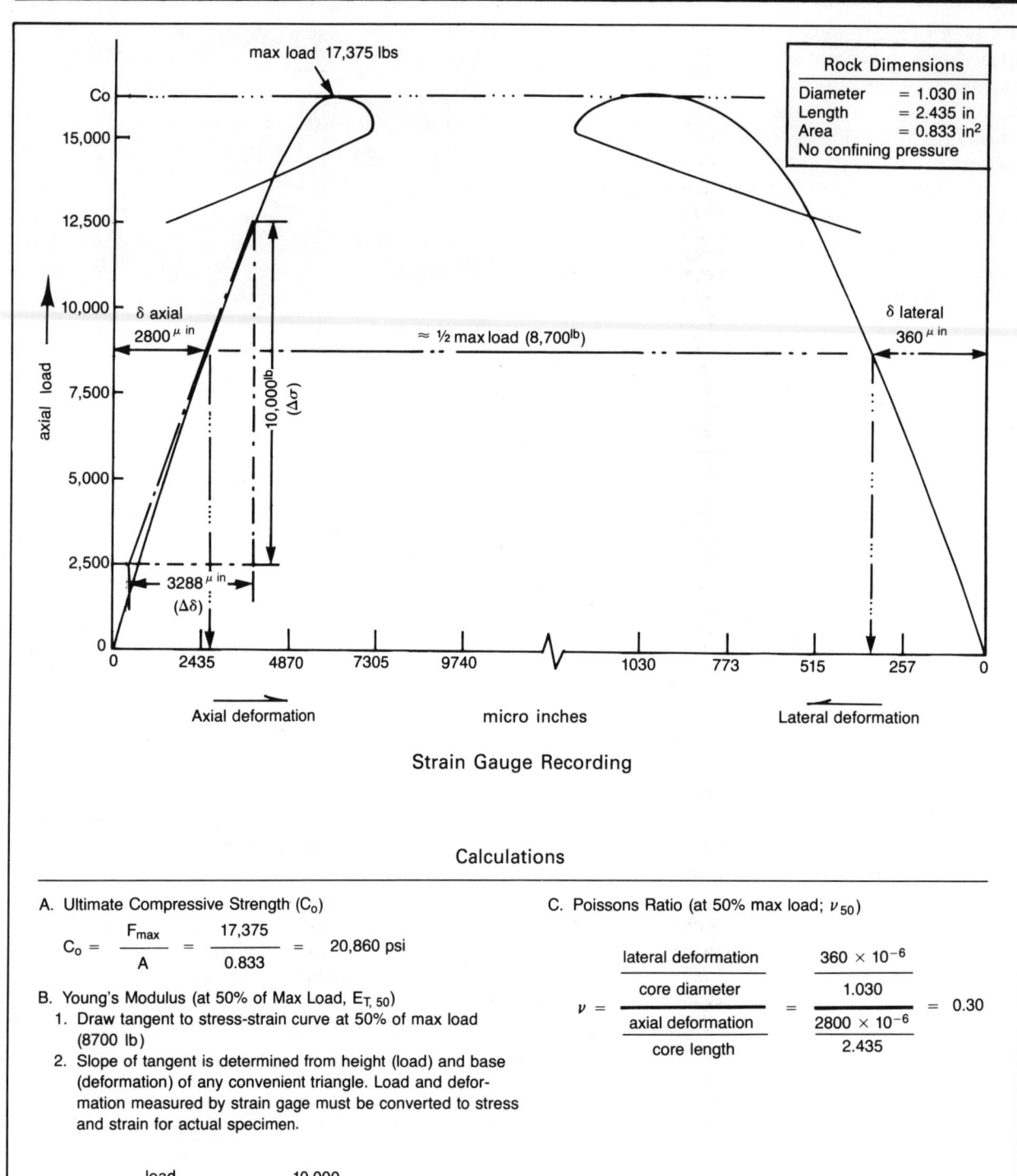

FIG. 8-45—*Static laboratory measurement of elastic constants. Courtesy of Dowell.*

**TABLE 8-12
Rock Properties of Typical Formations**

Formation	Depth ft	Location	Young's Modulus psi × 10⁶	Poisson's Ratio	Air Permeability md	Porosity %
Arbuckle Lime	3688	Kay Co., Ok	3.99	0.350	0.03	4.8
Austin Chalk	2319	Caldwell Co., Tx	2.38	0.276	0.03	17.6
Bartlesville	7900	Cleveland Co., Ok	6.80	0.124	1304.00	16.6
Bartlesville	8301	Canadian Co., Ok	4.80	0.144	7.70	11.8
Basal Quartz	4427	Calgary, Canada	2.08	0.147	0.86	13.4
Bromide	15,000	Grady Co., Ok	8.10	0.127	10.60	6.8
Clearfork	5541	Ector Co., Tx	5.81	0.340	<0.01	11.8
Clearfork Dolomite	6400	Lea Co., Nm	9.25	0.238	<0.01	2.8
Dakota	6076	San Juan Co., N.M.	2.66	0.163	0.24	7.5
Ellenburger	13,014	Upton Co., Tx	8.97	0.314	<0.01	0.7
Frio	9039	Aransas Co., Tx	0.95	0.150	1880.00	27.1
Grayburg	4262	Lea Co., N.M.	5.32	0.130	0.05	6.4
Mesa Verde	2330	Sublette Co., Wy	1.43	0.146	3.00	18.7
Mississippian	8859	Calgary, Canada	7.97	0.288	<0.01	4.3
Mississippian	8985	San Juan Co., N.M.	6.97	0.145	0.02	4.1
Muddy	6907	Campbell Co., Wy	1.96	0.143	10.70	18.3
Navarro	3733	Webb Co., Tx	7.57	0.230	<0.01	1.4
Oriskany	6000	Potter Co., Pa	4.50	0.262	0.03	3.5
Rodessa	9425	Jones Co., Ms	1.37	0.137	185.10	19.7
Second Frontier	7315	Sublette Co., Wy	1.52	0.110	0.10	12.4
Smackover	11,650	Quitman Co., Ms	1.96	0.320	8.60	14.1
Springer	12,320	Caddo Co., Ok	1.80	0.234	106.70	24.2
Tensleep	8782	Wyoming	5.88	0.268	2.40	9.0
Travis Peak	6401	Panola Co., Tx	4.08	0.100	0.45	11.8
Wilcox	8771	Bee Co., Tx	4.90	0.181	0.11	9.0

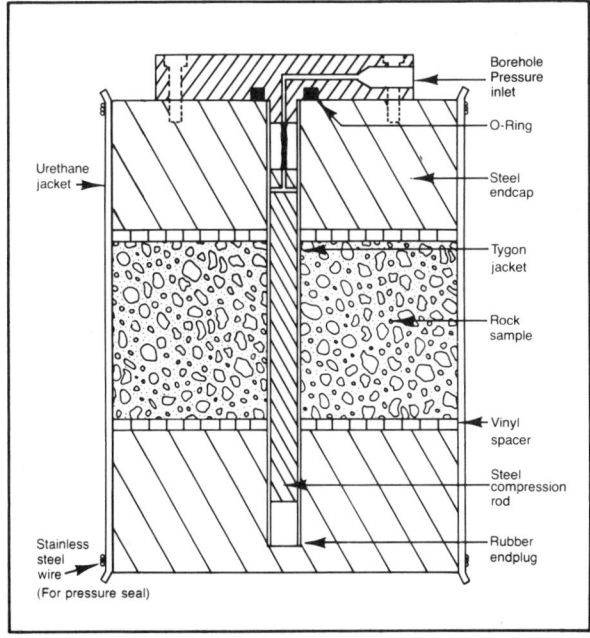

FIG. 8-46—*Measurement of tensile strength of rock.*

Measurement of Rock Properties Under Dynamic Conditions—Energy transmitted through rock moves in compressional, or primary, P-waves; and shear, or secondary, S-waves. P-waves cause longitudinal oscillatory particle motion along the wave path. S-waves cause transverse particle motion perpendicular to the wave path. P-waves travel faster than S-waves. The velocity of P-waves and S-waves depends on the density and the elastic properties of the rock as defined by the following characteristic equations:

$$V_p = \left[\frac{E_d (1 - v_d)}{\rho (1 + v_d)(1 - 2v_d)} \right]^{0.5}$$

$$V_s = \left[\frac{E_d}{2\rho (1 - v_d)} \right]^{0.5}$$

V_p = Compressional wave velocity
V_s = Shear wave velocity
E_d = Dynamic Young's modulus
v_d = Dynamic Poisson's ratio
ρ = Density of the rock

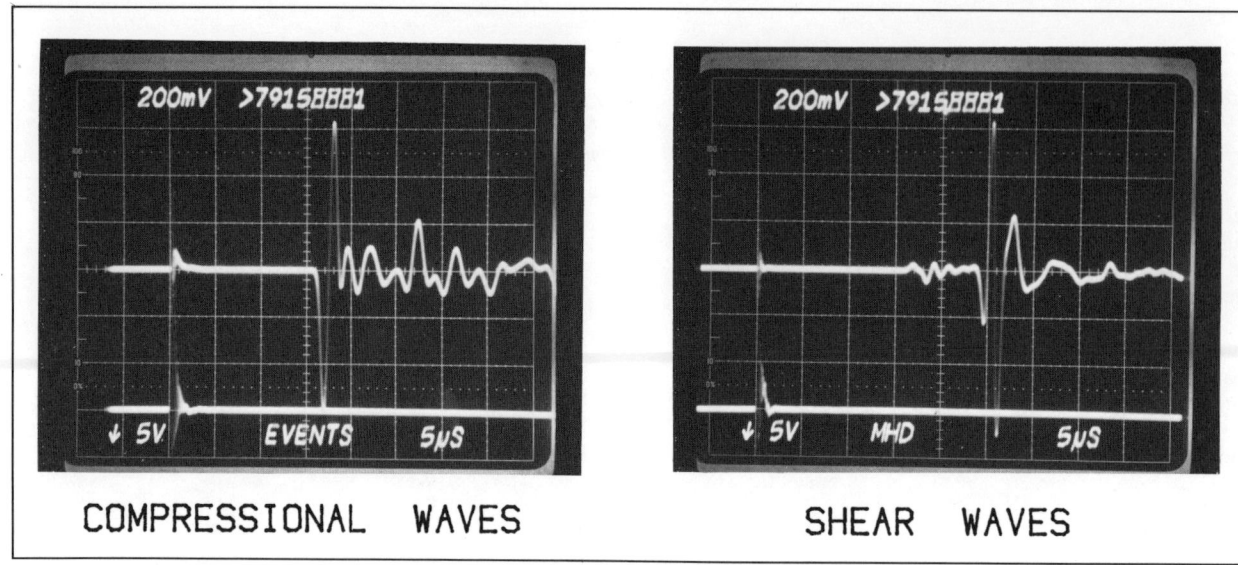

FIG. 8-47—*Laboratory measurement of compressional and shear wave velocities in Wilcox sandstone core. Courtesy of Dowell.*

In the laboratory, measurement of primary and secondary wave velocity (or more conveniently, wave travel time) can be made on formation cores using ultrasonic pulsing techniques. This can be done under triaxial loading conditions and at elevated temperatures. The elastic constants thus determined are dynamic values—and must be related to static values for use in hydraulic fracturing calculations. Figure 8-47 shows compressional and shear wave travel time measurements on a Wilcox sand core.

In the well, the combination of sonic log and density log measurements, with proper interpretation will provide similar in-situ dynamic measurements.

ELASTIC CONSTANTS—SUMMARY

Young's Modulus (E)

—Definition:

$$E = \frac{\text{stress}}{\text{axial strain}} = \frac{\sigma}{\varepsilon_{\text{axial}}} \quad \left[\frac{\text{lb/in}^2}{\text{in/in}}\right] = \text{lb/in}^2$$

For rocks, $E_{t,50}$ = Slope of tangent to stress-strain curve at 50% of ultimate strength

—In terms of other elastic constants:

$$E = \frac{9KG}{3K + G}$$

—With dynamic measurements (Downhole sonic & density):

$$E_d = \frac{\rho}{\Delta t_s^2}\left(\frac{3\Delta t_s^2 - 4\Delta t_c^2}{\Delta t_s^2 - \Delta t_c^2}\right) \times 2.15 \times 10^8$$

Poisson's Ratio (ν)

—Definition:

$$\nu = \frac{\text{lateral strain}}{\text{axial strain}} = \frac{\varepsilon_{\text{lateral}}}{\varepsilon_{\text{axial}}} \quad \left[\frac{\text{in/in}}{\text{in/in}}\right]$$

For rocks, measure lateral and axial strain at 50% of ultimate strength

—In terms of other elastic constants:

$$\nu = \frac{3K - 2G}{2(3K + G)} = \frac{E}{2G} - 1$$

—With dynamic measurement (Downhole sonic & density log)

$$\nu_d = \frac{1}{2}\left[\frac{\Delta t_s^2 - 2\Delta t_c^2}{\Delta t_s^2 - \Delta t_c^2}\right]$$

ELASTICS CONSTANTS—SUMMARY

Shear Modulus (G)

—Definition:

$$G = \frac{\text{shear stress}}{\text{deformation angle, radians}} = \frac{\sigma}{\theta}$$

$$\left[\frac{\text{lb/in}^2}{\text{radians}}\right]$$

—In terms of other elastic constants

$$G = \frac{3K}{2}\left(\frac{1-2\nu}{1+2\nu}\right) = \frac{3KE}{9K-E}$$

—With dynamic measurement (Downhole sonic & density log)

$$G_d = \frac{\rho}{\Delta t_s^2} \times 2.15 \times 10^8$$

Bulk Modulus (K)

—Definition:

$$K = \frac{\text{force/surface area}}{\text{change in volume/original volume}}$$

$$\left[\frac{\text{lb/in}^2}{\text{in}^3/\text{in}^3}\right] = \text{lb/in}^2$$

—In terms of other elastic constants:

$$K = G\frac{2(1+\nu)}{3(1-2\nu)} = \frac{EG}{3(3G-E)}$$

$$= \frac{E}{3(1-2\nu)}$$

—With dynamic measurements (Downhole sonic & density log)

$$K_d = \rho\left[\frac{3\Delta t_s^2 - 4\Delta t_c^2}{3\Delta t_s^2 \Delta t_c^2}\right] \times 2.15 \times 10^8$$

ρ = bulk density lb/ft^3
Δt_s = shear wave travel time, microsec/ft
Δt_c = compressional wave travel time, microsec/ft

Appendix C

Water Soluble Polymers

The term, polymer, is derived from Greek words meaning "many parts." The simple molecules from which polymers are formed are called monomers. Monomers react together under the proper conditions to form high molecular weight materials. For example, polyethylene film results from the joining, or polymerization, of thousands of ethylene molecules to form polyethylene molecules that have an atomic weight in the millions. For linear polymers, molecular weight is proportional to the degree of polymerization, i.e., the number of momomer units present in the polymer. Polymers that contain repeating units of only one compound are called homopolymers. Copolymers contain two or more different types of repeating units. They may be called random copolymers, block copolymers, graft copolymers, or branched copolymers depending on the arrangement of the repeating units.[33]

The majority of high molecular weight polymers used in the oil field to viscosify aqueous fluids are based on "natural" polymers. These polymers are obtained from products such as guar beans (guar and HP guar), cotton linters (CMC, HEC, and CMHEC), or by bacterial action (xanthan gum) and are not polymerized by man in order to obtain their high molecular weights.

Hydration of Polymers

In the solid state, natural polymers can undergo inter- and intramolecular hydrogen bonding. This capacity for hydrogen bonding is not completely satisfied because of disorganization of molecular chains. Thus, when natural polymers contact water, the water molecules quickly penetrate the solid polymer network and hydrogen bond to the available sites. This process hydrates the polymer, causing it to swell and expose new charge sites. Hydration continues until each polymer molecule is completely surrounded by partially immobilized water molecules.

Often, rapid wetting of the exterior of the polymer particles causes particles to swell and form a barrier to further water penetration. This results in the formation of "fish eyes" or "gel balls." The problem can be minimized by adding the polymer to the water slowly along with vigorous agitation. Also, hydration can be delayed, by treating the polymer surface with glyoxal, further aiding in mixing lump-free viscous fluid.

Viscosity

The most important property of a polymer solution for oil field use is its viscosity. Viscosity is defined as the ratio of shear stress to shear rate ($\mu = S_s/S_r$). In general, polymer solutions exhibit non-Newtonian, pseudo-plastic, shear-thinning behavior. See pages 137–138 for a discussion of non-Newtonian fluid rheology. Measurement of viscosity under conditions of shear rate, temperature, and polymer concentration appropriate to actual field use is important. Likewise, viscosity comparisons of polymer solutions must be made under similar conditions. Figure 8-48 shows the effect of polymer concentration on apparent viscosity at shear rate of $511 \ sec^{-1}$, and a temperature of 80°F, for HP guar, HEC, guar, and xanthan gum.

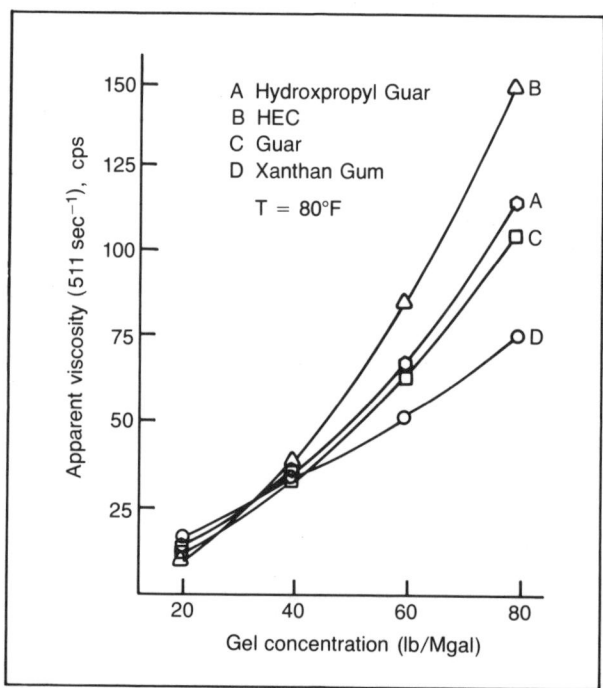

FIG. 8-48—*Effect of polymer concentration on apparent viscosity.[33] Copyright 1980, SPE-AIME.*

Choosing the best polymer for a given application depends both on economic and performance considerations. Factors that affect performance as far as viscosity is concerned include:

—sensitivity to metal ions of various salts in the mix water or formation fluid
—the effect of shear rate on viscosity
—temperature-time stability
—effect of solution pH
—the tendency of the polymer to adsorb on the formation
—compatibility with other solution additives
—tendency for bacterial degradation

Formation Damage

Formation damage is also an important concern in the choice of polymers for oil field use where these polymer fluids will be placed in contact with the producing formation. Examples are workover and completion fluids, and stimulation fluids. Key factors are the amount of polymer needed to provide the required viscosity, the ability to break, or degrade, the polymer when desired, and the amount of residue left after breaking. Figures 8-49, 8-50, 8-51, and 8-52 compare permeability loss in cores subject to unbroken guar gum, acid-broken guar, unbroken HEC, and acid-broken HEC.

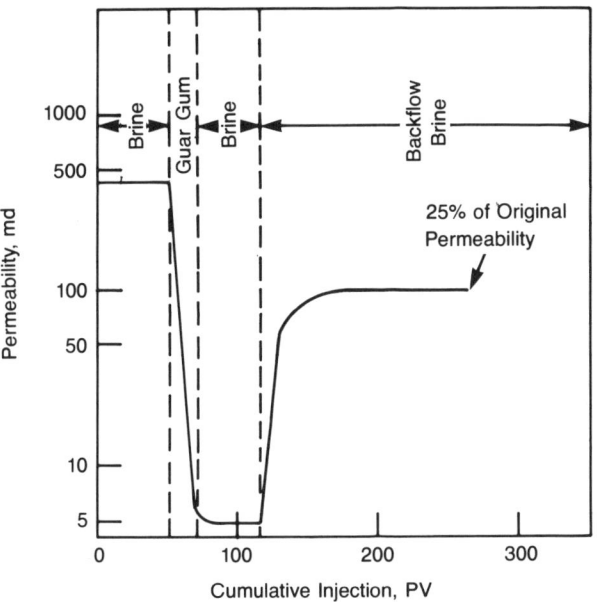

FIG. 8-49—*Unbroken guar gum. Copyright 1974, SPE-AIME.*

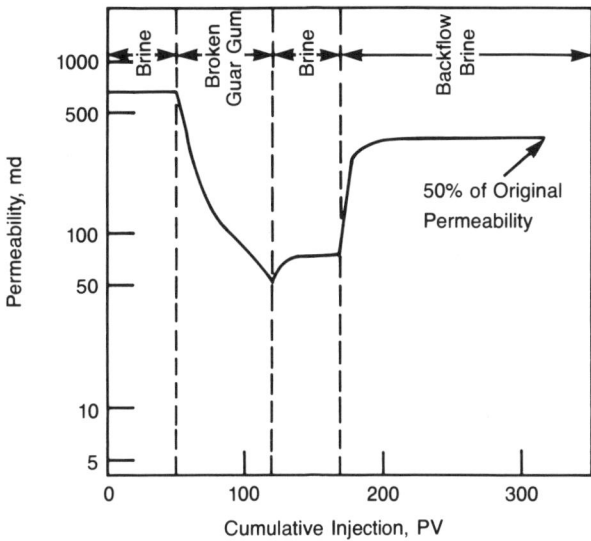

FIG. 8-50—*Acid broken guar. Copyright 1974, SPE-AIME.*

CHARACTERISTICS OF SPECIFIC OILFIELD POLYMERS
Guar Gum

Guar gum is derived from the seed of the guar plant. The guar seed endosperm is mechanically separated from the hull and embryo and ground to a pow-

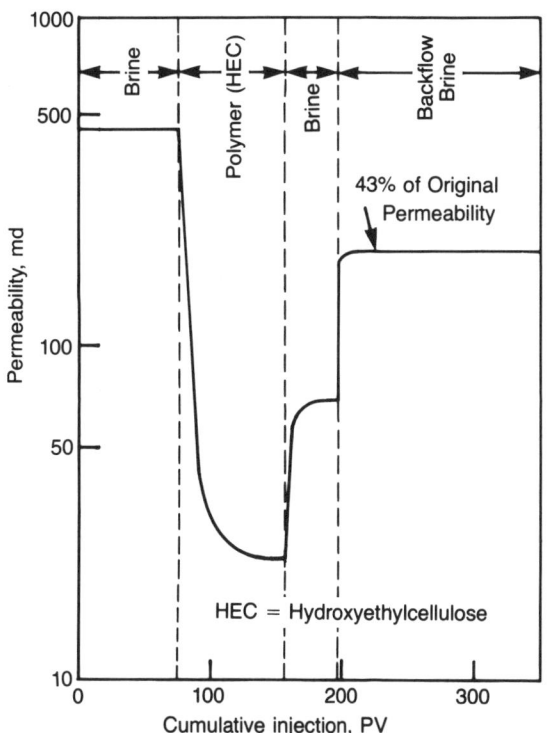

FIG. 8-51—*Unbroken HEC. Copyright 1974, SPE-AIME.*

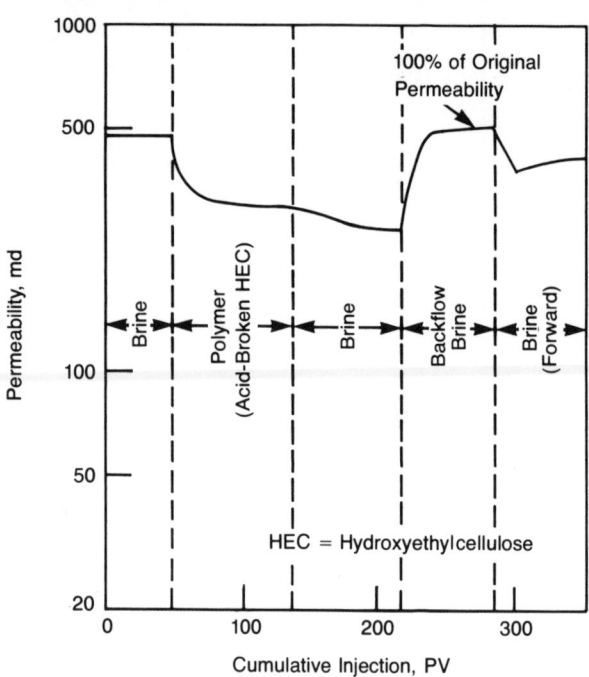

FIG. 8-52—*Acid broken HEC. Copyright 1974, SPE-AIME.*

der. Some of the hull and embryo (mostly protein) are unavoidably included in the powder. This residue (perhaps 5 to 12 percent) provides fluid loss control in guar gum fluids, but can also result in loss of permeability or formation damage. Figure 8-53 shows the chemical structure of guar gum. It is a branched copolymer of manose and galactose (two simple sugars) and is generally termed a *polysaccaride* or, specifically, a *galactomannan*.

Concerning sensitivity to metal ions, guar gum is nonionic and may therefore be hydrated in many types of mix waters. It is compatible with NaCl and KCl at salt concentrations below 5% by weight. Compatibility with polyvalent metal ions Ca^{++}, Mg^{++}, and Al^{+++} depends on salt concentration and solution pH.

Concerning the effect of shear rate on viscosity, Figure 8-54 shows shear stress versus shear rate relations for guar in comparison with HP guar, HEC, and xanthan gum in solutions containing 50 pounds of polymer per 1000 gallons of water. The flow behavior index or n' value for the guar fluid is approximately 0.32, showing that apparent viscosity is significantly reduced as flow velocity increases. Similar statements could be made for each of the other polymer fluids.

Concerning temperature stability, below 175°F, guar gum is fairly stable but degrades rapidly above this temperature.

Solution pH significantly affects guar hydration and degradation. The optimum rate of guar hydration occurs between a pH of 5.0 and 7.5. At a higher pH, hydration is somewhat inhibited. Once hydrated, guar is stable in the higher pH ranges but degrades rapidly at low pH. Even a weak hydrochloric acid solution (5%) will cause rapid loss of viscosity.

Compared to cellulose or biopolymer materials, guar tends to adsorb more rapidly on formation clays, sands, or on frac sand grains. This results in reduced solution viscosity—and perhaps contributes to for-

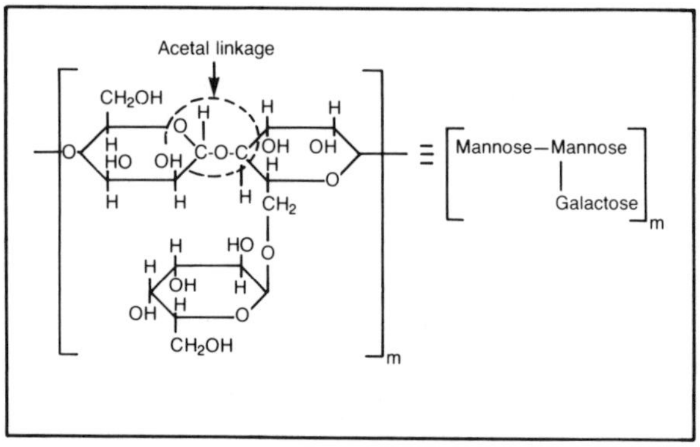

FIG. 8-53—*Chemical structure of guar gum. Courtesy of Dowell.*

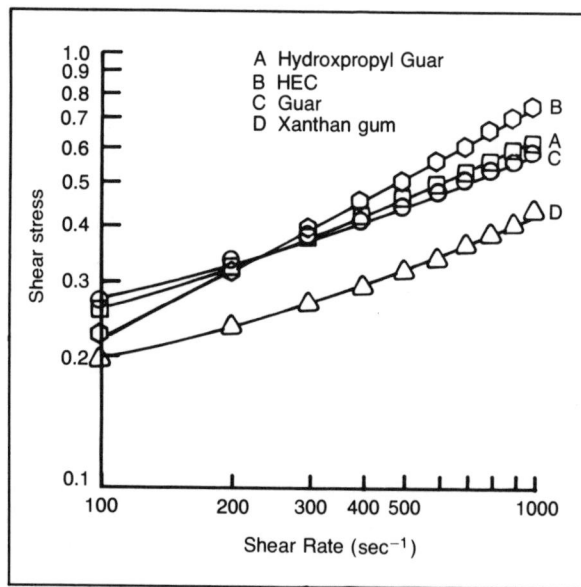

FIG. 8-54—*Shear stress vs. shear rate for polysaccharide gels.*[33] *Copyright 1980, SPE-AIME.*

mation damage or loss of fracture conductivity.

Guar being nonionic (or slightly anionic), is normally compatible with other additives that are anionic or nonionic in nature. Cellulose viscosifiers are more compatible with cationic materials than is guar.

Hydroxypropyl Guar

Hydroxypropyl guar (HP guar) is prepared by reacting highly purified guar with propylene oxide. This process consists of putting guar seed endosperm splits through a series of acid and water soaks to remove most of the embryo and hull before grinding the splits.

Broken HP guar contains no more than 2% insoluble materials. HP guar hydrates more rapidly than guar in cold water. HP guar is more soluble in water-miscible solvents such as methanol, ethanol, and ethylene glycol. HP guar has somewhat better high temperature stability than guar. HP guar is more resistant to enzymatic degrading but can be degraded with ammonium persulfate in a chain reaction. Ammonium persulfate, $(NH_4)S_2O_8$, is needed only in catalytic amounts.

In summary, HP guar fluids are cleaner, less damaging, more alcohol compatible, and more stable than guar fluids, but usually have higher fluid loss than guar fluids.

Crosslinking of Guar or HP Guar

In some hydraulic fracturing situations, high viscosity fluids are needed to provide wide fractures or to permit carrying large concentrations of frac sand. To obtain these high viscosities, greater concentrations of polymers could be used. However, the higher cost, and the negative aspects of formation damage from increased polymer residue, usually make this approach unattractive. The chemical structure of guar or HP guar is such that under proper conditions, the molecular chains can be crosslinked to provide much higher viscosity for a given concentration of guar or HP guar. With cellulose viscosifiers, the chemical structure is such that traditional crosslinking methods are ineffective.

Borate Crosslink—Each of the sugar residues in guar or HP guar has two hydroxyl groups positioned in the cis-form, as opposed to the trans-form positioning of hydroxyls in cellulose or biopolymers. Crosslinking takes place through these cis-hydroxyl groups. Figure 8-55 shows this crosslinking with dissociated borate ions. Crosslinking is fully reversible

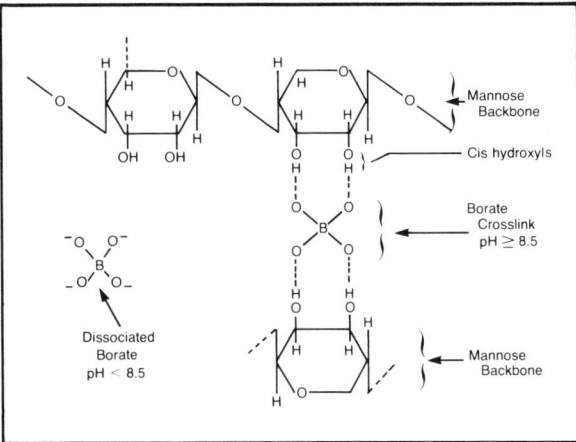

FIG. 8-55—*Chemical structure of guar gum with borate crosslink. Courtesy of Dowell.*

through changes in pH, and is also affected by dissolved salts. Guar or HP guar fluids will not crosslink in the presence of borate until the pH of the fluids is raised above a critical number (8.5 with 2% KCl water). This pH value depends on the concentration of dissolved salts. In field mixing of dry powder, pH must be dropped below 7.0 to hydrate the polymer, then raised above 8.5 to crosslink with borate.

The borate crosslink is ionic and, if broken due to shearing action, will reform when shear rates are reduced. The crosslink bond is also sensitive to temperature. It will break and not reform if the system is subjected to temperatures above 210°F.

Organo-Metallic Crosslink—Under higher temperature conditions where borate fails, organically complexed metallic ions can be used to crosslink guar as shown in Figure 8-56. This organo-metallic crosslink forms a tight covalent bond that will not reform read-

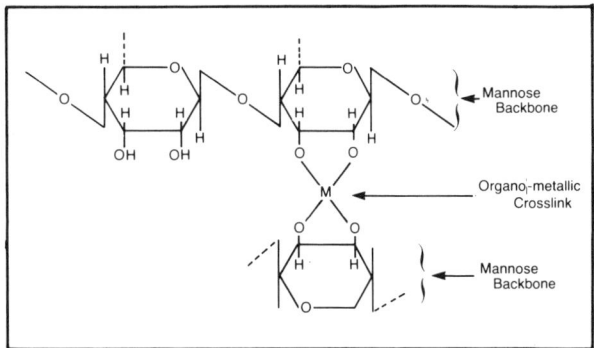

FIG. 8-56—*Chemical structure of guar gum with organo-metallic crosslink. Courtesy of Dowell.*

ily upon shearing but is insensitive to temperatures up to approximately 325°F. The temperature limitation appears to be caused by cleavage of ions in the guar backbone rather than to the crosslink itself.

In summary, the borate crosslink should be used below 210°F because it will reform after shearing. The organo–metallic crosslink should be limited to conditions above 210°F.

Cellulose

Cellulose is a linear molecule formed by polymerization of a simple sugar glucose by natural plant processes. It is insoluble in water and must be chemically modified to make it water soluble for oil field use. Usually, cellulose is treated with sodium hydroxide, which results in separation and swelling of the polymer particles and allows penetration of chemical reagents such as ethylene oxide.

Sodium Carboxymethyl Cellulose (CMC) is prepared by the reaction of cellulose with chloracetic acid in the presence of sodium hydroxide.

Hydroxyethyl Cellulose (HEC) is prepared by the reaction of alkali cellulose with ethylene oxide.

Sodium Carboxymethylhydroxyethyl Cellulose (CMHEC) is formed by the treatment of sodium cellulose with ethylene oxide and subsequently with chloroacetic acid.

Carboxymethyl Cellulose (CMC) is a water-soluble anionic polymer. It is compatible with salts of monovalent metals such as NaCl and KCl. Tolerance for salts of divalent metals such as $CaCl_2$ depends on the degree of substitution (D.S. values from 0.4 to 1.4 are commercially available), the solution pH, and the manner of salt addition. Trivalent metal ions can cause an insoluble floc to form which may plug formation pore channels. CMC is extensively used in drilling fluids to suspend weighting materials and to reduce fluid loss.

Hydroxyethyl Cellulose—HEC is a non-ionic polymer that is compatible with salts of monovalent metals (NaCl, KCl), and also with many divalent metals ($CaCl_2$, $MgCl_2$, etc.). HEC finds wide application in completion and workover fluids. Most fluid loss additives used in cementing are based on HEC.

HEC can be split at the acetal link by acid. The resulting depolymerized solution has very little residue to cause formation damage. HEC is also susceptible to cellulase enzyme degradation, but not to the extent of guar gum.

Carboxymethylhydroxyethyl Cellulose—CMHEC has a unique anionic-non-ionic character. It is used in fluid loss additives and retarders for cementing and, to a limited extent, as a viscosifier in frac fluids.

Xanthan Gum

Often called "biopolymer" or XC polymer, xanthan gum is exuded by the microorganism *Xanthamonas Campestris*. Water solutions of xanthan gum are highly pseudoplastic or shear thinning. Under high shear conditions, xanthan gum solutions have little apparent viscosity, but at reduced shear, viscosity regain occurs rapidly. Below a concentration of 1000 ppm (.35 lb/bbl), xanthan gum solutions lose viscosity significantly in the presence of low salt concentrations. Xanthan gum is quite stable in acid or in the presence of enzymes. It is used as the primary thickener in gelled acid. Strong oxidizing agents such as sodium hypochlorite, hydrogen peroxide, or ammonium persulfate are needed to degrade xanthan gum. The combination of xanthan and guar gum exhibits enhanced viscosity development over either material alone. This combination is used to viscosify frac fluids on a limited basis. Xanthan gum is used extensively to viscosify low solids drilling fluids.

Starch

Starch is a high-molecular weight polymer of glucose molecules formed during the growth of plants (seeds contain some 70% starch; tubers, up to 30%). Compared to other polysaccharides, starch does a poor job of building viscosity. Starch and starch derivatives such as hydroxyethyl starch and carboxymethyl starch have been used for fluid loss control in drilling fluids and to a limited extent in workover fluids.

POLYMER BREAK MECHANISMS

Frac fluids require significant viscosity to produce the desired fracture geometry and to provide the needed proppant-carrying capacity. After the proppant has been placed, however, this viscosity must be reduced to permit recovery of the fluid from the formation. Gravel-packing fluids have somewhat similar requirements.

The widely-used guar and cellulose based polymers can be broken (depolymerized) by (1) mild acid solutions, (2) enzymes, (3) free radical degradation by materials such as ammonium persulfate, or (4) free oxygen. In addition, these polymers have temperature limitations above which they depolymerize quite rapidly. Treatment conditions must be considered in selecting the proper type and concentration of breaker.

For example, enzymes lose effectiveness above approximately 130°F. For fluid temperatures above 130°F, breakers such as ammonium persulfate can be used. At very high temperatures (>400°F.), no breaker may be necessary.

Guar

Guar gum undergoes rapid degradation as the pH of the solution is reduced. Even in 5% hydrochloric acid, guar solution viscosity declines rapidly as shown in Figure 8-57. Guar gum is extremely susceptible to enzymatic attack as shown in Figure 8-58. The presence of bacteria that produce either cellulase or hemicellulase enzymes causes rapid depolymerization.

These enzymes, or bacteria that produce these enzymes, can cause significant problems in field mixing and storage. Mix water and mix tanks must be free both of bacteria and enzymes. Bactericides are usually required even for batch mixing and certainly for extended storage. Enzymes, which may remain in the mix water even though the bacteria are eliminated, can be removed by reducing pH below approximately 2.

The effect of temperature on viscosity reduction or depolymerization is shown in Figure 8-59. Above approximately 150°F., with guar gum, free polymer radicals form spontaneously and act to split the guar molecule by removing protons from the guar backbone. This effect occurs rapidly above 175°F.

HP Guar

HP guar is somewhat more stable than guar to high temperature depolymerization and to enzymatic attack. Oxidizing agents, such as ammonium persulfate may be needed to form free polymer radicals. As in the case of guar, these free radicals then extract protons from the guar backbone, split the molecule, and reduce viscosity. These fragmented molecules in turn become free radicals that extract more protons and split more guar molecules. At lower temperatures (below approximately 125°F), oxidizing agents must be activated chemically to form the initial free radicals.

If high temperature stability is desired, spontaneous formation of free radicals must be inhibited. This can be done through the use of free radical scavengers (Dowell J353 and K46) that quickly satisfy the negative charge formed when the polymer splits due to thermal activity.

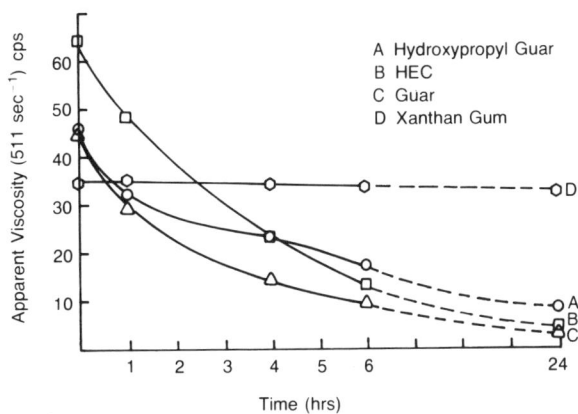

FIG. 8-58—*Viscosity vs. time at 80°F: 50 lb/1000 gal of various gelling agents with 2% KCl and 0.3 lb enzyme/1000 gal.*[33] *Copyright 1980, SPE-AIME.*

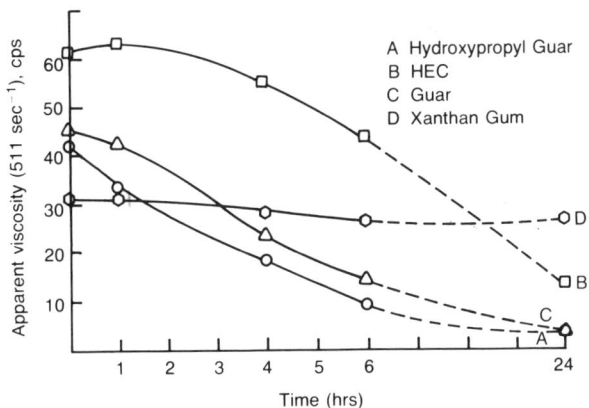

FIG. 8-57—*Viscosity vs. time at 80°F: 50 lb/1000 gal of various gelling agents with 5% HCl.*[33] *Copyright 1980, SPE-AIME.*

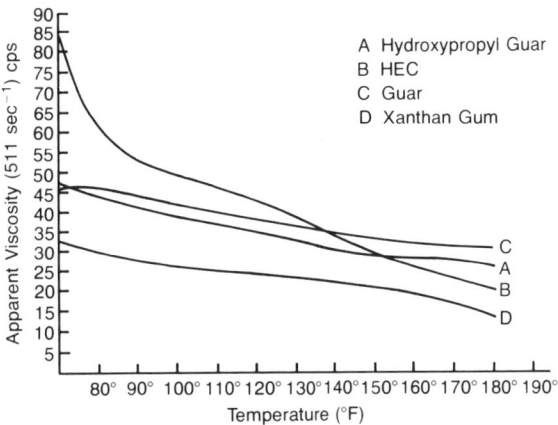

FIG. 8-59—*Effect of temperature on apparent viscosity.*[33] *Copyright 1980, SPE-AIME.*

HEC

The break mechanisms described for guar and HP guar also apply to HEC. Unlike guar or HP guar, however, once broken, no residue remains from HEC.

Xanthan Gum

The structure of xanthan gum is such that the break site is protected, and therefore is not readily susceptible to acid, enzyme, or oxydation degradation. This property limits its use in frac fluids or gravelpacking fluids.

POLYMER COST COMPARISON

A number of factors must be considered to properly represent the cost of a polymer-thickened fluid in oil field use. Table 8-13 gives a relative cost comparison of commonly used frac fluid viscosity builders. Prices are based on Dowell Midcontinent U.S. April 1981 charges. Depending on the needed fluid characteristics, cost of crosslinking materials and stabilizers, fluid loss materials, surfactants, breakers, and mixing charges must also be considered. A factor that is sometimes difficult to evaluate, but perhaps is even more important than the fluid cost, is the effect of the fluid characteristics on the overall success of the job.

Table 8-14 compares typical frac fluids on the basis of viscosity (@ 160 sec^{-1} shear rate) versus polymer concentration and temperature. It also provides a reasonable comparison of cost per gallon, and cost per unit of viscosity (@ 160 sec^{-1} and 150°F). For the high temperature fluids, the cost per unit of viscosity is based on the viscosity after five to six hours at 250°F. It should be noted that the fluid loss characteristics are not considered in this comparison. Likewise, formation damage or flow restriction in the packed fracture due to polymer plugging is not considered. Other factors, such as compatibility with CO_2 may also warrant consideration. For example, metallic crosslinking is more stable under lower pH condition of CO_2 fluids than is borate crosslinking.

One obvious conclusion from the comparison of Table 8-14 is that the crosslinked guar or HP guar fluids provide viscosity at a lower cost and with less polymer (from a formation damage standpoint) than do noncrosslinked fluids.

TABLE 8-13
Cost Comparison of Oilfield Polymers

Polymer[a]	Cost per lb[b]
Guar (J111, J133)	2.85
HP Guar (J266)	3.85
HP Guar (J337, J347)	3.65
HEC (J160, J301F, J303D)	5.20
Xanthan Gum	5.85

[a] Dowell trade names
[b] April 1981 Midcontinent U.S. prices

TABLE 8-14
Comparison of Frac Viscosity and Cost

Fluid System Polymer Type (Dowell Trade Name)	Polymer Concentration lb/1000 gal	Typical Viscosity Cp @ 160 sec^{-1} Temperature °F					Cost	
		100	150	200	250	300	$/1000 gal	$/Cp @ 150°F and 160 sec^{-1}
Linear Aqueous Gels								
Guar (J111)	20	10	4	2	—	—	$ 60.00	$16.70
	40	49	29	16	—	—	120.00	4.00
HP Guar (J337)	20	12	14	—	—	—	85.00	6.30
	40	39	37	20	—	—	165.00	4.50
	80	230	189	120	—	—	330.00	1.75
HEC (J160)	20	11	6	—	—	—	110.00	18.50
	40	68	36	—	—	—	220.00	6.00
	80	300	175	90	—	—	440.00	2.50

TABLE 8-14 (cont.)

Fluid System Polymer Type (Dowell Trade Name)	Polymer Concentration lb/1000 gal	Typical Viscosity Cp @ 160 sec^{-1} Temperature °F					Cost $/1000 gal	$/Cp @ 150°F and 160 sec^{-1}
		100	150	200	250	300		
HEC (J301F + J303D)	60	164	90	49	18	5	310.00	3.50
	80	317	217	138	70	23	415.00	1.90
	100	475	316	215	143	66	515.00	1.65
	120	778	560	405	299	176	620.00	1.10
Crosslinked Aqueous Gels								
Guar with Borate Crosslink (J111 or J133)	20	97	97	4	—	—	90.00	0.90
	40	864	667	396	2	—	175.00	0.33
	60	1836	1750	1390	127	—	265.00	0.15
HP Guar with Borate Crosslink (J347)	20	177	107	—	—	—	90.00	0.85
	40	633	471	58	—	—	195.00	0.45
HP Guar with Organo-metallic Crosslink (J347)	30	[1 hr] [5 hr]		304 38	— —	— —	150.00	—
	40	[1 hr] [6 hrs]		455 320	— —	— —	195.00	—
	60	[1 hr] [6 hrs]		— —	80 26	48 10	260.00	10.00$^{(a)}$
HP Guar with Organo-Metallic Crosslink and High Temperature Stabilizer (J347 + J353)	25	[1 hr] [5 hrs]		320 90	184 33	— —	160.00	$ 4.85$^{(b)}$
	40	[1 hr] [5 hrs]		301 121	79 48	— —	$220.00	4.60$^{(b)}$
	60	[1 hr] [5 hrs]		— —	211 79	— —	285.00	3.60$^{(b)}$
Polyemulsion 2/3 Diesel, 1/3 NaCl Brine with Guar (J133)	8	176	101	50	—	—	85.00$^{(c)}$	0.85
	16	307	163	100	67	48	105.00$^{(c)}$	0.65
Gelled Oil (J292)	NA	176	165	170	—	—	150.00$^{(c)}$	0.90
Aerated Fluid Foam Frac 70% Quality with Nitrogen	@ 1000 psi bottom hole treating pressure						74.00	—
	@ 10,000 psi bottomhole treating pressure						460.00	—

(a) after six hours @ 250°F
(b) after five hours @ 250°F
(c) assumes full recovery of diesel

Chapter 9 Scale Deposition, Removal, and Prevention

Causes of scale deposition
Prediction of scaling tendency
Identification of scale
Scale removal
Scale-prevention methods

INTRODUCTION

Scale is deposited in formation matrix and fractures, wellbore, downhole pumps, tubing, casing, flowlines, heater treaters, tanks, and salt water disposal and waterflood systems. Scale deposits usually form as a result of crystallization and precipitation of minerals from water.

The direct cause of scaling is frequently pressure drop, temperature change, mixing of two incompatible waters, or exceeding the solubility product. Scale sometimes limits or blocks oil and gas production by plugging the formation matrix or fractures, perforations, wellbore, or producing equipment.

The composition of scales is as variable as the nature of the waters that produce them. The most common oil field scale deposits are calcium carbonate ($CaCO_3$), gypsum ($CaSO_4 \cdot 2H_2O$), barium sulfate ($BaSO_4$) and sodium chloride ($NaCl$) (See Table 9-1). Calcium sulfate ($CaSO_4$) or anhydrite does not usually deposit downhole but may be deposited in boilers and heater treaters. A less common deposit is strontium sulfate ($SrSO_4$).

Scale deposited very rapidly may have gas channels, by very porous, and be easy to remove with acid. Scale deposited slowly may be very hard and dense, and may be difficult to remove with acid or other chemicals.

Loss of Profit

The greatest loss of profit each year in the U.S. from scale deposits is probably caused by the loss of oil and gas production. It is estimated that a majority of the 700,000 oil, gas, and service wells in the U.S. have appreciably reduced productivity or injectivity because of scale deposited in the wellbore, perforations, the formation matrix, or in formation fractures. Scale causes a large number of costly pulling jobs, fracturing of wells to bypass scale, and other remedial work each year in both production and injection wells, such as the two examples cited below. Many oil and gas fields have probably been prematurely abandoned because of scale.

1. Mobil reported gypsum (gyp) scale caused by mixing incompatible waterflood waters plugged a 4-in. flowline in four days.

2. In 1957, Earlougher reported that 187 remedial fracturing jobs were performed to bypass scale plugging of the formation flow channels and wellbores in a single waterflood project in West Texas.

CAUSES OF SCALE DEPOSITION

Primary factors affecting scale precipitation, deposition, and crystal growth are: supersaturation; mingling of two unlike waters having incompatible compounds in solution; change of temperature; change of pressure on solution; evaporation (affects concentration); agitation; contact time; and pH.

Tendency to Scale—$CaCO_3$

In oil wells, calcium carbonate precipitation is usually caused by pressure drop releasing CO_2 from bicarbonate ions (HCO_3^{-1}). When CO_2 is released from solution, the pH increases, the solubility of dissolved carbonates decreases, and the more soluble

TABLE 9-1

Chemical Name	Chemical Formula	Mineral Name
	Water Soluble Scale	
Sodium Chloride	NaCl	Halite
	Acid Soluble Scales	
Calcium Carbonate	$CaCO_3$	Calcite
Iron Carbonate	$FeCO_3$	Siderite
Iron Sulfide	FeS	Trolite
Iron Oxide	Fe_2O_3	Hematite
Iron Oxide	Fe_3O_4	Magnetite
Magnesium Hydroxide	$Mg(OH)_2$	Brucite
	Acid Insoluble Scales	
Calcium Sulfate	$CaSO_4$	Anhydrite
Calcium Sulfate	$CaSO_4 \cdot 2H_2O$	Gypsum
Barium Sulfate	$BaSO_4$	Barite
Strontium Sulfate	$SrSO_4$	Celestite
Barium Strontium Sulfate	$BaSr(SO_4)_2$	

bicarbonates are converted to less soluble carbonates. As an illustration of the severity of the problem, loss of only 100 milligrams of bicarbonate per liter of water can result in deposition of 28.6 lb of calcium carbonate per 1,000 bbl of water.

Scale precipitation also varies with calcium ion concentration (common ion effect—such as from $CaCl_2$), alkalinity of water (concentration of bicarbonate ion), temperature, total salt concentration, contact time, and degree of agitation:

—Scaling will increase with increased temperature. Figure 9-1 shows the effect of temperature on solubility in fresh water of calcium carbonate.

—Scaling increases with an increase in pH.

—Scaling increases and becomes harder with increased contact time.

—Scaling decreases as total salt content (not counting Ca ions) of water increases to a concentration of 120g NaCl/1000g of water. Further increases in NaCl concentration decrease $CaCO_3$ solubility and scaling increases.[31]

—Scaling increases with increase in turbulence.

Mixing of two incompatible waters will cause precipitation of $CaCO_3$ scale. An example is the mixing of saline water and fresh water highly charged with bicarbonate.

Tendency to Scale—Gypsum ($CaSO_4 \cdot 2H_2O$) or Anhydrite ($CaSO_4$)

The most common form of calcium sulfate scale deposited downhole is hydrous calcium sulfate or gypsum ($CaSO_4 \cdot 2H_2O$).

A reduction in pressure decreases solubility and causes scaling. Pressure drop from 2,000 psi to atmospheric pressure may precipitate as much as 900 ppm (0.3 lb/bbl of water) calcium sulfate from a typical Permian basin saline water. Figure 9-2[26] illus-

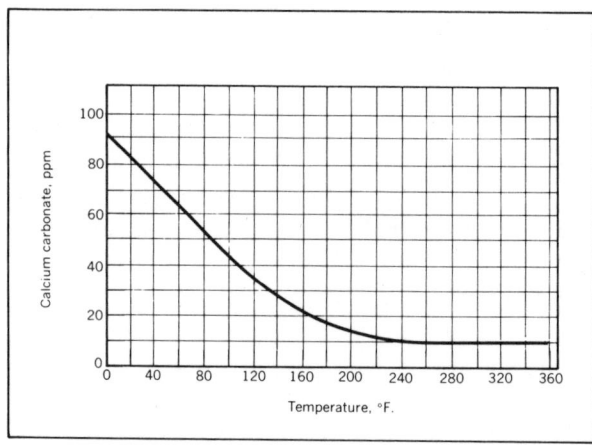

FIG. 9-1—Effect of temperature change on solubility of $CaCO_3$ (After Carlberg).

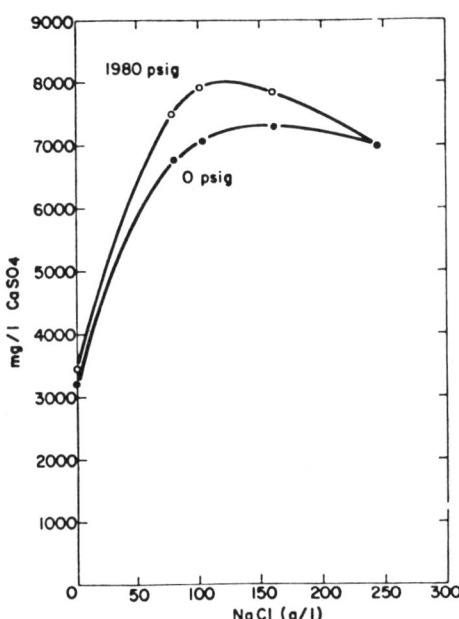

FIG. 9-2—*Gyp scale solubility at 95°F.*[26] *Permission to publish by Society of Petroleum Engineers.*

trates solubility of gyp scale at zero and 1980 psig at 95°F. with various concentrations of NaCl. These curves should be particularly useful in estimating downhole scaling of gypsum.

Mixing of two waters, one containing calcium ions and the other containing sulfate ions, often causes gyp scaling, particularly in waterflooding.

Casing leaks or poor cement jobs are frequent causes of scaling due to downhole mixing of water from the producing zone with water from other porous zones.

An increase of magnesium ions in the range of 24,400–36,600 mg/liter may increase the solubility of $CaSO_4$ up to several times the solubility of $CaSO_4$ in distilled water and thereby decrease scaling.

Agitation increases scaling tendency.

Within the pH range of 6 to 8, pH has very little effect on solubility and scaling.

Evaporation of water due to evolution of free gas near or in the wellbore may cause supersaturation and gyp scaling. Hydrates in gas wells frequently become supersaturated due to evaporation, with resultant scaling.

The effect of temperature on solubility of gyp scale and anhydrite is illustrated in Figure 9-4. A change in temperature will change the solubility of calcium sulfate or gyp and the tendency to precipitate.

In wells having anhydrite ($CaSO_4$) stringers in the producing zone, water flowing in the reservoir is saturated with (and in equilibrium with) anhydrite. The same water, at disturbed flow conditions near the wellbore, is supersaturated with respect to gyp and will precipitate gyp scale ($CaSO_4 \cdot 2H_2O$) near or in the wellbore.[24]

Figure 9-3 Gypsum Solubility Curves by Case[25] shows the effect of Ca and SO_4 ions in waters with varying chloride ions. These curves should be particularly useful in predicting gyp scale when mixing waters in water injection systems. Since this figure was originally charted by Case only for relative values of SO_4 as Na_2SO_4, Ca as $CaCO_3$, and Cl as NaCl, values for these ions, which are obtained from a water sample from the well or water system, must be multiplied by a conversion factor (see Fig. 9-3) in order to obtain values for $NaSO_4$, $CaCO_3$, and NaCl. Once values for these compounds are obtained, plot them on Figure 9-3. If the relative values of Ca and SO_4 intersect above and to the right of the relative value for Cl, gypsum scale will form. If relative values for Ca and SO_4 intersect below and to the left of the Cl line, $CaSO_4$ will not precipitate.

This system can be illustrated by assuming 2,000 ppm of calcium, 4,000 ppm of sulphate, and 18,000 ppm of chlorine. After conversion, the value of $CaCO_3$ is 5,000, $NaSO_4$ is 5,920, and NaCl is 29,700. These values, when plotted on the chart, indicate that the intersections of the values for $NaSO_4$ and $CaCO_3$ are well above and to the right of the NaCl line and that gypsum scale will form.

Tendency to Scale—$BaSO_4$ and $SrSO_4$

For a given NaCl concentration, $BaSO_4$ scaling increases with decreases in temperature as a result of decreasing $BaSO_4$ solubility.

Both $BaSO_4$ and $SrSO_4$ scales are usually caused by mingling of two unlike waters, one containing soluble salts of barium or strontium and the other containing sulfate ions.

Pressure drop may decrease the solubility of $BaSO_4$ in a given NaCl solution and cause scaling.

Barium sulfate is often precipitated in gas wells as hydrates are evaporated.

Solubility of $BaSO_4$ is noted in Table 9-2 with changes in percent NaCl and temperature.[28] Solubility of $BaSO_4$ in many of the high salinity oil-field brines may average 85 to 100 mg/liter.

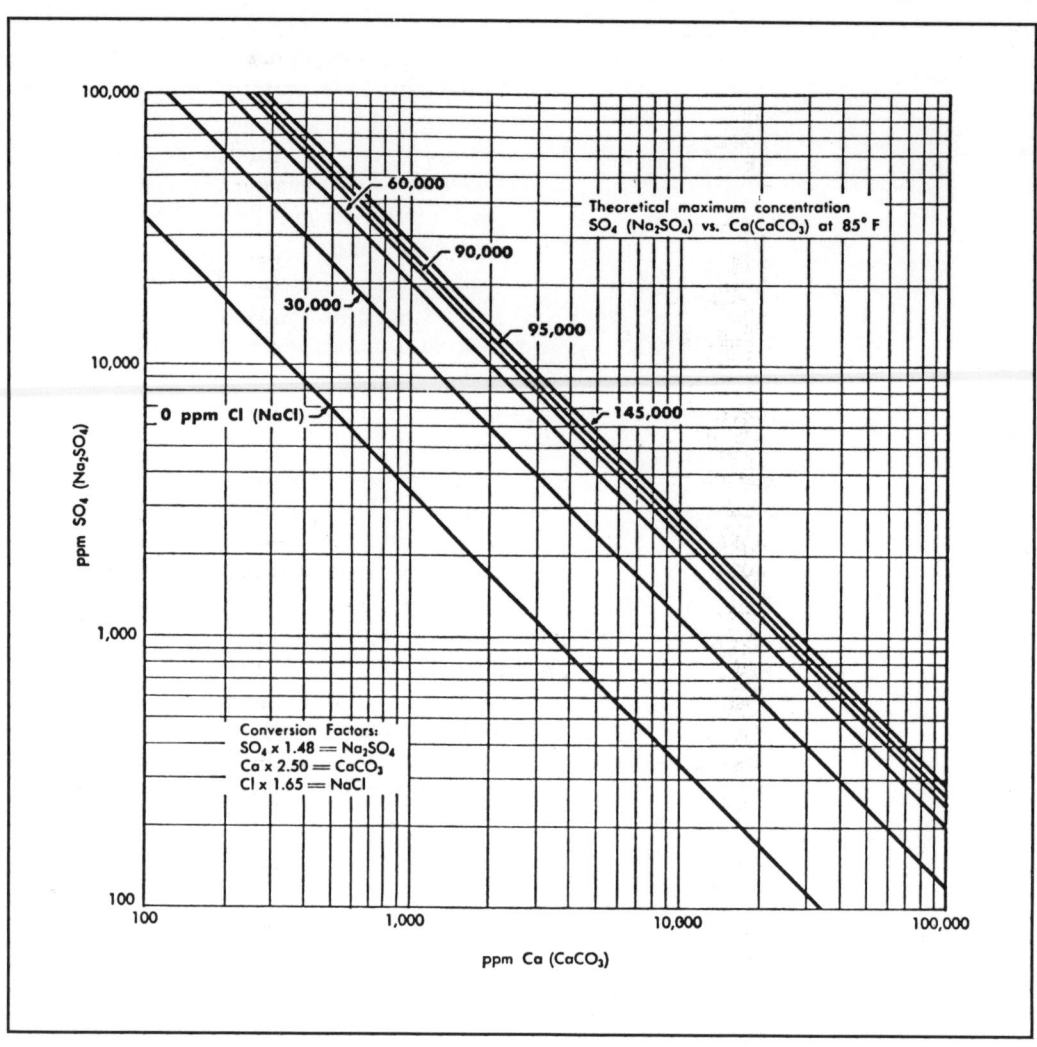

FIG. 9-3—*Gypsum solubility curves.*[25] Permission to publish by The Petroleum Publishing Company.

An example of $BaSO_4$ precipitation occurred in the mixing of two incompatible waters in a 16-in. gathering system in an Illinois waterflood. Bottomhole well temperature was 100°F. Soluble barium in water was 13 to 310 mg/liter and the soluble sulfate range was from 50 to 85 mg/liter. Scale deposition was so severe that it was necessary to run *a pig twice daily* to keep lines free of scale.

Tendency to Scale—NaCl

Precipitation of sodium chloride is normally caused by supersaturation—usually due to evaporation or decreases in temperature. For example, from Table 9-3, it may be noted that 4,000 mg/l of NaCl will be precipitated from saturated salt water if temperature drops from 140°F to 86°F.

Salt precipitation may be quite severe near bottom in gas wells or high GOR oil wells producing very little or no water at the surface. Precipitation may result from both drop in temperature and drop in pressure through perforations and into the tubing. Dry gas will evaporate water, leaving the salt as a precipitate.

Table 9-4 shows the great difference in solubility of NaCl, Gypsum, $CaCO_3$, and $BaSO_4$ in distilled water.

Tendency to Scale—Iron Scales

Iron scales are frequently the result of corrosion products such as various iron oxides and iron sulfide.

Scale Deposition, Removal, and Prevention

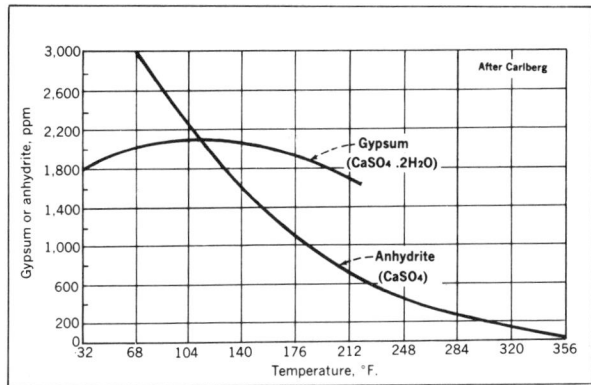

FIG. 9-4.—Effect of temperature change on solubility of gypsum and anhydrite in fresh water (After Carlberg).

Sulfate-reducing bacteria can be a source of hydrogen sulfide, which then reacts with iron in solution or with steel surfaces to form iron sulfide. If oxygen is introduced to a system, it can react with iron to form a precipitate and with steel surfaces to form an oxide coating.

PREDICTION AND IDENTIFICATION OF SCALE
Prediction of Scaling Tendencies

Techniques discussed under "Tendency to Scale" are very helpful in predicting various types of scale. The Stiff and Davis[29] method has been used for many years to show scaling tendencies. However, the age and method of collecting samples may have a bearing on the water analysis values obtained. For example, an aged sample of water may show different values than a fresh sample for pH, bicarbonate content, and CO_2. The best procedure is to measure water properties immediately after sampling.

Analysis of waterflood water will provide a reliable basis for estimating scaling in injection lines and in injection wells.

TABLE 9-2
Solubility of BaSO₄ in NaCl

Temperature °F	NaCl-mg/liter	BaSO₄ solubility mg/liter
77	0	2.3
203	0	3.9
77	100,000	30.0
203	100,000	65.0

TABLE 9-3
Effect of Temperature on NaCl Solubility

| Temperature °F | Salt in solution | | |
	Wt. percent	PPM	Mg/liter
32	25.9	259,000	310,000
86	26.8	268,000	323,000
140	27.1	271,000	327,000
176	27.7	277,000	335,500
212	28.5	285,000	346,500

Analysis of produced brine is an aid in predicting scaling in surface facilities, but may *not* provide a reliable basis to estimate downhole scaling in producing wells. Downhole deposition of scale, frequently due to release of CO_2 from bicarbonate ions in water as pressure declines tends to cause an error in predicting scaling tendencies from produced brine.

If bottomhole pressure is near original, bottomhole samples brought to the laboratory under subsurface pressure and temperature conditions may provide reliable information on both downhole and surface scaling tendencies under original reservoir conditions.

To determine calcium carbonate supersaturation, take a wellhead sample of water and run test on water at the time of sampling. If the calcium carbonate supersaturation is more than 10 percent of the bicarbonate alkalinity content, then the water will usually have a scaling tendency.

Identification of Scale

X-ray diffraction is the most used method for scale identification. This involves directing a beam of X-rays onto a powdered sample of scale crystals.

Each crystalline chemical compound in the scale diffracts X-rays in a characteristic manner, which permits its identification. It is the fastest method and requires the least amount of sample.

TABLE 9-4
Solubility of Various Scale in Distilled Water

Scale	Temperature °F	Solubility mg/liter
Sodium chloride	77	318,300.0
Gypsum	77	2,080.0
Calcium carbonate	77	53.0
Barium sulfate	77	2.3

TABLE 9-5
Analyses of Field Brine and Scale

	Analyses of brines		Analyses of scale*	
	Milligrams/liter			
Chemical	Bartlesville	Arbuckle	Type scale	Percent
Sodium, Na	52,000	58,400	—	—
Calcium, Ca	10,700	13,900	$CaCO_3$	0.65
Magnesium, Mg	1,807	2,182	$MgCO_3$	8.12
Barium, Ba	250	nil	$BaSO_4$	80.10
Sulfates, SO_4	nil	194		
Chlorides, Cl	104,750	120,750	—	—
Alkalinity, HCO_3^{-1}	44	50		
			$SrCO_3$	4.45
			$SrSO_4$	3.60
			SiO_2	0.11
			Fe_2O_3	0.22
			Water soluble	
			Salts	0.87
			Oil	1.02
			Moisture	0.46
			Total	99.60%

*Scale in interior and exterior of tubing and on rods, with pellet types of scale in the bottom of pumping wells.

Chemical analysis may also be used for scale identification. Samples of scale are decomposed and then dissolved in chemical solution. Chemical elements are then analyzed by standard techniques of titration or precipitation.

Scale compounds will usually not be identified unless the analysis is made for each specific chemical cation and anion. By comparison, all chemical compounds can readily be identified from an X-ray analysis.

After adding HCl to the scale sample, effervescence usually indicates $CaCO_3$, especially if the sample does not contain iron sulfide or iron carbonate. The odor of H_2S will indicate the presence of sulfide scale.

Analysis of field brine can show scaling tendencies. Table 9-5 shows an example of $BaSO_4$ scale caused by mixing of Bartlesville and Arbuckle brines from wells in the Boston Pool, Oklahoma.

Table 9-6 shows an example of predictable $CaCO_3$ precipitation due to pressure drop and release of CO_2 from HCO_3^{-1}, in the Grayburg formation, Hobbs field, New Mexico.

Table 9-7 shows water analyses from two flow stations and from an individual well in a specific field.

As a general rule, water should be taken from individual wells for analysis related to well scaling problems. Although water from the several locations varied considerably, service company analysis suggested scaling of $CaCO_3$ at all temperatures above 0°F. Most wells from Flow Station B also showed gyp scaling.

TABLE 9-6
Analyses of Brine and Scale, Hobbs, New Mexico

Analyses of brine		Scale after extraction of water and oil	
Chemical	Milligrams per liter	Type scale	Percent
Chlorides, Cl	4,755		
Sulfates, SO_4	54	$CaSO_4$	1.8
Alkalinity, HCO_3^{-1}	2,335		
Calcium, Ca	476	$CaCO_3$	95.5
Magnesium, Mg	291		
Sodium, Na	2,150		
Sulfides, H_2S	345	FeS	1.2
		SiO_2	0.9
		Total	99.4

TABLE 9-7
Water Analysis from Seven Rivers—Queen Formation, New Mexico

	Milligrams/liter at various locations		
Chemical	Batt. A	Batt. B	Well #1
Chlorides, Cl	36,600	3,900	12,100
Sulfates, SO_4	800	2,200	300
Alkalinity, HCO_3^{-1}	1,950	1,075	1,695
Calcium, Ca	1,000	760	320
Magnesium, Mg	1,530	440	218
Iron, Fe	nil	nil	nil
Sodium, Na	20,720	2,255	7,820
Sulfides, H_2S	60	120	60

SCALE REMOVAL

Scale is classified by methods of removal. Chemically inert scales are not soluble in chemicals. Chemically reactive scales may be classified as (a) water soluble, (b) acid soluble, and (c) soluble in chemicals other than water or acid.

Mechanical Methods

For perforated casing, reperforating is a most effective method of bypassing perforations sealed with scale.

Mechanical methods such as string shot, sonic tools, drilling, or reaming have been used to remove both soluble and insoluble scales from tubing, casing, or open hole.

Scale may be removed from surface lines with "pigs" or by reaming out.

Chemical Removal

Water-Soluble Scale—The most common water-soluble scale is sodium chloride which can be readily dissolved with relatively fresh water. Acid should not be used to remove NaCl scale.

If gyp scale is newly formed and porous, it may be dissolved by circulating water containing about 55,000 mg/liter NaCl past the scale. At 100°F, 55,000 mg/liter NaCl will dissolve three times as much gypsum as fresh water.[28]

Acid-Soluble Scale—The most prevalent of all scale compounds, calcium carbonate ($CaCO_3$), is acid-soluble. Hydrochloric acid (HCl) or Acetic acid can be used to remove calcium carbonate. Formic acid and sulfamic acid have also been used.

Acetic acid has special application downhole in pumping wells when it is desired to leave chrome-plated or alloy pump in a well during acid treatment. Acetic acid will not damage chrome surfaces at temperatures below 200 F; however, HCl may severely damage chrome.

Acid-soluble scales also include iron carbonate ($FeCO_3$), iron sulfide (FeS), and iron oxides (Fe_2O_3).

HCl plus a sequestering agent is normally used to remove iron scale. The sequestering agent holds iron in solution until it can be produced from the well. A sequestered Fe acid, such as 15% HCl containing Acetic and Citric acid, may provide over 15 days of sequestering time.

Normally 15% sequestered HCl is used, but 20% may be necessary because of slow reaction with iron compounds.

A 10% solution of Acetic acid may be used to remove iron scales without an additional sequestering agent; however, Acetic acid is much slower acting than HCl.

Calculation of required acid treatment is based on type and amount of scale as indicated in Table 9-8.

Scales are frequently coated with hydrocarbons, thus making it difficult for acid to contact and dissolve the scales. Surfactants can be added to all types of acid solutions to develop a better acid-to-scale contact. Surfactant selection for this use should be tested to determine that the surfactant will prevent acid-crude oil emulsions and will also leave rock surfaces water-wet.

Acid-Insoluble Scales—The only acid-insoluble scale which is chemically reactive is calcium sulfate or gypsum. Calcium sulfate, though not reactive in acid, can be treated with chemical solutions which can convert calcium sulfate to an acid soluble compound, such as $CaCO_3$ or $Ca(OH)_2$. Table 9-9 shows the relative solubility of gyp in some of the chemicals normally used for gyp conversion. Test conditions

TABLE 9-8
Acid Required for $CaCO_3$ and Iron Scales

Type of acid-soluble scale	Gallons of 15% HCl/cu ft of scale
$CaCO_3$	95
Fe_2O_3	318
FeS	180

TABLE 9-9
Gyp Solubility Tests

Type of solution	Percent gyp dissolved	
	24 Hours	72 Hours
NH_4HCO_3	87.8	91.0
Na_2CO_3	83.8	85.5
Na_2CO_3–NAOH	71.2	85.5
KOH	67.6	71.5

were 200 cc of solution and 20 grams of reagent grade gyp.

Most of the chemicals shown in Table 9-9 convert gyp to acid-soluble $CaCO_3$. KOH converts gyp to $Ca(OH)_2$, which is soluble in water or a weak acid; however, only 68 to 72% is converted to $Ca(OH)_2$, leaving an unconverted scale matrix.

After converting gyp, the residual fluid is circulated out. $CaCO_3$ can then be removed with either HCl or Acetic acid.

Scale-removal procedure if waxes, iron carbonate and gyp are present is:

—Degrease with a solvent such as kerosene, or xylene, plus a surfactant.
—Remove iron scales with a sequestered acid.
—Convert gyp scales to $CaCO_3$ or $Ca(OH)_2$.
—Remove converted $CaCO_3$ scale with HCl or Acetic acid. Dissolve $Ca(OH)_2$ with water or weak acid.

Compounds such as EDTA (Ethylenediamine-tetraacetic acid) and DTPA (diethylenetriamine-pentaacetic acid) can dissolve gyp without the necessity of conversion to $CaCO_3$ or $Ca(OH)_2$. But EDTA and DTPA are not used extensively because of the high cost. EDTA is manufactured and sold by Dow Chemical under the trade name of Versene.

Halliburton's LSD (Liquid Scale Disintegrator) is a converter and disintegrator of anhydrite ($CaSO_4$) and gypsum ($CaSO_4 \cdot H_2O$) scale. LSD produces a water dispersible sludge which is acid soluble, although acid may not be required. Normal well treatment is:

1. Degrease prior to LSD treatment. In a pumping well, place lower end of tubing near the bottom of producing zone, and circulate degreasing solvent for 12 to 24 hours and then pump from well.
2. Dump LSD into annulus and let it soak for several hours, then circulate with pump for a minimum of 24 hours. A surfactant may be added to LSD to water wet scale and to prevent emulsions.
3. For flowing or gas lift wells, degrease and then soak with LSD for a minimum of 24 hours.

Chemically Inert Scales—The most common chemically inert scales are barium sulfate ($BaSO_4$) and strontium sulfate ($SrSO_4$). Barium sulfate scale on the formation face or in perforations may be removed by mechanical methods such as string shots, drilling out, or under-reaming, or bypassing by reperforating. The best approach is to prevent deposition.

SCALE PREVENTION

Inhibition of Scale Precipitation by Inorganic Polyphosphates

Inhibiting scale with a few parts per million of molecularly dehydrated polyphosphates is called a "threshold treatment." When a crystal nucleus is formed, polyphosphate is adsorbed on the surface and slows further crystal growth.

For about 25 years, sand-grain size polyphosphate particles have been injected as a part of regular fracturing treatments. Also, wells have been fractured specifically to inject the phosphate particles. The polyphosphate slowly dissolves in produced water and prevents scale precipitation.

Reversion to Orthophosphate—When placed in solution all polyphosphates tend to hydrolyze into orthophosphates. In the presence of calcium ions, insoluble calcium precipitates may be formed. Polyphosphates will also revert to orthophosphates in the presence of acid.

Rate of reversion to orthophosphates depends on temperature, acidity, mineral content, phosphate content, and nature of polyphosphates. Reversion does not start until polyphosphate goes into solution.

Sodium-calcium phosphates are normally used in well treatment because of their slow dissolving rate. This property assures a near constant phosphate concentration in the water over a long period of time. The objective is for the phosphate-containing water to be produced from the well before reversion can occur.

Inhibition of Scale with Polyorganic Acid

During the mid 1960's, a very stable polyorganic acid, LP-55, was field-proven by Halliburton.[17] This liquid material inhibits scale formation through "threshold treatment" approach. It has no temperature limit and does not precipitate solids.

Placement during fracturing—If placed during a conventional frac treatment, the LP-55 liquid should precede the introduction of sand. If injected during a minifrac, carried out only to place the chemical, the well is fraced and then LP-55 is pumped in the well at rates of 0.25 to 1.0 bbl/minute to permit rapid leakoff.

The feedback rate is controlled by pressure drop when the well is produced from formation capillaries or from minute fractures.

It may be combined with a corrosion inhibitor if desired; however, lab tests of compatibility should be made before injecting both a corrosion inhibitor and scale inhibitor into a well.

The polyorganic acid is placed on the basis of 10 gals of chemical per 1,000 gals of water or alcohol.

The polyorganic acid can also be injected in a water solution down the annulus to reduce scale deposition in the casing, tubing, pump, and surface equipment. It can also be used in the power oil of hydraulic lift type of downhole pumps.

Inhibition of Scale with Organic Phosphates and Phosphonates

Various organic phosphates are now offered to inhibit against calcium sulfate, barium sulfate, and to a lesser degree, against calcium carbonate scale. Many of these water soluble liquid organic phosphates are suitable for squeeze treatments into the formation.

Calgon[23] has reported on lab and field results of its organic inhibitors S-31 and S-51. S-31, an aminoethylenephosphonate (AMP), is designed to inhibit against $CaSO_4$, $BaSO_4$ and $CaCO_3$.

Some of the most successful treatments with S-31 appear to be effective for six to twelve months; however, Calgon reported treatments in Lake Maracaibo, Venezuela to have inhibited scale for 15 months in wells producing 2,000 barrels of water per day.

Calgon's most recent development is S-51, an organic phosphate, designed to inhibit against calcium sulfate scaling.

Exxon Inhibitors—Exxon Chemical Company offers several organic phosphate inhibitors, two of which, COREXIT 7640 and COREXIT 7641, have been used for several years. COREXIT 7640 is primarily designed to inhibit calcium sulfate and barium sulfate scaling by formation squeeze treatments. It may also be used in surface water distribution systems.

Exxon Production Research Company[24] reported average effective inhibiting of nine months against very severe gyp scaling in one West Texas field, with retreatment being considered when the inhibitor in the produced water decreases to about 7 ppm. With this approach, some wells have been relatively scale free with retreatments every 15 months depending on water production.

COREXIT 7640 treating procedure for downhole squeeze is:

1. Preflush with 30 barrels fresh water.
2. Squeeze 165 gallons COREXIT 7640 dissolved in 30 barrels of fresh water.
3. Overflush with 100 barrels fresh water.
4. Shut-in well for 24 hours.
5. Return to production.

COREXIT 7641 is a water soluble organic phosphate, especially designed to prevent deposition of $CaCO_3$, $CaSO_4$ and $BaSO_4$. It may be applied either by continuous injections or by the formation squeeze technique.

Visco Inhibitors—Visco Division of Nalco Chemical Co. offers three organic phosphate scale inhibitors—VISCO 950, 953, and 959.

VISCO 950 inhibits against deposition of $BaSO_4$, $CaSO_4$, and $CaCO_3$. VISCO 950 may be added to surface water systems or may be used for continuous feed into down the casing-tubing annulus. As an effective formation squeeze treatment scale inhibitor, Visco 950 is best used with a 1:9 ratio of produced water, with 200–500 barrels of overflush into the formation.

Visco 953 controls $CaCO_3$, $CaSO_4$, and $BaSO_4$ scales at very low dosages in oil and gas wells, water injection systems, and salt water disposal systems. For water injection systems, continuous treatment is best; for downhole usage, batch treatment or semicontinuous treatment is most effective. Visco 953 also is used as a formation squeeze treatment scale inhibitor, mixed at a 1:9 ratio with overflush into the formation. However, the most effective treatment is to mix a 1–3% solution of Visco 953 with produced water, displacing the solution into the formation.

VISCO 959 is designed for formation squeeze jobs to inhibit $CaCO_3$ and sulphate deposits in oil and gas wells and producing equipment. Typical treatment is to preflush with 5 to 10 bbl of produced water, pump in VISCO 959 mixed with formation water at a ratio of one to ten, overflush with 50 to 250 bbl of formation water, depending on daily water production. Usual treatment is about 165 gal of VISCO 959.

Tretolite Inhibitors—Tretolite SP-223, an organic phosphorate, has been reported as very successful in preventing calcium carbonate and calcium sulfate scaling. It may be used on a continuous basis to circulate into a well or squeezed into the formation. Wells or lines are protected from scaling as long as 5 ppm of chemical can be detected in the fluid stream in contact with the pipe or equipment.

Tretolite SP-219, a liquid scale inhibitor, which is a combination of a phosphate and a polymer, prevents deposition of barium sulfate and calcium carbonate. It may also be used to protect against $CaSO_4$ deposition. Concentrations of SP-219 for protection against $BaSO_4$ and $CaCO_3$ are from 5 to 50 ppm, depending on the severity of the problem.

Baroid Inhibitors—Surflo H-35 is an anionic organic phosphonate designed to prevent the deposition of Calcium Carbonate, Calcium Sulfate, and Barium Sulfate. It may be injected as a continuous treatment in wells or used as a periodic squeeze treatment into the formation. When squeezed, 4 to 12% of Surflo H-35 is mixed with fresh water or produced water and then overflushed with 25 to 100 bbl of water.

Inhibiting Scale with Polymers

ARCOHIB S-223,[27] a salt of polyacrylic acid, developed by Atlantic-Richfield, is marketed by major service companies under various trade names. Procedure for use in prevention of $BaSO_4$, $CaSO_4$ and $CaCO_3$ is as follows:

1. Pump in 100–500 gal HCl spearhead to insure $CaCO_3$ scale cleanup.
2. Pump in a slug consisting of 45 bbl produced water, 100 gal 15% HCl, and 100 gal ARCOHIB polymer.
3. Follow up with enough $CaCL_2$ to raise the $CaCL_2$ content in the 45 bbl slug of water to 10,000 ppm. This raises the pH from about 1.2 to about 4.5 and causes the polymer to crosslink and precipitate as a gel in the formation.
4. The crosslinked polymer is then pushed into the formation and dispersed with 100 to 200 bbl of produced water. The polymer is slowly dissolved by produced water to provide inhibition against $BaSO_4$, $CaSO_4$, and $CaCO_3$ scale.

Exxon Chemical's COREXIT 7647, a low molecular weight polymer, has been effectively used for continuous injection and circulation of wells and surface systems and also for squeeze treatments into the formation. It is a stable inhibitor up to 500°F and controls deposition of both carbonate salts of calcium and magnesium, strontium and barium.

$CaCO_3$ Scale Prevention by Pressure Maintenance

If calcium carbonate scale can be predicted as a result of drop in reservoir pressure, pressure maintenance should be considered as a means of reducing scaling.

Summary

Steps to be taken in solving scale problems are:

1. Identify the scale and the reason for its deposition.
2. Remove deposit by chemical or mechanical means.
3. In perforated completions, it may be more satisfactory to bypass scaled perforations by reperforating.
4. Inhibit against further scale deposition.

REFERENCES

1. Burcik, E. J.: "The Inhibition of Gypsum Precipitation by Sodium Polyphosphate," *Producers Monthly,* 19, No. 1 (Nov. 1954) 42.

2. Case, L. C.: "Chemical Treatment of Oil Wells for the Prevention of Corrosion and Scale," U.S. Patent No. 2,429,593, Oct. 28, 1947.

3. Case, L. C.: "Prevention and Removal Methods for Scales in Oil-Producing Equipment," *Oil & Gas Journal,* 45, No. 22 (Oct. 5, 1946) 68.

4. Case, L. C.: "Scales in Oil-Producing Equipment . . . Their Occurrence and Causes," *Oil & Gas Journal,* 45, No. 21 (Sept. 28, 1946) 76.

5. Crawford, P. B.: "Scale Forming Waters, Part 1," *Producers Monthly,* 21, No. 12 (Oct. 1957) 15.

6. Crawford, P. B.: "Scale Forming Waters, Part 2," *Producers Monthly,* 22, No. 1 (November 1957) 32.

7. Crawford, P. B.: "Scale Forming Waters, Part 4," *Producers Monthly,* 22, No. 4 (January 1958) 6.

8. Crawford, P. B.: "How a Corrosive Water Becomes Scale Forming Between the Wellhead and the Sandface," *Producers Monthly,* 22, No. 4, Part 3 (Feb. 1958) 16.

9. Crawford, P. B.: "Scale Forming Waters, Part 5," *Producers Monthly,* 22, No. 5 (Mar. 1958) 10.

10. Earlougher, R. C. and Love, W. W.: "Sequestering Agents for Prevention of Scale Deposition in Oil Wells," J. Pet. Tech., April 1957, p. 17.

11. Featherston, A. B., Mihram, R. G. and Waters, A. B.: "Minimization of Scale Deposits in Oil Wells by Placement of Phosphates in Producing Zones," J. Pet. Tech., March 1959, p. 29.

12. Jones, E. N.: "Corrosion and Scale Control," *World Oil*, 133, No. 3 (Aug. 1951) 204.

13. Morgan, Z. V.: "Calcareous Depositions from Formation Waters," *Oil & Gas Journal*, 49, No. 46 (March 22, 1951) 102.

14. Plummer, F. B.: "Treatment of Oil Wells to Remove Carbonate Scales," *Oil & Gas Journal*, 44, No. 1 (July 7, 1945).

15. Bell, W. E.: "Chelation Chemistry—its importance to water treatment and chemical cleaning," *Materials Protection*, February, 1965.

16. Ostroff, A. G.: "Compatibility of Waters for Secondary Recovery," *Producers Monthly* (March 1963).

17. Tinsley, J. M., Lasater, R. M., and Knox, J. A.: "Design Technique for Chemical Fracture Squeeze Treatments," J. Pet. Tech., Nov. 1967, p. 1,493.

18. Weintritt, D. J., Cowan, J. C.: "Unique Characteristics of Barium Sulfate Scale Deposition," J. Pet. Tech., Oct. 1967, p. 1,381.

19. Carlberg, B. L., Casad, B. M., Brace, R. L., and Chambers, F. B. Jr.: "Stabilize Water for High-Temperature Flooding," *Oil & Gas Journal* (July 2, 1962), pages 120–122.

20. Charleston, James: "Scale Removal in the Virden, Manitoba Area," J. Pet. Tech., June 1970, p. 701.

21. Bezemer, C., Bauer, K. A.: "Prevention of Carbonate Scale Deposition: A Well-Packing Technique with Controlled Solubility Phosphates," J. Pet. Tech., April 1969, p. 505.

22. Lasater, R. M., Gardner, T. R., Glasscock, F. M.: "Scale Deposits are Controlled Now With Liquid Inhibitors," *Oil & Gas Journal* (January 15, 1968).

23. Ralston. P. H.: "Scale Control with Aminomethylenephosphates," J. Pet. Tech., Aug. 1969, p. 1,029.

24. Kerver, J. K. and Heilhecker, J. K.: "Scale Inhibition by the Squeeze Technique," The Petroleum Society of C.I.M., Calgary, Alberta, Canada, (May 1968).

25. Case, L. C.: *Water Problems in Oil Production, an Operator's Manual*. The Petroleum Publishing Company, August, 1977.

26. Fulford, R. S.: "Effects of Brine Concentration and Pressure Drop on Gypsum Scaling in Oil Wells," J. Pet. Tech., June 1968, p. 559.

27. Miles, Leon: "A New Concept In Scale Inhibitor Formation Squeeze Treatments," API Division of Production, Paper No. 851-44-J, April, 1970.

28. Patton, Charles C.: "Oil Field Water Technology Manual," International Petroleum Institute, Ltd., 1971.

29. Stiff, H. A., Jr. and Davis, L. E.: "A Method for Predicting the Tendency of Oil Field Waters to Deposit Calcium Carbonate," SPE of AIME Transactions, Vol. 195 (1952), pp. 213–216.

30. Vetter, O. J. G.: "How Barium Sulfate Is Formed—An Interpretation," SPE of AIME Trans., Vol. 259, pp. 1515–24.

31. Englander, H. E.: "Conductometric Measurement of Carbonate Scale Deposition and Scale Inhibitor Effectiveness," J. Pet. Tech., July 1975, p. 827.

32. Knowles, T. C.: "Wellbore Scale Control Restores Production in Means Field," The Oil and Gas Journal (Dec. 1974), pp. 89–93.

33. Ostroff, A. G.: "Introduction to Oil Field Water Technology," Prentice-Hall, Inc., 1965.

34. Wheeler, D. W., and Weinbrandt, R. M., "Secondary and Tertiary Recovery with Sea Water and Produced Water In The Huntington Beach Field," SPE 7464, Oct. 1978.

35. Hausler, R. H.: "Predicting and Controlling Scale From Oil-Field Brines," The Oil & Gas Journal, Sept. 18, 1978, p. 146.

36. Metler, A. V., and Ostroff, A. G.: "The Proximate Calculation of solubility of Gypsum in Natural Brines from 28 to 70 C," *Environmental Science and Technology*, vol. 1, no. 10, p. 815, Oct. 1967.

Chapter 10 Corrosion Control

Why and how metals corrode
Types of corrosion
Detection and measurement of corrosion
Corrosion control—materials and design
Corrosion control—coatings and inhibitors
Corrosion control—removal of corrosive gases
Corrosion control—cathodic protection
Corrosion control—non-metallic materials

THE CORROSION PROCESS

Corrosion in the oil field appears as leaks in tanks, casing, tubing, pipelines, and other equipment. Base metal disappears as corrosion changes it to another type of material.

Metal ores are mostly oxides and sulfides, which are more stable than pure metals. Energy is required to reduce the ore and produce metal, as in the blast furnace. Most metals are like fuels; they tend to combine with oxygen, sulfur, and other elements. These combining reactions of metals produce heat or some other form of energy and change the metals back to more stable compounds.

Metals such as zinc and magnesium are more reactive and release more energy when they corrode than metals such as silver and gold. Silver and gold do not corrode readily but are usually too expensive to be used as a means of reducing corrosion.

Four elements are necessary for corrosion to occur: (1) an anode, (2) a cathode, (3) an electrolyte (water), and (4) a metallic path for electron flow. Figure 10-1 illustrates a corrosion cell and shows these four elements. The anode and cathode may be located at two different locations on the same metal surface as shown in Figure 10-1. An example of this is when a grain boundary corrodes because it is anodic with respect to an adjacent grain body. The anode and cathode may be two separate pieces of metal; for example, if the threads of a steel nipple are made up in a bronze valve body, which becomes the cathode, the steel will be anodic and corrode. Examples of electrolytes are oil field brines, moisture in soil, and sea water.

The metallic path is usually through the object that is undergoing corrosion. When the anode and cathode are separated by some distance, it is frequently possible to stop corrosion by interrupting the metallic path. Thus, in the previous example of the steel nipple and bronze valve, it may be possible to electrically isolate the two metals with a plastic bushing or to isolate the threads with teflon tape. Corrosion caused by this dissimilar metal or galvanic couple will cease unless there is a continuous path for electron flow from the valve to the nipple.

In the corrosion of iron in a weak acid solution such as many oil field brines, iron goes into solution as ferrous ions at the corroding or anodic surface, leaving two negative electrons in the metallic iron:

$$Fe^\circ \rightarrow Fe^{++} + 2\,(\text{electrons})^-$$
(Iron)   (Ferrous ion)   (Remain in iron)

The negative electrons in the metal move toward noncorroding or cathodic areas, and hydrogen ions in the solution react with these electrons at the cathode, forming hydrogen gas that bubbles off:

$$2\,H^+ + 2\,(\text{electrons})^- \rightarrow H_2^\circ$$
(Hydrogen ion)   (Hydrogen gas)

A metal surface is often covered with electrolytic cells, with some areas being anodic and others being cathodic. Impurities or imperfections in the metal contribute to the formation of anodic and cathodic areas on a metal surface.

Corrosion is associated with electric current flow, which can be predicted by Ohm's Law. The amount of current, I, is directly proportional to the applied voltage, E, and inversely proportional to the resistance, R:

$$I = E/R$$

In the electrochemical corrosion circuit, current flows from the anodic areas into the electrolyte and flows from solution into metal at cathodic areas. All corrosion inhibitors and coatings are designed to reduce current flow, usually by increasing resistance in the electrochemical corrosion circuit.

Since corrosion is related to flow of electricity through many materials, it is necessary to consider electrical resistivity of each.

Resistance (R) to current flow through a material or electrolyte is inversely proportional to the cross-sectional area (A) of the object and directly proportional to the length (L) of the current path. This is represented by the equation

$$R \propto \frac{L}{A}$$

or, using the symbol, ρ, for a proportionality constant,

$$R = \frac{\rho L}{A}$$

This proportionality constant (ρ) is unique for each material or electrolyte. If the units of L and A are cm and cm^2 respectively, and R is expressed in ohms, the above equation shows the units of ρ will be ohm-cm. Resistivity is the name frequently used to refer to the proportionality constant ρ.

Materials in Table 10-1 are arranged in order of increasing resistivity. Note the extreme range between metals and the best insulators.

Sea water or brine is usually considered a good

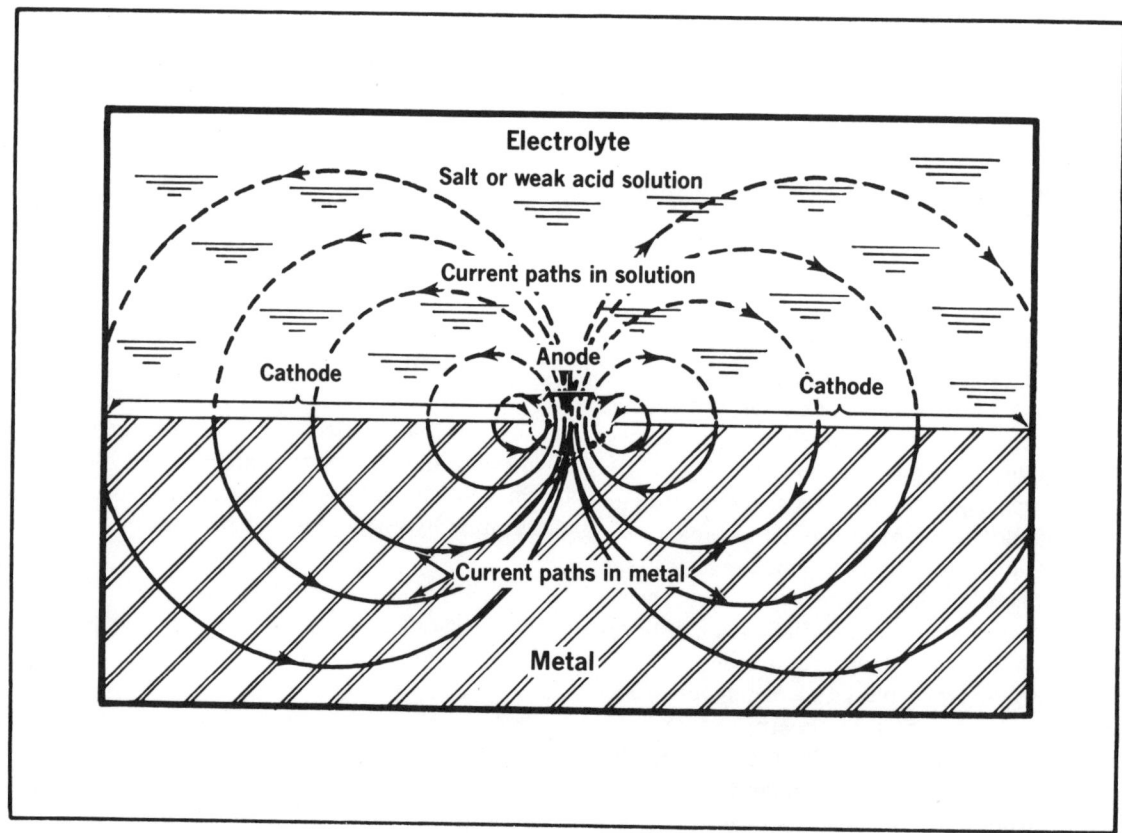

FIG. 10-1—*Flow of current from a corroding pit.*

TABLE 10-1
Typical Values of Electrical Resistivity

Class	Material	Resistivity in ohm-cm	Resistivity in comparison with silver
Metals	Silver	1.6×10^{-6}	1
	Copper	1.7×10^{-6}	1+
	Iron (pure)	$10. \times 10^{-6}$	6+
	Steel (for pipelines)	12 to 24×10^{-6}	7 to 15
	Nichrome	100×10^{-6}	63
Semi-Conductors	Graphite	.0008	500
	Carbon	.006	3,750
	Magnetite (Fe_3O_4)	.05	31,500
	$CaCl_2$ (fused)	.86	538,000
	FeS (fused rod)	1.5	790,000
Electrolytes	Saturated NaCl Brine	6.0	3,750,000
	Sea Water	20.0	12,500,000
	Soils and Formations	500 to 100,000	312 million to 62.5 billion
	Tap Water	500 to 10,000	312 million to 6.25 billion
	Very Pure Water	17.2×10^6	10.7 trillion
Insulators	Porcelain	5 to 30×10^{13}	31,200,000 trillion to 187,500,000 trillion
	Glass	2 to 9×10^{13}	12,500,000 trillion to 56,200,000 trillion
	Fused Silica	5×10^{18}	3.12 trillion trillion

conductor but its resistivity is roughly a million times that of iron. However, sea water is a good conductor compared to distilled water. Some corrosion products such as magnetite or ferrous sulfide are better conductors than brine or sea water.

Many questions regarding the flow of cathodic protection currents can be explained through consideration of Table 10-1. For example, an insulating joint in an uncoated well flow line containing produced water will almost completely stop the flow of current to the well, although the pipe is filled with material of relatively low resistivity when compared to good insulators.

At the cathodic spot where the electrical current enters the metal (Fig. 10-1) bubbles of hydrogen gas form on the metal surface, but no metal loss occurs. Hydrogen gas clinging to the cathode increases resistance to current flow; so the current flow and the corrosion rate are simultaneously reduced. This is termed polarization and plays an important role in reducing corrosion. There are other well known means of increasing resistance to current flow and reducing corrosion.

Pounds of iron loss from an anode are proportional to the current flow from the anode. One ampere of current flowing for a one-year period will remove 20 lb of iron. For perspective, a ⅛-in diameter hole in tubing represents about 1/100 ounce of metal loss.

Types and Causes of Corrosion

It is seldom possible to determine the cause of corrosion simply by visual or even laboratory examination of the corroded piece. Thorough knowledge of the environment, location in the system, operating history, and metallurgy of the corroded item are important in diagnosing the causes and prescribing a cure for the corrosion problem.

Factors in the environment that must be determined by field or laboratory measurement include electrolyte properties (resistivity, chlorides, pH, etc.), temperature, pressure, CO_2, H_2S, O_2 content, and bacteria presence and type.

In the case of pipeline or well casing corrosion, location information would include proximity to stray current sources. Depth in the well is important location information in the case of tubing or sucker rod corrosion.

Operating history includes such information as previous inhibitor treatments, acid treatments, producing rates, water oil ratio, and handling damage.

Metallurgy examination usually reveals that environmental factors are the principal cause of corrosion. In a few cases, probably less than 10%, metallurgy factors may be the principal cause of failure in oil field operations.

In the absence of CO_2, H_2S and O_2, water is usually not corrosive at normal oil field temperatures. Unfortunately, one or more of these gases is almost always present in varying quantities in production operations.

Various treatments of the metal such as scratches, tong marks, hammer marks, greases, and paint, as well as original heat treatment or imperfections in the basic metal, control the location of the cathodic and anodic areas. Physical damage appears in the form of pits, holes, cracks, general metal loss, and loss of strength or ductility.

Carbon Dioxide (CO_2) or Sweet Corrosion—Carbon Dioxide dissolves in water to form carbonic acid. The solubility is directly proportional to the pressure and inversely proportional to temperature. A rule of thumb that has evolved from field experience with corrosion in sweet gas condensate wells making only condensed water (i.e. no "bottom" water) is shown in Table 10-2. The CO_2 partial pressure is calculated by multiplying the mole % CO_2 (from a gas analysis) by the pressure at the point of interest in the system.

This rule-of-thumb does not necessarily apply in gas-condensate wells making salt water or in oil wells. CO_2 corrosion can be very rapid where CO_2 partial pressures are high. When small quantities of water are condensing or being produced, a thin film of water may exist on tubing surfaces. Diffusion of CO_2 through such a thin film to the corroding metal surface is very rapid under these conditions. Corrosion rates of several hundred mils per year have been reported under the worst of these conditions.

CO_2 also causes corrosion in oil wells once the water-oil ratio becomes high enough to cause the steel to switch from oil-wet to water-wet. This almost always occurs before the water cut reaches 80% and sometimes when it is as little as 20%. Heavier oils tend to keep the steel oil-wet at somewhat higher water-oil ratios than higher gravity crudes.

Hydrogen Sulfide (H_2S) or Sour Corrosion—Two types of damage can occur in systems containing H_2S: normal weight loss corrosion and hydrogen-induced damage, which can take the form of blistering or cracking. Black scale (iron sulfide) on a steel surface is indicative of hydrogen sulfide attack. The action on iron or steel is indicated by:

$$Fe° + H_2S + Moisture = FeS + H_2°$$

The black corrosion product that is formed, having a higher resistivity than steel or the electrolyte, can sometimes be a barrier to further corrosion, particularly where the FeS layer tightly adheres to the steel. If FeS is discontinuous and loosely bonded to the steel, the FeS layer can accelerate corrosion by galvanic action since the FeS is cathodic to steel.

Hydrogen-induced blistering or cracking is a complex function of many variables. Strength of the metal, composition of the alloy, hardness, stress level, pressure, temperature, and pH all have an effect on the tendency of metals to blister or crack. Because hydrogen-induced cracking is usually associated with hydrogen sulfide in production operations, it is frequently called sulfide stress cracking (SSC). Under conditions of high stress, SSC can be catastrophic and occur after only a few hours of exposure to the environment with little or no evidence of weight loss due to corrosion.

The most widely accepted theory of the mechanism of SSC holds that the atomic hydrogen (H), formed at cathodes on the metal surface, is able to enter the metal. Hydrogen atoms normally react to form molecular hydrogen (H_2) before entry into the metal has time to occur. In the presence of sulfide ions, the rate at which these hydrogen atoms combine is slower. Given this additional time, the hydrogen atoms move into the metal lattice along dislocations until they reach an imperfection or void in the metal. Here they combine to form much larger hydrogen molecules. This gaseous hydrogen continues to collect until the pressure inside the metal and the external stresses combine to exceed the tensile strength of the metal, and failure occurs.

Sulfide stress cracking is not a problem in all systems containing H_2S. The National Association of Corrosion Engineers (NACE) standard (MR-01-75)[17]

TABLE 10-2
Carbon Dioxide (CO_2) Corrosion in a Sweet Gas Condensate Well (Condensate Water Only)

CO_2 Partial Pressure	Corrosiveness
30+	Corrosion almost certain
7–30	Corrosion possible
0–7	Corrosion not likely

should be used to (1) determine whether the system of interest is one where SSC can be expected and (2) select steels and alloys that will resist SSC at system conditions.

This standard is continually being updated, principally through field experience with various metals and alloys. Representatives of producing companies, equipment manufacturers, steel mills, alloy producers, and research institutions from around the world serve on the NACE committee T-1F, which is charged with approving materials and revisions to the MR-01-75 standard.

Oxygen Corrosion—Oxygen dissolved in water causes very rapid corrosion. This corrosion forms a scale that may vary from dense and adherent to loose, porous, and thick.

Downhole oxygen corrosion in producing wells is usually caused when air enters the casing-tubing annulus. In a water injection system, air may enter the fluid stream at numerous places. In pumping wells, oxygen corrosion can occur in casing, tubing, pump, and the lower section of the rod string.

Corrosion rate increases three-fold when oxygen dissolved in water increases from less than 1 ppb to 0.2 ppm. Serious corrosion has been indicated in water that contains H_2S and as little as .09 ppm oxygen, with corrosion problems eliminated after oxygen is removed. Frank[4] reported in 1972 that when carbon dioxide and oxygen are present in equal amounts in water, corrosion of steel is about ten times as fast as when the same water contains no oxygen. The oxygen content should be below 50 ppb for good corrosion control.

Trace amounts of oxygen—the amounts that get in through leaky pump packing, faulty gas blankets, faulty flanges, and leaky control valves—can cause a special type of corrosion known as concentration cell corrosion. This can destroy equipment in a very short time.

In less than three days oxygen can penetrate a 3-ft layer of oil on top of water in a storage tank. Trace amounts of oxygen get through a gas blanket and set up a concentration cell, the most severe type of corrosion.

One of the most corrosive environments in oil field operations is caused by trace amounts of oxygen entering a sour brine system. This type environment has destroyed major items of equipment in less than six months.

Differential Aeration Cell—Corrosion attack will occur where steel is subject to variations in the environment. This is responsible for much of the metal loss in underground steel. Examples are pipelines passing through different soils, well casings penetrating strata of various compositions, or flow lines crossing abandoned saltwater pits.

In an environment where the level of oxygen varies from one location to another, steel experiences very rapid corrosion. This type of corrosion is called differential aeration cell, differential oxygen cell, or oxygen concentration cell. The area of steel starved of oxygen is anodic to the adjacent area of steel and corrodes, as illustrated in Figure 10-2.

The bottom side of a pipeline will corrode because it is contacted by lesser amounts of air. Pipe under the road at road crossings with less access to air corrodes near the edge of the road. Tanks resting on soil foundations will corrode underneath and near the outside edge, due to differential aeration. Corrosion occurs in crevices, lap joints, butt joints, bolt heads, gaskets, scale, debris, and moist insulation because these areas can be starved of air while an adjacent area is not. Trace amounts of air in any system can trigger this serious corrosion problem.

Erosion Corrosion—Corrosion is reduced in many cases by thin films that form on the surface of the metals and alloys. These films can be oxide films, adherent corrosion products, thin scale, or inhibitors. If such films are partially removed or destroyed, localized corrosion can result.

High velocity or turbulent flow of gases and fluids will mechanically remove these surface films and increase corrosion rates. This action is accelerated if

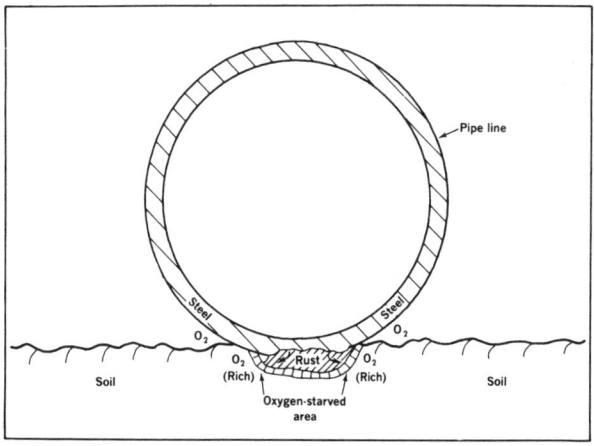

FIG. 10-2—*Area of steel starved of oxygen is anodic to adjacent area of steel, and corrodes.*

suspended solids or entrained gases are present in the fluid streams. Localized corrosion attack will appear as parallel pits or grooves. A special form of erosion corrosion is cavitation, which frequently occurs in centrifugal pumps. This problem, which is usually curable by increasing the pressure on the suction of the pump, is caused when gas bubbles are formed in the lower pressure area created by the pump suction, then violently collapsed as the impeller continues to rotate, which increases the pressure.

Corrosion Resulting From Bacteria—A typical oil field brine may contain many different kinds of bacteria. These can be separated into two groups: (1) aerobic bacteria where oxygen is present in the water and (2) anaerobic bacteria where no oxygen is present. If the water contains oxygen, the major problems will be molds, algae, fungi, and other slimey growths that plug formations and foul equipment. Chlorine treatments will kill any organism living in air. Oxygen is usually excluded or removed from water to be injected downhole to reduce corrosion. This action prevents the growth of the green molds, fungi, and algae.

Sulfate-reducing bacteria, which are found in many oil field brines, are anaerobic and can cause corrosion. These bacteria digest sulfates in water and convert the sulfates to corrosive hydrogen sulfide. The resulting iron sulfide corrosion product, particularly in combination with small amounts of oil, is a very troublesome plugging agent in water injection and disposal systems.

Slug treatments with bactericide at fairly high concentrations (100 ppm minimum) are usually effective in controlling sulfate reducers. The interval between treatments should be determined by monitoring with culture bottles.

Corrosion Fatigue—Steel will fatigue and break when alternately loaded and unloaded many times. Such failures will occur more frequently, even at lighter loads, if the steel has a nick, dent, notch, imperfection, or mashed area. Loads are concentrated in the damaged area, with the stress in the metal being increased up to ten or more times normal stress. Corrosion pits cause stress concentrations resulting in metal fatigue or corrosion fatigue.

One oil field study found that 90% of rod breaks occur at dents, nicks, or bends, with the cause of breaks being attributed primarily to improper handling.[10] Although a majority of sucker rod breaks are due to fatigue, a rod string can give many years of trouble-free service if handled properly and not overloaded. The load limit or endurance limit is defined by API RP 11BR.[29] If a rod string is allowed to corrode, higher stresses are set up because of pits. Proper use of a corrosion inhibitor will prevent pitting.

DETECTION AND MEASUREMENT OF CORROSION
Finding Corrosive Environments

Corrosion can be prevented or reduced if a corrosive environment is recognized. Early detection of active corrosion will allow initiation of control measures and usually will prevent serious damage.

How can active corrosion be identified before a catastrophic failure or irreparable damage occurs? Look for factors that increase corrosion rates. If any of these factors are present, measure the corrosion rate. If metal loss is not significant, it may be cheaper to allow the corrosion to continue than to combat it.

Identify Potential Sources of Corrosion—The first step in control is the identification of corrosion-causing factors. Visual inspections may be adequate. Chemical, bacteriological, or electrochemical tests are usually made to verify conclusions. Here are some of the factors that contribute to corrosion:

1. Water must be present before corrosion can start. The water may be oil field brine, fresh water, water spray, vapor, or condensation.

2. In oil or gas wells, acid gases, hydrogen sulfide, and carbon dioxide form acids when dissolved in water. Hydrogen sulfide may be generated by sulfate-reducing bacteria in waters containing dissolved sulfates.

3. Air (oxygen) enters oil field systems by "breathing" of vessels, through the casing-tubing annulus, from packing leaks at pumps and valves, from faulty gas blankets, or from malfunctioning vapor recovery systems, and other sources. Trace amounts of air can cause severe pitting.

4. Dissimilar metals in physical contact can cause galvanic corrosion. Some common occurrences are brass valves in steel lines, brass polished rod liners, and bronze pump impellers in steel cases.

5. High fluid velocities or turbulence can remove protective films. This can be a problem in high-capacity gas wells, high-speed centrifugal pumps, throttling valves, chokes, and heat exchangers.

6. *Concentration cell corrosion* occurs in many situations:

Pipelines: External corrosion of lines crossing cultivated fields, brine-polluted areas, caliche beds, and roads. *Internal* corrosion of lines handling fluids with considerable suspended corrosion products, loose scale, or other solids.

Well Casings: External corrosion caused by faulty electrical insulation between wellhead and flow line or gas gathering line. Oil wells with known casing leaks should be checked for external corrosion. *Internal* corrosion is frequently caused by "breathing" of air if the casing-tubing annulus is open to the atmosphere.

Vessels: External corrosion of tanks set on soil foundations, buried drain lines beneath tanks, and crevices where moisture can collect between tank and supports. *Internal* corrosion caused by "breathing" air into vessels, faulty gas blankets, and solids accumulation on tank bottoms.

Tests for Corrosive Conditions

Chemical Tests—The type and amount of acid gases or oxygen dissolved in water or in gas streams, vapor zones, and gas blankets can be determined by chemical tests that provide a clue to the type and severity of corrosion. Dissolved iron analysis can indicate the severity of downhole corrosion.

Scale, such as iron oxide formed by corrosion, can be analyzed chemically. The composition will usually indicate probable cause.

Tests for Bacteria—Most oil field waters contain either aerobic or anaerobic bacteria. Sulfate reducers, anaerobic bacteria that digest sulfate in water to produce hydrogen sulfide, are the primary cause of bacteria-related corrosion in production operations. If black iron sulfide suddenly appears in water or a "rotten egg" odor is detected, test for sulfate reducing bacteria.

Electrochemical tests—These are used to check surface lines, well casings, or other buried steel structures.

Tests for Pipelines—Soils contain moisture and will conduct electricity. Corrosion cells develop along a pipeline, and metal loss occurs as current flows from anodic to cathodic areas. Low electrical resistance of soil, as in brine polluted areas, will allow rapid corrosion.

Resistivities of soils can be measured along proposed routes to locate areas of low soil resistivities and high corrosivity. Pipe can then be either protected or laid in the least corrosive areas. Plastic pipe is frequently more economical than steel where operating pressures and temperatures allow its use.

The tendency of sections of steel pipe in contact with soil to corrode can be predicted by the use of a voltmeter and copper-copper sulfate electrode. Surveys conducted using these electrodes generate data that can be interpreted by corrosion specialists. Such tests are used to moniter systems that are under cathodic protection to determine adequacy of the protection. For existing lines without cathodic protection, corroding sections or "hot spots" can be located by these tests.

Test of Current Flow in Well Casings—Electrical current flowing in the casing may be corrosion current or current from a cathodic protection system. Whatever the current source, a voltage or IR drop occurs along the outside of the casing due to current flow and resistance of the pipe. If the grade and weight of the casing are known, voltage drop can be used to calculate current flow.

Voltage changes in the casing are measured with a logging tool with two sets of contactor rollers at a known spacing, separated by an electric insulator. Polarity of the voltage reading between the two logging tool contacts indicates whether current is flowing up or down the casing. A series of readings provides a casing potential profile (Fig. 10-3). The slope of the profile indicates whether current is entering or leaving the casing through the interval of interest. A negative slope indicates current leaving, i.e. an anodic area undergoing corrosion (from about 1,250 to 2,200 ft between points A and B in Figure 10-3).

In curve No. 2. (shown as a dashed line), cathodic protection has been applied and current is no longer leaving the pipe at any point. The casing is fully protected.

The potential profile is one tool used to find active corrosion on the outside of the casing and to show effectiveness of cathodic protection.

Many high-resistivity formations such as anhydrites will not permit cathodic current to pass from near-surface anodes, so lower portions of a casing string may not be protected. The casing potential profile log will detect such conditions. A deep groundbed with anodes placed below such a high-resistivity zone in a hole drilled through the zone is better than a shallow groundbed under these conditions.

Measurement of Corrosion Rate

Where factors favorable for corrosion are found, the next step is to determine corrosion rates. Rate measurements are usually made over an extended

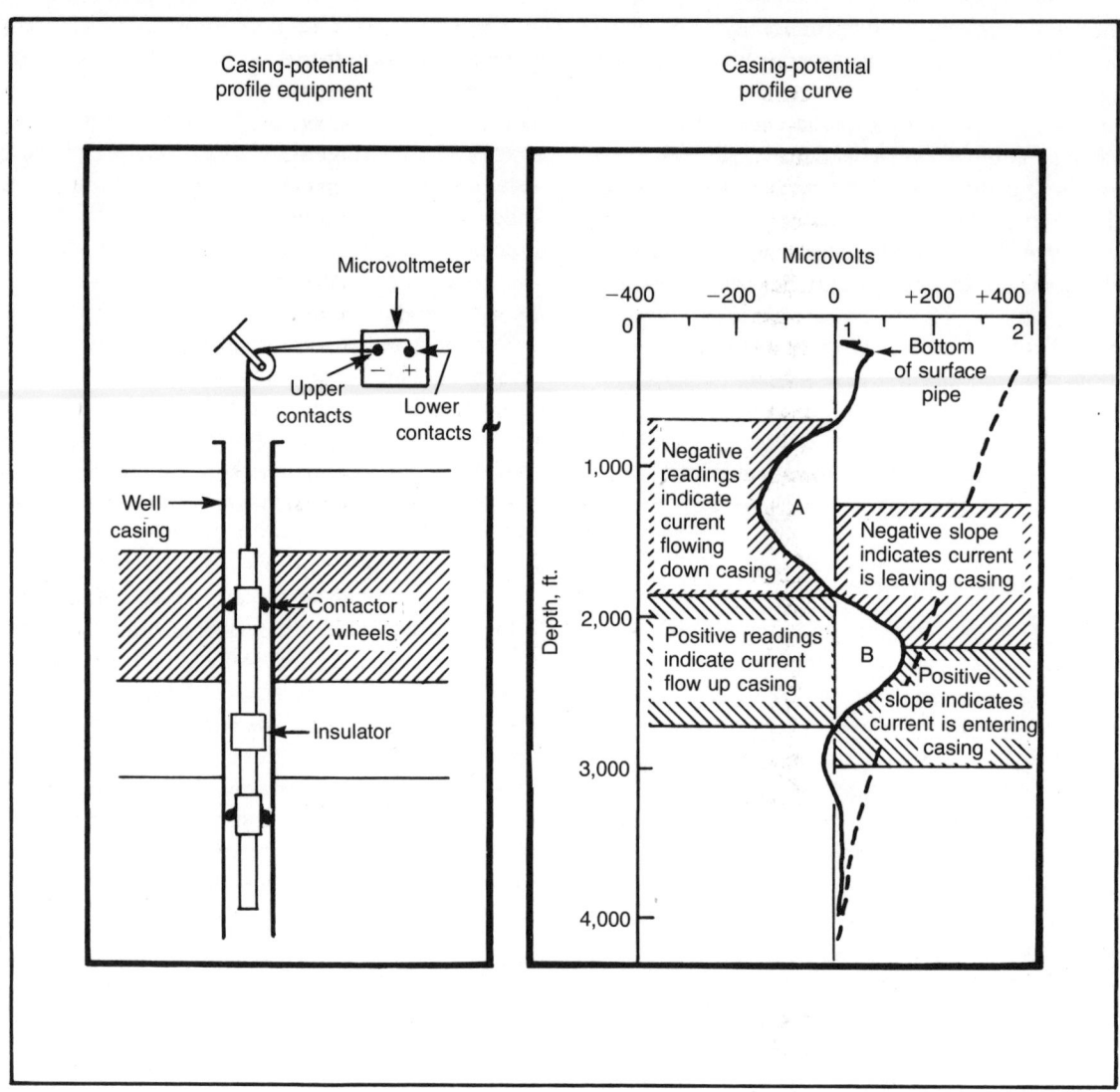

FIG. 10-3—*Casing-potential profile.*

time because single tests do not provide absolute values of damage. A comparison of several tests gives a more accurate appraisal of whether corrosion control can be economically justified.

Four techniques are used commonly for evaluating active corrosion: inspection, corrosion-rate tests, chemical-rate tests, and equipment-performance history. A specific technique may be more suitable, depending on the cause of the corrosion, the equipment involved, and operating conditions. Frequently, a combination of techniques will provide the most useful information. The following are the principal means of determining corrosion rate.

Visual Inspection—Out-of-service equipment can be inspected to determine corrosion damage. All of the equipment can be scanned rather than the local areas, as is the case with most other techniques. Good records and descriptions are essential for future reference and comparison.

Caliper Surveys—Caliper surveys are run on wire lines to inspect the internal surface of tubing or casing. Mechanical feelers contact the inside metal surface and will detect metal loss due to pitting, metal thinning, or rod wear. Better detail is obtained with instruments where all feelers record, such as the Kinley tool. This allows determinations at any point of pits, general metal reduction, mashed areas, kinks, or "dog-legs."

When tools are run through internally-coated tubing, precautions are necessary with some types of caliper to prevent pipe coating damage. Caliper surveys are most useful if they are conducted periodically to determine the progression of pits or area metal loss. Periodic surveys are useful in determining the effectiveness of corrosion inhibitor treatment. Caliper "tracks" have been responsible for corrosion damage where a sound inhibitor film was not re-established immediately following caliper surveys.

Casing Inspection Log—Magnetic flux leakage detection tools are available from a variety of service companies. These tools use the distorted magnetic field around an anomaly in the pipe wall, such as a corrosion pit, to create a signal in a pick-up coil, which is transmitted to the logging truck where it is recorded as a "kick" on a strip chart. Such devices have been used for many years for inspection of new tubular goods at manufacturing facilities. In the oil field, they are employed for inspection of new tubulars to be used in critical service and for grading the condition of used tubulars.

The principal of magnetic flux leakage was adapted by Shell Development Company for use on a downhole logging device, and tools have since evolved that are capable of detecting fairly small metal loss and discriminating between internal and external metal loss.

Comparisons by a number of companies of actual downhole inspection logs with recovered casing have

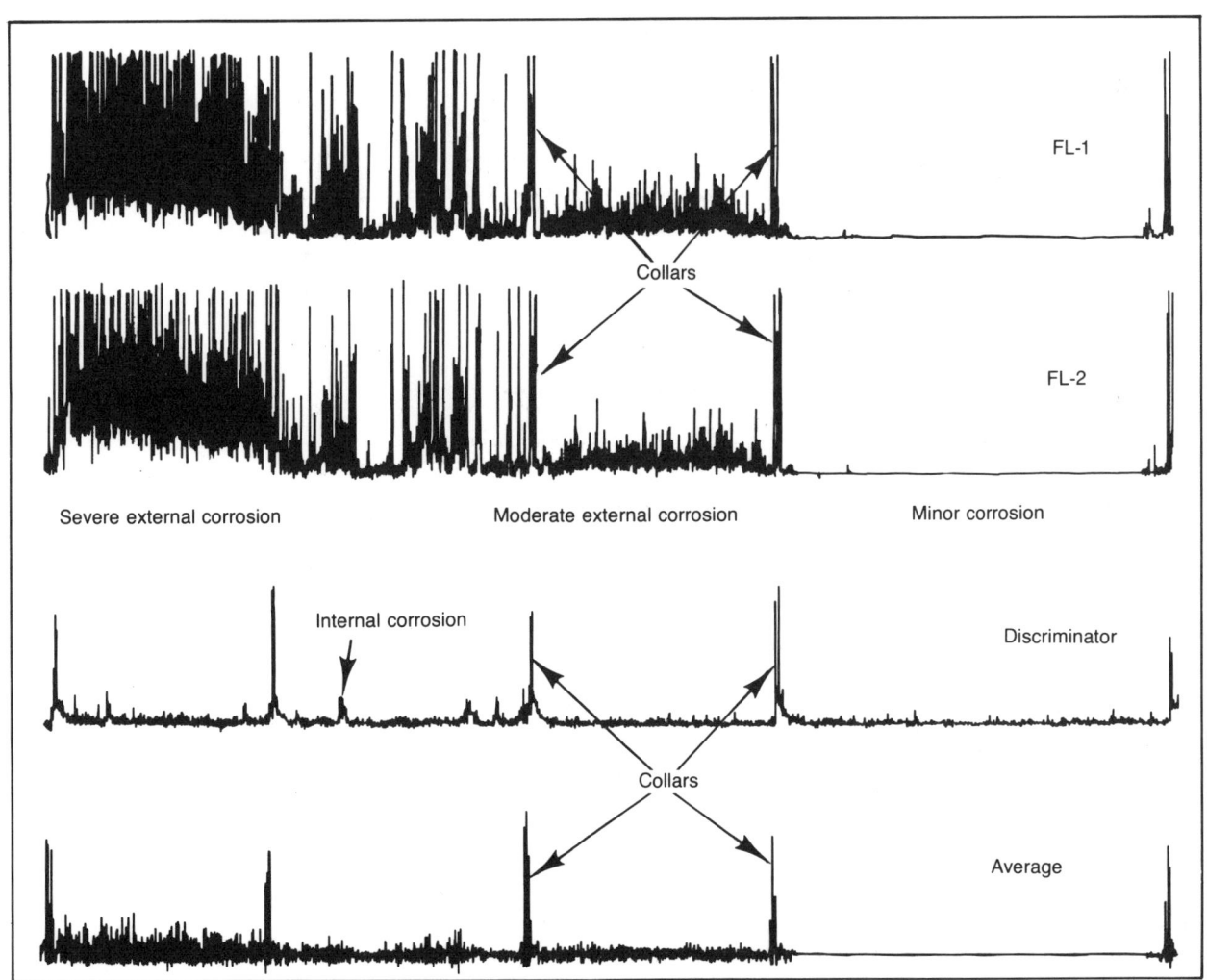

FIG. 10-4—*Four channel vertilog output.*

shown the "Vertilog" or "CAT" log currently licensed to Dresser Atlas to be capable of detecting the smallest amount of metal loss. A sample Vertilog is shown in Figure 10-4. Other types of casing inspection logs are available and may be adequate or more economical in particular circumstances.

Casing inspection logs are frequently used in conjunction with casing potential profile logs to verify anodic areas and assess the extent of the metal loss over the corroding interval.

Ultrasonic Thickness Tests—Ultrasonic thickness measuring instruments use the principle that the speed of transmission of a sound wave through a material is a constant characteristic of the material. The time required for an ultrasonic pulse introduced at one surface to traverse the metal thickness, reflect off the other surface, and return to the detector is measured and converted to metal thickness. Such measurements have the attraction of requiring access only to one side of an in-service pipe or vessel. The biggest disadvantage is that each individual measurement only inspects a tiny area of the total metal surface. Thus, concentrated pitting is very easily missed and the reliability of ultrasonic measurements in detecting pitting is directly proportional to the number of measurements taken. Ultrasonic measuring instruments are very useful where erosion, sand cutting, or erosion-corrosion are problems. Since these phenomena generally cut a relatively wide swath in the pipe or vessel surface, a few well-placed measurements, particularly at ells or reducers, are very helpful in assessing metal loss caused by these forms of attack.

Metal Loss Rate Tests Using Coupons—Weight-loss tests are the most common of all corrosion rate measurement tests. Corrosion coupons are strips of mild steel of various sizes with a hole in one end so that coupons can be mounted on an insulating (plastic) rod and inserted in a pipe line or tank through a threaded fitting. Arrangements can also be made, as shown in Figure 10-5, to insert coupons into a pipe or vessel under pressure through a valve. Ring coupons are used in tool joints to measure drill pipe corrosion rates.

After preparation, usually by sandblasting and degreasing, coupons must be accurately weighed before exposure. Corrosion of the coupon must be prevented while it is stored or being transported to and from the test location. Grease, oil, or fingerprints on the uncorroded coupon will prevent the coupon from corroding properly when exposed. After being exposed to the corrosive fluid for two to four weeks, the coupon must be removed and cleaned of all corrosion products without any attack of the metal by the cleaning technique.

Weight-loss, area of coupon, and exposure time are used to calculate corrosion rate, which is reported in mils per year (MPY) of metal loss (one mil equals .001 in). Pitting penetration rate and physical appearance of the coupon should be included in reporting corrosion of coupons.

Two types of tools are used to insert coupons into systems under pressure. The "lubricator" tool, such as the Fincher Engineering coupon lubricator, uses a packing ring to seal around a push rod, which is attached to the coupon. For higher pressures, the more expensive extractor tools, such as those manufactured by Cosasco or McMurry-Hughes, couple to a special full-opening valve. The inside chamber of the entire tool is then exposed to system pressure by opening the valve; coupon insertion takes place by turning a gear and screw arrangement. The fitting shown in Figure 10-5 can be used for installing other corrosion monitoring devices, pressure gauges, thermometers, or a fluid sampling "dip tube." Coupons were the earliest corrosion rate measuring device and continue to be widely used because they are simple to use and provide visual evidence of corrosion or effectiveness of an inhibitor treating program.

Determination of corrosion rates through coupons can be inexpensive if the number of coupons used in a system is large and personnel for processing coupons are available. Some inhibitor suppliers provide coupons and then clean and weigh them after exposure. The principal disadvantages of coupons are (1) the time required to obtain results and (2) their ability to show corrosion only at the point of installation.

Metal in the coupon is seldom identical with metal in the pipe line; therefore, corrosion rate of coupon may not be identical with that of the pipe. However, if corrosion is negligible on a properly installed coupon, corrosion of the pipe may also be negligible. Similarly, if injection of an inhibitor reduces coupon corrosion rate to a negligible value, similar results may be expected on the vessel or pipe. Corrosion rates are initially very high on the sandblasted coupon but decline as the corrosion product film forms on the coupon.

A low MPY rate may be serious if concentrated pitting corrosion is occurring, while a high MPY loss with a general area type of metal loss may be relatively insignificant. Thus, pit depth should be measured, and pitting penetration rate should be reported

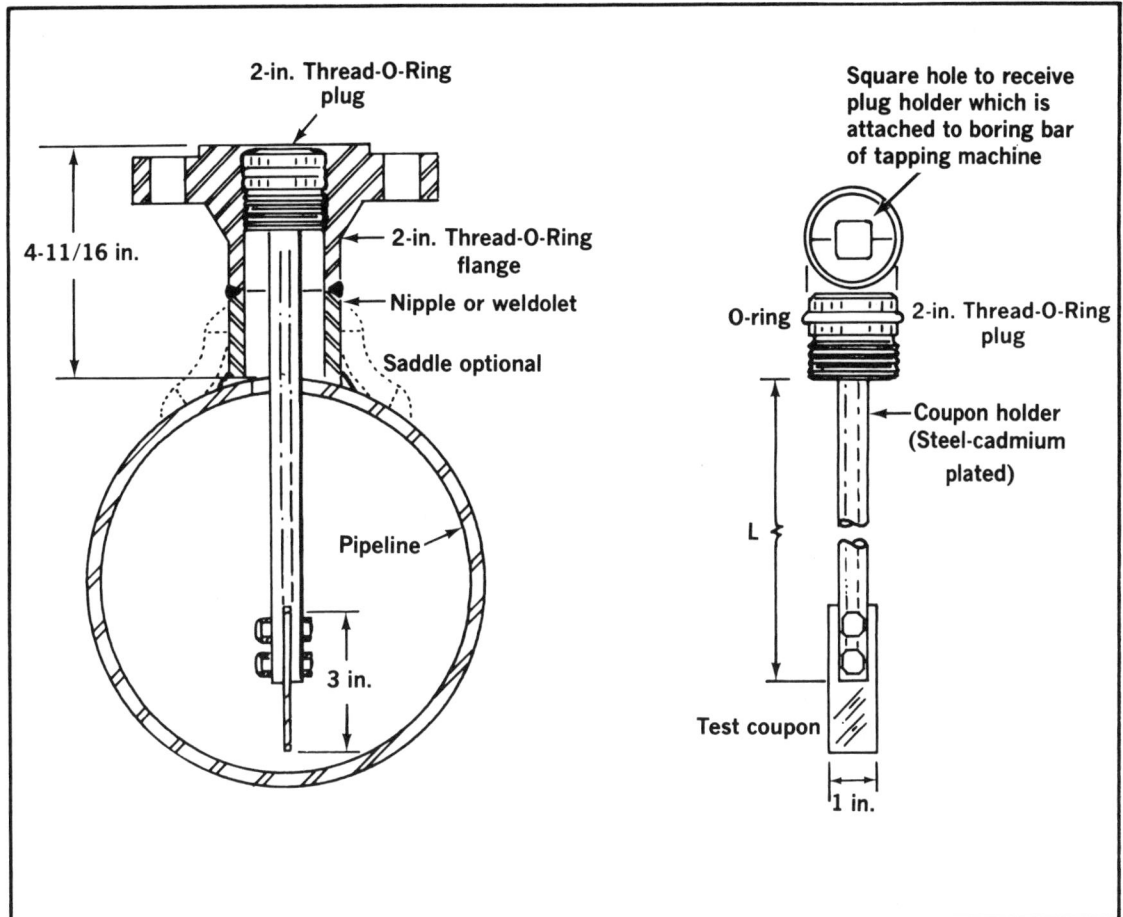

FIG. 10-5—*Fitting for in-stream corrosion test coupon.*

as well as average corrosion rate based on the entire surface area of the coupon. Where pitting is too shallow to measure directly, an estimate of the corroded area may be used to calculate pitting penetration rate.

Pipe test nipples with metal identical to that in the line being checked for corrosion are installed directly in the line or in a bypass. The nipples must be cleaned and weighed before and after the test interval, usually three months to one year, to determine loss of metal. Such nipples may also be used to monitor scale or paraffin deposition.

Hydrogen Probes—Hydrogen is generated at the cathode of a corrosion cell. The hydrogen probe takes advantage of this phenomenon to give a semi-quantitative indication of corrosion rate. Figure 10-6 shows a simple pressure-type hydrogen probe. The probe is inserted in the system through a tee or other threaded fitting. As corrosion occurs on the body of the probe, atomic hydrogen, in the presence of H_2S, permeates the metal probe and collects in the sealed cavity. Here hydrogen atoms combine to form H_2 and increase the pressure reading on the gauge. Attempts to correlate rate of pressure buildup with an exact corrosion rate have not been successful because the rate of hydrogen penetration is dependent on so many variables. The hydrogen probe is most useful in monitoring the effectiveness of corrosion control programs, such as inhibitor treatments. The principal attractions of the hydrogen probe are (1) that it can be left in the system for an indefinite period of time, requiring only occasional inspection to assure the probe is not badly pitted so as to be unsafe and (2) that it provides a day-to-day qualitative indication of corrosion activity, inhibitor effectiveness, or recommended interval between inhibitor batch treatments.

Pressure buildup rates of 20 psi per day in systems

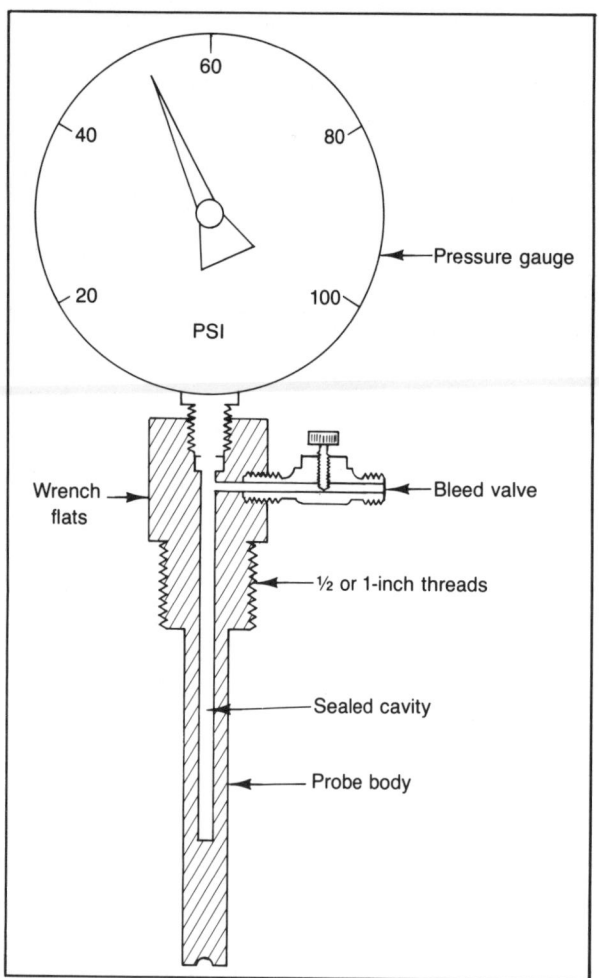

FIG. 10-6—*Section of hydrogen probe.*

these devices use the hydrogen generated from corrosion on the pipe or vessel wall, with a pool of electrolyte on the outside of the system. Hydrogen entering the electrolyte after permeating the pipe wall causes a change in potential between two special electrodes. This change in potential is monitored on a meter or recorder; corrosion activity is inferred from these readings. A major advantage of such a device is that nothing needs to be inserted into the system. Additional information on various types of hydrogen probes and their relative advantages and disadvantages are given in an NACE state-of-the-art report on corrosion monitoring with hydrogen probes.[27]

Electrical Resistance Method (Corrosometer)—The Corrosometer is used principally in gas streams because it does not have to be submerged in water to function.

The Corrosometer measures electrical resistance and is an adaptation of the Wheatstone bridge. This bridge consists of four resistances, with two of the resistances in the probe and two in the instrument box. One of the probe resistances, a small strip, wire, or tube of steel or other metal, is exposed to the corrosion environment. Changes in resistance as corrosion reduces the cross-section of the probe provide a very precise measure of corrosion rate. Probe elements may be obtained in various sizes and thicknesses and may be used in gas streams where the corrosive electrolyte is not continuous.

By taking two readings of the instrument and applying characteristic corrections for the particular probe, corrosion rate may be determined to the nearest microinch (millionths of an inch). The corrosion rate can then be expressed in mpy or other convenient units. As an extreme example, using a thin (4 mil) probe element, a moderate corrosion rate can be measured with fair precision in one hour. A less sensitive and more rugged probe and a longer time interval are usually preferred.

The Corrosometer is connected through a cable to a probe installed in a pipe or vessel through a threaded fitting. Any number of probes may be installed in various locations. All probes can be observed at convenient time intervals with one instrument, which may be supplied with power from 110v AC or from internal batteries.

The Corrosometer, like the hydrogen probe, appreciably reduces total time required to obtain reliable results. For example, if a corrosion inhibitor treatment is to be evaluated, several probes can be installed at various places in the system and observed

with active corrosion are not uncommon. With effective corrosion control, pressure buildup may drop to less than 1 psi per day.

The principal shortcomings of the hydrogen probe are (1) it indicates corrosion only at the point of insertion into the system, (2) it is not quantitative, and (3) it does not function in all systems. Theoretically, H_2S or other agents that enhance hydrogen permeation must be present in the system for the probe to function. Hydrogen probes, however, have been used successfully in systems where H_2S could not be detected. This may be caused by the presence of H_2S in minute amounts below detectable levels.

Electronic versions of the hydrogen probe have been used in research for many years, and efforts continue toward the building and marketing of a reliable electronic hydrogen probe for field use. Some of

until each one shows a definite straight line corrosion rate. Then without moving or changing any of the probes, inhibitor injection is started. Continued measurement of corrosion rates at each probe will show the time required for the inhibitor to reach each probe and inhibitor effectiveness at each location in percent reduction of corrosion. Inhibitor injection can then be stopped and the persistence of the inhibitor observed at each location.

The effect of other operating procedures on corrosion rate can be similarly evaluated. Several months would be required to obtain such information with coupons.

Disadvantages of the Corrosometer are:

1. It is somewhat expensive.
2. It is a sensitive and delicate instrument not easily repaired. However, the instrument and measurements can be checked easily for accuracy.
3. Untrained operators may have difficulties in operating the instrument.
4. Misleading results can be given if a deposit forms on the probe.

Linear Polarization Resistance Method—(Corrosion Rate Meter)—The corrosion rate meter uses a three-electrode probe that is inserted into the system. The corrosion rate meter measures corrosion current and thereby measures corrosion rate, because metal loss is directly proportional to current flowing from the test electrode. With the corrosion rate meter, an external current is applied from the auxiliary electrode to the test electrode to change its potential with respect to the reference electrode by exactly 10 millivolts. The applied current is related to metal loss and corrosion rate.

The instrument gives direct readings in MPY, which represent instantaneous corrosion rates. Very low rates are detected. Versions of the instrument are available that can record data for multi-test points on a continuous basis. This is useful in studying changes throughout a system caused by introduction of oxygen scavengers, inhibitors, air leaks, or other changes.

The test probes must be submerged in electrolyte, and probe positioning must be done with care in a flowing stream to avoid "shadowing" of one electrode by another. Short-circuiting of electrodes with corrosion products or solids must be avoided. Because the corrosion rate meter is very sensitive to fouling with paraffin and iron sulfide, it is most useful in relatively clean, oil-free water systems. As with weight-loss coupons, the sandblasted probe surfaces typically indicate high corrosion rates soon after insertion into the system. While these high rates may be acceptable for monitoring inhibitor effectiveness, the probes are usually left in the system for several days before attempting to determine corrosion rates representative of actual rates occurring on the pipe or vessel wall. Uniform corrosion is indicated although some progress has been made in predicting pitting.

Chemical Test for Corrosion Rate—The measurement of iron dissolved in a produced water stream can indicate a metal-loss rate. The corrosion product must be water soluble. Therefore, the test is applicable primarily to CO_2 corrosion in which water-soluble ferrous bicarbonate $Fe(HCO_3)_2$ is the corrosion product. This type of corrosion is usually associated with gas wells and wells producing sweet crude. Insoluble corrosion products such as iron sulfide and iron oxide are less reliable in evaluating corrosion rates.

The iron content is measured in parts per million and then converted to iron loss in pounds per day using the water production rate of the test well and the Nomograph shown in Figure 10-7. An iron loss of less than 0.02 lb per sq ft per year of exposed metal is a low corrosion rate in a producing or injection system. Iron loss rates may not correlate with equipment failures.

With uniform corrosion, the loss may be high and

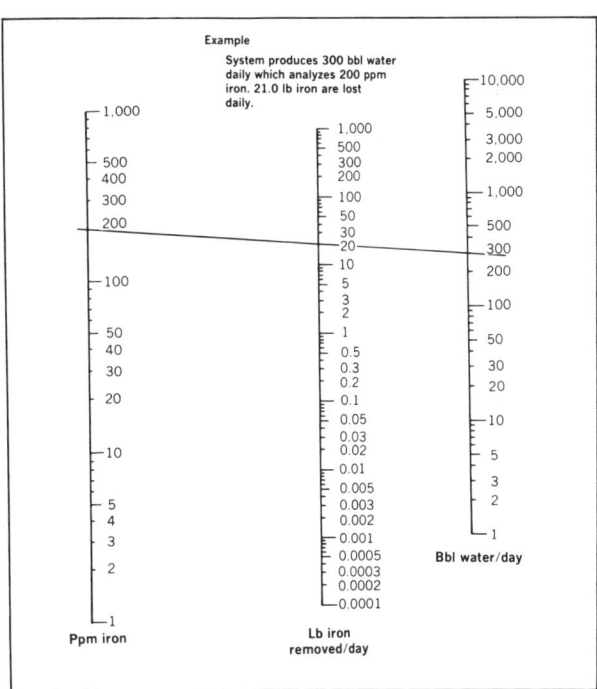

FIG. 10-7—*Nomograph for determining iron loss.*

damage small. With pitting corrosion, the loss could be low and damage severe. A common practice is to supplement iron count data with tubing caliper surveys run in key wells at intervals of one to two years.

Iron is frequently present in formation water. Thus iron measured in wells making formation water may be "background" iron, corrosion product, or some of each. In such wells, *changes* in iron content are used to monitor inhibitor treatment effectiveness and life. Iron measurements are an easy and inexpensive corrosion monitoring tool that gives a direct indication of *downhole* corrosion.

Corrosion Records—In any production operation, a study of corrosion records can be started at any time if records of purchases and repairs due to corrosion are available.

A preliminary study may show that leaks and repairs are confined to certain parts of the system. For example, most flow line leaks may occur at road crossings or near tank batteries or wells. Control measures might then be limited to vulnerable places and most leaks might be prevented at minimum cost.

If records of repair and replacement costs due to corrosion are available, the operator can determine how much can be spent to prevent corrosion. The record of casing leaks provides a basis for estimation of future leaks and determination of economic incentives to reduce the leaks.

When very extensive systems are involved, corrosion surveys and records can be kept in computer files. Computers can be employed effectively in studying any corrosion problem where a large mass of data is involved.

Corrosion failure records may be conveniently made part of more comprehensive equipment failure records, which include mechanical, electrical, and all other failures associated with production operations.

Failure plots of rods, tubing, and pumps greatly enhance the analysis of problems in rod-pumped wells. These plots of failure rate vs. depth will permit the recognition of failure patterns, which can be used to differentiate between corrosion and mechanical problems. A "shot-gun" failure pattern indicates corrosion. Split barrels, wallowed ball and seat, and a concentration of failures near the top or bottom of rods, or at a rod size change are due to mechanical conditions.

Plots of leaks or downhole failures versus time on semi-log paper or log-log paper (Fig. 10-8) provide a good method of predicting future failure frequencies and evaluating the effectiveness of measures to reduce corrosion. Figure 10-8 shows the casing leak history in one large field, before and after installation of insulating plastic nipples between the wellheads and the flowlines. The insulated nipples were installed during the fourth year, and by the sixth or seventh year, it was apparent that casing leaks had been appreciably retarded.

Based on a 30-year projection with and without insulated nipples, casing leaks will be reduced about 95%. This great reduction will occur after isolating the well casing from the flowline if most of the external casing corrosion is caused by current flowing from the flowline into the well casing.

In this illustration (Fig. 10-8) a definite trend of casing leak frequencies had been established before application of insulated nipples between the flowlines and the wellheads. Afterwards, the trend was reduced, indicating the effectiveness of the protection. Also, pinpointing line leaks on an aerial map will indicate "hot-spots" that may need attention. Generally, other measurements such as pipe-to-soil potential surveys are made in conjunction with this mapping to define the problem.

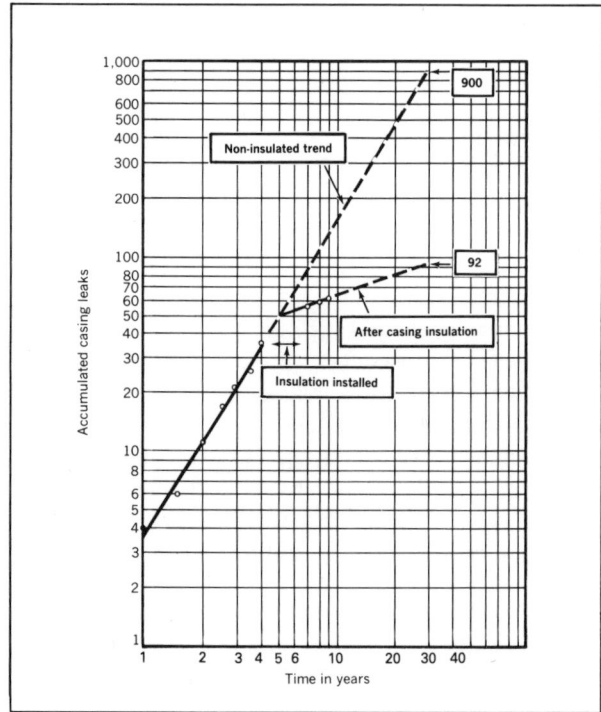

FIG. 10-8—*Comparison of casing leaks with and without insulated joints.*

CORROSION CONTROL

It is usually impossible or too expensive to stop all corrosion. Corrosion may sometimes be allowed to proceed at an acceptable rate if the projected economic loss from corrosion is less than the cost of corrosion control. For example, in many areas, lease production equipment may not be coated for corrosion control if damage occurs at a slow rate. Various vessels should be internally coated where corrosion is causing serious damage. Corrosion inhibitors are used widely but are generally less than 90% effective.

The degree of corrosion control is also influenced by safety aspects, governmental regulations, and environmental considerations. There are a number of ways to minimize corrosion in oil field operations including materials selection, engineering design, inhibitors, coatings, removal of corrosive gases, cathodic protection and use of nonmetallic materials.

Select Proper Materials to Reduce Corrosion Rate

Metals and Alloys—Iron and steel are the most commonly used metals in oil field operations because of their lower cost, ease of fabrication, and strength. However, there are numerous applications where high-priced alloys are more economical than the use of steel. For example, various expensive alloys are used in sucker rod pumps because other means of corrosion control are relatively ineffective.

Materials that have been used successfully in sucker rod pumps in sour service are categorized for use by the degree of system corrosiveness and the level of sand production in NACE Standard MR-01-76.[18]

Selection of metals is affected by corrosive environment as well as physical requirements. When hydrogen sulfide is present, the effect of hydrogen embrittlement on the strength and durability of a metal is a frequent concern. Acceptable metals to resist embrittlement in an H_2S environment are shown in NACE Standard MR-01-75.[17]

In carbon dioxide and oxygen environments, embrittlement is not a problem, and metals are selected for metal loss control. NACE Standard RP-04-75 lists materials that have been successfully used in aerated and nonaerated salt water handling systems.[21] Tables in this standard show components of various items of oil field equipment, such as centrifugal pumps and turbine meters, and usually list a number of alloy choices for each component part. Economics usually favor the use of low carbon steel tubulars with suitable protective measures for carbon dioxide exposure.

Aluminum bronze, stainless steel, and other alloys are used extensively for valves and smaller pieces of equipment in CO_2 injection projects where wet CO_2 is present. This includes injection wellhead equipment, where CO_2 and water are injected alternately, and producing well valves and other components. CO_2 surface injection lines are usually separate from water injection lines. Dry CO_2 is noncorrosive and easily handled in bare carbon steel lines, although Schremp and Roberson have found that certain elastomers are subject to failure in this service.[12] For oxygen problems, low carbon steels and air exclusion are the cheapest approach. If oxygen cannot be excluded or removed, the higher priced alloys may be justified.

Galvanic corrosion is primarily a metals selection problem. A simple solution is to use similar metals. Other means of controlling this problem are to (1) select metals close together in the galvanic series, (2) use inhibitors, (3) use proper coatings, (4) electrically insulate, (5) use cathodic protection, or (6) select metals so that anodic area is large compared to cathodic area.

Caution must be exercised in using coatings to mitigate galvanic corrosion. A pinhole or holiday in the coating on the anodic metal can result in a very small effective anode area coupled to a comparatively large cathode. This coated anodic component may be penetrated in a relatively short time by very concentrated galvanic action at the holiday in the coating.

There is a tendency to overemphasize galvanic corrosion. Many couples of metals that are relatively close together in the galvanic series (Table 10-4) rarely cause trouble.

Corrosion Control Through Original Design

Large savings in future repair and maintenance are usually possible through proper planning for corrosion control when structures and other equipment are being designed and installed. Because all metals have an inherent tendency to corrode, corrosion is normal rather than abnormal and will usually occur unless preventive measures are taken.

A new installation should be designed to last the life of the project, but over-design should be avoided. Many times, future failures are inadvertently built into system design.

Many types of corrosion can be eliminated or minimized by proper engineering design. Here are some of the more prevalent corrosion problems resulting from improper design.

1. Crevices cause concentration cell corrosion.
2. Poor drainage of lines and equipment may cause concentration-cell corrosion.
3. Dissimilar metals coupled together cause galvanic corrosion.
4. Improper selection of metal may result in sulfide embrittlement.
5. If fluid velocity is too slow, solids settle and shelter bacteria. Trace amounts of oxygen set up concentration cells. If it is too high, protective films are eroded. Recommended velocities are shown in Table 10-3.
6. Air-exclusion equipment is undersized.
7. Pump suction conditions may promote cavitation.
8. Flow lines or gas-gathering lines are either not insulated from the well, or insulation is not maintained.

These are only a few examples of corrosion problems resulting from poor design. The more obvious corrosive conditions should be avoided. Installation of insulating flanges or fiberglass-reinforced plastic nipples should be standard procedure at the wellhead in flowlines and gas-gathering lines.

TABLE 10-3
Flow Velocities in Oil Field Lines to Minimize Corrosion Recommended Maximum and Minimum

		Flow Velocities, Ft/sec	
Type service	Type fluid	Minimum	Maximum
Injection lines—cement lined	Liquid	—	5
Injection lines—plastic lined	Liquid	—	10
Injection tubing	Liquid	2	10
Heat exchangers	Liquid	5	10
Flow lines & tubing	Gas*	—	*

*API recommends a maximum velocity equal to $100/\sqrt{\rho}$ for dry gas where ρ is the flowing gas density in lb/ft^3. Experience shows $75/\sqrt{\rho}$ is a practical upper limit for wet gas in a mildly corrosive system. For highly corrosive systems, $60/\sqrt{\rho}$ has been used.

TABLE 10-4
Galvanic Series of Metal and Alloys in Flowing Sea Water

Magnesium and magnesium alloys
Zinc
Commercially pure aluminum (1100)
Cadmium
Aluminum 2024 (4.5 Cu, 1.5 Mg, 0.6 Mn)
Steel or iron
  Cast iron
Chromium stainless steel 13% Cr (active)
Ni-Resist Cast Iron (high Ni)
18-8 stainless steel (active)
  18-8 Mo stainless steel (active)
Lead-tin solders
Lead
Tin
Nickel (active)
  Inconel (active)
Hastelloy B (60 Ni, 30 Mo, 6 Fe, 1 Mn)
  Chlormet 2 (66 Ni, 32 Mo, 1 Fe)
Brasses (Cu-Zn)
  Copper
    Bronzes (Cu-Sn)
      Cupronickels (60-90 Cu, 40-10 Ni)
        Monel (70 Ni, 30 Cu)
Silver solder
Nickel (passive)
  Inconel (passive) (80 Ni, 13 Cr, 7 Fe)
Chromium stainless steel 11–30% Cr (passive)
  18-8 stainless (passive)
  18-8 Mo stainless (passive)
Hastelloy C (62 Ni, 17 Cr, 15 Mo)
  Chlorimet 3 (62 Ni, 18 Cr, 18 Mo)
Silver
Titanium
Graphite
Gold
Platinum

Note: Materials at the upper end of the list are the most active and anodic to those below. Materials below in the Table are more passive and cathodic to those above.

Galvanic Cells and Corrosion-Resistant Metals—Two different metals, immersed in the same electrolyte (salt or acid solution), will have an electrical potential difference between them, and, if the two metals are connected together, current will flow from one metal to the other. This system is called a galvanic cell; it is the basic arrangement of the dry-cell battery.

The galvanic series of metals in aerated sea water is shown in Table 10-4. Metals such as magnesium, zinc, or aluminum shown in the upper part of the table are called "active" metals; those on the lower end of the table, such as nickel and monel, are called "passive" metals. Definite values of the electric potentials are not given because metals and alloys do not have definite and fixed potentials in sea water.

Metals shown in Table 10-4 will not have the same relative electrical potentials or be in the same order in the galvanic series in other corrosive media. As an extreme example, serious pitting of admiralty brass tubes in a cooling system was observed adjacent to steel baffles. It would be expected that steel would be anodic, but observation of corrosion showed that in this particular cooling fluid, which contained sulfides, ammonia, and other material, the admiralty brass was anodic and the corroding electrode.

On immersing in aerated sea water, each metal or alloy will corrode at the relative rate characteristic of that metal as shown in Table 10-4, with metals higher in the table being anodic and corroding at a faster rate than those lower in the table.

If any of the two metals in the series are connected together and immersed in sea water, this action will occur:

1. Electric current will flow in such a direction between the two metals as to reduce corrosion on the less active metal and increase corrosion on the more active metal.

2. Flowing current is greater and corrosion rate increases when metals are more widely separated in the galvanic series.

3. Corrosion rate per unit area of the corroding metal is almost proportional to the total area of the noncorroding metal.

As an example of galvanic corrosion, if iron or mild steel and copper are immersed in aerated sea water, steel, the more active metal, will corrode at the rate of about 5 mils per year. If steel is joined to an equal area of copper, corrosion rate of steel will be nearly doubled to about 10 mils per year. If the relative area of the steel compared to the copper is very small, the metal loss per year per unit area of steel is very high. The copper is cathodically protected and does not corrode when joined to steel.

Destructive galvanic cells are sometimes built into waterflood plants, injection systems, tank farms, and other installations. With improper design, these galvanic cells may actually accelerate corrosion. For example, at one time many states in the U.S. required that metallic structures be adequately grounded for protection against lightning. Rods or steel pipes, ¾- to 1-in. in diameter and about 8 ft long, are driven into the ground and connected to the structure.

These rods are usually copperclad to prevent corrosion. It may appear that steel tanks, resting on the earth, should be grounded in this manner. Calculations have shown that the electrical resistance to ground of the required number of rods in parallel is much higher than the electrical resistance to ground of the tank bottom. Therefore, the rods do not protect the tanks against lightning but do introduce a galvanic cell, which increases the corrosion rate on the outside of steel tank bottoms.

When an important structure, such as a large saltwater pump, is being designed or purchased, a qualified corrosion engineer should check the design so as to avoid harmful galvanic cells. It is usually possible (1) to provide adequate grounding for protection against lightning and accidental failures of electrical insulation, (2) to eliminate harmful galvanic cells, (3) to provide the most suitable protective coatings to reduce corrosion, and (4) to arrange circuits so that cathodic protection can be installed economically if and when required.

Galvanic cells contribute to the corrosion of many structures. But, either through design or accidental choice, galvanic couples are often an advantage. The gates and seats of valves are often faced with monel or stainless steel, both of which are strongly cathodic to the surrounding steel valve body. In addition to erosion resistance, the critical seat and gate surfaces are cathodically protected by the steel body.

The relative area of the steel is large, so the galvanic couple causes only a slight increase in the corrosion rate per unit area of the steel, and the steel body is so heavy that it seldom fails. In these applications where dimensions must remain exact, where fluid velocities and mechanical abrasion are problems, and where corrosion cannot be controlled with inhibitors or coatings, the more expensive corrosion resistant metals can be used to advantage.

In cathodic protection through the use of sacrificial anodes including sacrificial metallic coatings, corrosion is transferred from the steel to the more active metal whose only function is to corrode and protect the steel.

Insulating Flanges or Nipples—Insulating flanges, couplings, or nipples in a pipe line can be installed and tested during construction at a much lower cost than after the pipe line is in operation. During construction, it is not always possible to determine where the insulating joints will be needed; when in doubt, insulating flanges or nipples should usually be installed. They can be easily shorted out if not needed.

Insulating threaded bushings, unions, or flange

joints can be purchased. Insulating materials for most flange joints are also available from the corrosion service companies. Figure 10-9 shows a cross section of a typical insulating flange joint.

Reinforced fiberglass insulating joints, usually 10 in. in length, are being used currently in preference to flanges because flange washers or insulating sleeves often become lost or damaged if the flange is broken out frequently. A pressure tight insulating nipple is an indication of satisfactory insulation. However, the use of insulated nipples is frequently limited by pressure ratings, particularly in the larger sizes.

CO_2 Enhanced Oil Recovery Projects—Corrosion problems can be so severe that they become the limiting factor in determining whether CO_2 injection projects are economical. Materials that will be contacted by wet CO_2 or water containing CO_2 must be carefully selected.

Wellhead equipment where water and CO_2 are alternately injected are generally constructed of special alloys. Valves and meters are commonly type 316 stainless steel or aluminum bronze. Low carbon stainless (e.g. 316-L) should be used for welded components to avoid sensitization to rapid corrosion in the heat-affected zone. Threaded stainless steel connections should be avoided because of frequent galling problems during make-up.

Thick-film epoxy coatings have blistered severely in high pressure CO_2 service. Thin-film phenolics and epoxy-modified phenolics have given good service in injection well tubing.

The most severe corrosion occurs downhole in producing wells after CO_2 breakthrough. Coatings have been used in wells with hydraulic or electric submersible pumps, but rod wear makes them ineffectual in beam-pumped wells. Frequent batch treatments, usually 2 to 5 per week, or continuous inhibitor treatments should be provided for in planning well completions.

Separators and water tanks should be internally coated and have cathodic protection. Continuous inhibitor treatments may be required in the water injection system. Special precautions are usually taken to separate the CO_2 and water injection piping. Blind flanges or removable spools are used to prevent back-mixing of CO_2 and water in surface piping.

Adequate fittings for installation of corrosion monitoring devices should be provided. Scale is frequently encountered in early stages of production as carbonated water mixes with formation or previously injected water. Nonmetallic materials find much application in lower pressure portions of CO_2 enhanced recovery projects, such as water gathering systems.

Metallurgy for CO_2 recovery plants must be carefully selected to avoid severe corrosion. Corrosion inhibitors are difficult to apply and only partially effective in such recovery systems.

Corrosion problems and solutions in CO_2 projects are covered by a rather extensive body of literature, which is cataloged in a report issued by NACE T-1-3 committee.[28] A report issued by the U.S. Department of Energy also provides an overview of CO_2 corrosion with an extensive list of references.[13]

Coatings

Coatings prevent corrosion by isolating the substrate metal from the corrosive environment. A few coatings, such as galvanizing or zinc-rich paints, have the additional effect of protecting the steel at pinholes, or holidays, in the coating by sacrificing themselves through galvanic action. The importance of surface preparation, which is the most critical step in any coating application, must be stressed in any discussion of coatings. Many good coatings systems have failed in service because they were applied over poorly cleaned substrate. In the case of internally coated tubulars, probably more than 75% of all "coating" failures are attributable to inadequate cleaning of the substrate. NACE (see TM-01-70) and the Steel Structures Painting Council (SSPC) both have visual comparator standards for surface preparation, which inspectors use to assure that the specified surface preparation is attained by the coating contractor.[26,6] The NACE standards are steel coupons sealed in plastic, while the SSPC provides photographs of

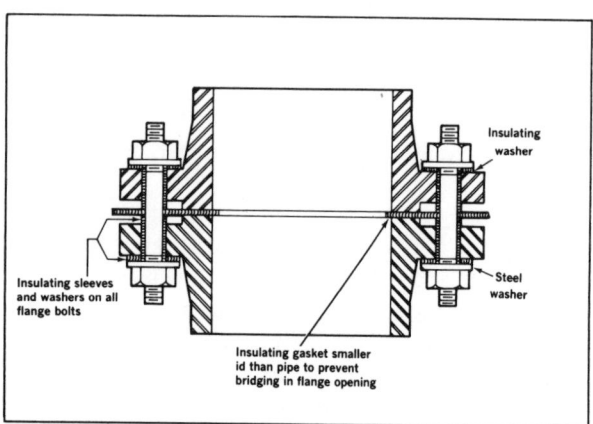

FIG. 10-9—*Cross section of insulating flange joint.*

various degrees of substrate cleanliness using different methods of surface preparation. Most coating suppliers have technical representatives who will help select a specification for surface preparation, as well as all other phases of coating application, based on the particular conditions at a specific work site and anticipated service environment of the coating. The corrosion or coating department of some companies have their own specifications. Some provide inspectors, particularly on large jobs, or can usually recommend a reliable third-party inspector.

Coatings in production operations can be conveniently categorized for discussion as (1) internal coatings and liners for tubulars, (2) immersion coatings such as used in oil field tanks, (3) external pipeline coatings, (4) atmospheric coatings or paints, and (5) ceramic and metallic coatings for smaller parts including those coatings used for purposes other than corrosion resistance, e.g. chrome hard-facing.

Coatings are seldom used as the sole method of preventing corrosion, except in the case of paints or where corrosion is rather mild. Inhibitors and cathodic protection are usually used in combination with coatings to approach 100% protection as closely as economically possible.

The concept that coatings with holidays are worse than no coating at all because corrosion is "accelerated" at holidays has been widely discussed and researched. This research has proven that in the absence of galvanic corrosion and stray or interference currents, organic coatings do *not* cause accelerated corrosion at holidays when applied in the normal thicknesses encountered in production operations. This is obvious when one considers the lack of conductivity of organic coatings and their inability to participate in the electrochemical corrosion reactions that occur with metals.

Internal Coatings and Liners for Tubulars—Plastic coatings, plastic liners, and cement lining for tubulars have found the most application in production and drilling operations. By preventing corrosion, less problems may be expected with plugging of injection and disposal systems from corrosion products such as iron sulfide and iron oxide. Friction losses with plastic coatings and liners are approximately one-half the friction loss in bare steel pipe.

The term "plastic-coated" is frequently used to refer to all internal coating systems. However, these systems, methods of application, and their suitability for different services vary considerably. Table 10-5 shows various service environments encountered with recommendations for a plastic coating system.

The epoxy thick film coatings are applied to a thickness of around 15 mils by heating the pipe to a temperature above the melting point of the epoxy powder and then introducing the powder into the hot pipe where it fuses to the pipe. Most of these systems are applied after a liquid primer has been sprayed on the pipe. After applying the coating, the pipe is baked to cross-link or "cure" the plastic.

The thin-film coatings are usually applied at less than 9 mils and as little as 5 mils in some cases. They are sprayed on as a liquid, usually in more than one coat and sometimes over a primer, with partial curing by baking between coats.

Surface preparation is critical to the survival of any coating system, and nowhere is this more important than in the case of internal plastic coating for tubulars. The most common method of surface preparation is accomplished by pulling a blast lance through the pipe as it rotates. Flint or other very hard, sharp abrasive is used, and multiple passes of the blast lance are usually required to remove hard mill scale from new pipe.

New coating systems are being developed each year. Most of these fall into one of the three generic categories shown in Table 10-5, but some companies have recently introduced a thick-film powder-applied phenolic or phenolic-like coating. These were developed partly in response to requests for a thick-film coating that would stand the severe environment of

TABLE 10-5
Internal Plastic Coating Systems for Tubulars

Service	Generic Coating System Recommendation
Salt water gathering disposal and injection systems	Epoxy thick film to 150°F*
Gas condensate wells	Phenolic thin film to 350°F*
Drill pipe	Epoxy modified phenolic thin film to 225°F*
CO_2 (alternated w/water) injection tubing and producing tubing and flow lines in CO_2 injection projects	Epoxy modified phenolic thin film to 225°F* or phenolic thin film to 300°F*

*Many manufacturers claim their coating systems are good to considerably higher temperatures than these. The temperatures shown are conservative; however, corrosion or coating specialists should be consulted about the specific service before using these systems at temperatures higher than shown here.

CO_2 tertiary recovery projects. The powder-applied epoxy thick films are generally unacceptable in this service. While these newer coatings hold some promise, they have some application problems and will not receive wide acceptance until favorable field experience is gained.

One other generic system, urethane, is used to some extent interchangeably with the epoxies and epoxy phenolics in water service. Urethane is not included in Table 10-5 because performance of urethane has been inconsistent.

Plastic liners are thin-walled (40–90 mils) tubes that are inserted into the steel pipe. A grout or cement is then pumped behind the liner. The liner may be PVC or polyethlene, or fiberglass for somewhat higher temperatures. Other more expensive plastics such as teflon are available, but their extra cost is not justified for oil field use since the more conventional liner materials are sufficiently resistant to normal oil field fluids. The plastic liner insertion technique is generally done in a plant. A similar technique is used with coiled polyethylene for in-place restoration of corroded gas lines.

Cement lining is accomplished by pumping a cement-pozzolon or cement-sand slurry into the pipe, then spinning it to cast a somewhat uniform layer of cement inside the pipe. The pipe is then set aside until the cement hardens. Although the use of plastic coatings is increasing, cement, at roughly one-half the cost of plastic coating, continues to be an economically attractive alternative where the reduction in inside diameter and increased surface roughness are not major objections.

The joint has always been a weak link in corrosion resistance for internally coated pipe. Threaded and coupled pipe should use couplings that have been coated in the stand-off area with the same plastic used on the pipe or with Ryton. Mechanically coupled, grooved end pipe (victaulic pipe) is suitable for lower pressure line pipe applications where the outside surface from the pin to the groove has been coated. Care must be taken to use masks when applying coating to tubing with special connections, such as Hydril or Atlas Bradford, to prevent the coating from being applied to metal-to-metal seal areas. Special interference fit connections have gained acceptance for use with internally coated line pipe. Most of these systems require a special end preparation where one end is slightly crimped and the other belled by hydraulic powered equipment. The joint is made by forcing the pin into the bell with field-joining hydraulic equipment. The joint strength is achieved by undersizing the inner diameter (ID) of the bell to approximately .050 in. less than the outer diameter (OD) of the pin.

The interference fit attained has passed laboratory tests to pressures equivalent to the minimum yield strength of API grade B line pipe (35,000 psi). Eight-inch pipe with one of these joint systems has been used successfully in water injection service to well over 2000 psi.

Some companies that apply internal coatings to tubular goods now have their own version of this interference fit joint or have a contract with a company that performs the end preparation and joining. The joint system is used quite extensively in the United States with smaller diameter (to 12-in.) uncoated pipe as well as coated pipe, since even with the additional cost for end preparation, installed cost is considerably less than for welded pipe. One company offers an interference joint system that uses a machined coupling with ID less than the pipe OD. This overcomes the objection to special end preparation, but the machining is expensive.

Special weld couplings are available for cement-lined or plastic-coated line pipe. In addition, one company offers a bell-and-sleeve arrangement. These systems isolate the lining or coating from the heat of welding. The sleeve used with the bell-and-sleeve system is internally coated, providing a theoretically continuous coating after welding. The belling operation distorts the bevel end, especially with seamless pipe, and makes proper welding very difficult with the bell-and-sleeve joint. Seamless pipe should always be rebeveled after belling and prior to coating.

Plastic-coated, plastic-lined, and cement-lined pipe all require special handling procedures. Cement-lining is especially vulnerable to damage from flexing since the cement has very little tensile strength and no bond to the steel. Some operators lay cement-lined pipe by joining it in the ditch rather than risk damage by lowering in longer segments. Bending internally coated or lined pipe can damage the coating. Bend radii are calculated for plastic coated line pipe such that the yield strength of the steel is not exceeded. For shorter radius bends, internally coated fittings or shop bends should be used whenever possible.

Plastic-coated tubing should be made up with the aid of a stabbing guide, a hinged funnel-like device that guides the pin into the box. Other precautions should be observed, such as leaving thread protectors on the pin until the joint is suspended above the rig floor by the elevators and using nonmetallic drift mandrels.

Tubing with a plastic liner uses an elastomer seal

ring in the coupling stand-off area between the pins. Make-up distance is important with this system. Insufficient make-up will not compress the ring and will allow fluid to migrate behind the seal and corrode the pins and coupling. Too much make-up will distort the ring and lead to the same type damage, as well as possibly creating an obstruction through which wire line tools will not pass. The lining company will furnish a representative on location to measure each coupling and pin to assure proper make-up. The representative should also drift each joint after make-up to assure the seal ring is not obstructing the tubing ID.

Caution must be exercised in acidizing through plastic coated tubing. Acidizing should be avoided entirely in cement-lined tubing. Table 10-6 gives some general guidelines for acidizing through plastic-coated tubing. Damage is cumulative if more than one acid job is done, and as the table shows, allowable exposure times decrease with increasing temperature.

Thin-film coatings are generally not purchased 100% holiday free. In fact, most applicators charge extra if thin-film coatings are specified to be 100% holiday free. Most specifications require that no more than 5–10% of the joints on a given order have holidays. Holidays are detected by pulling a wet sponge on a lance through the coated pipe, which is wired through a fixed voltage source and buzzer. The buzzer sounds when the electrolyte contacts any bare metal.

In spite of exercising great care in handling and installing internally coated pipe, damage inevitably occurs to the coating. All coatings are also vulnerable to wire-line damage. Thus inhibitors are frequently required to protect holidays and damage to pipe where bare substrate is exposed to the corrosive media.

The combination of installing internally coated pipe and using corrosion inhibitors may appear to be difficult to justify. Operators who utilize both usually use the following arguments as justification: (1) corrosion is so difficult to control that all practical means should be used to prevent it; (2) coating is cheap compared to a fishing job or junked hole where corrosion reduces the thickness of the pipe wall (with isolated pitting at holidays in the coating, there is usually sufficient tensile strength, even after the tubing is penetrated, to allow the tubing to be pulled without parting); (3) even though the coating does not provide 100% protection from corrosion on every joint, there will be enough well-coated joints that can be rerun after a pulling job to more than pay for the coating, and (4) hydraulic efficiency and injection water quality improvement are further justification for coating in addition to corrosion protection.

Nevertheless, many operators maintain that if inhibitor is required anyway, plastic coatings are a waste of money. Indeed, bad experiences with plastic coatings have caused some operators to avoid the use of plastic coating under any conditions. However, it has been proved that proper specifications, quality control, and inspection can overcome the problems that led to most of these bad experiences.

Immersion (Tank) Coatings—Tank coatings are used in salt water handling and crude storage vessels. The most widely used are catalyzed coal tar epoxies, straight epoxies or epoxy-phenolics, polyesters, and vinyls. Fiberglass flake fillers are sometimes used with the basic resins, particularly with polyesters. Fiberglass systems using successive layers of mat and resin are used for restoring corroded tanks, particularly tank bottoms. Most immersion coating systems used in production operations are two-part catalyst cured, although at least one company offers a baked-

TABLE 10-6
Guidelines for Acidizing Through Plastic-Coated Tubing

Type coating	Acid	Maximum Recommended Cumulative Exposure Time
Thin-film Phenolic	15% HCl (Inhibited)	80 hrs. at 100°F 40 hrs. at 150°F 12 hrs. at 200°F
Thin film Phenolic	12% HCL/3% HF* (Inhibited)	20 hrs. at 100°F 8 hrs. at 150°F **NR at 200°F or above
Thin-film Phenolic	9% HCL/6% HF* (Inhibited)	4 hrs. at 100°F **N.R. at 150°F or above

*TK-7, a modified thin-film phenolic manufactured by AMF Tuboscope is not recommended for exposure to HF at any temperature.
**Not Recommended

on epoxy. This baked-on system is applied to steel panels in the plant; then the panels are transported to the field and bolted together using rubber washers and special plastic covers to protect the bolts.

The inability to achieve a perfect holiday-free coating frequently makes it prudent to supplement the coating with other corrosion control techniques. It is common practice to use cathodic protection in conjunction with immersion coatings in water vessels. There are certainly valid arguments in favor of using cathodic protection alone in salt water vessels; one of the strongest is the frequently lower cost of cathodic protection when compared to coating plus cathodic protection.

Coatings remain the only viable method for protection from vapor space corrosion in crude and water handling vessels. Where tank bottoms or sediment are persistent problems, cathodic protection is difficult to achieve below such deposits.

Most organic coatings that cure at ambient temperature conditions are limited to maximum immersion service temperatures of 150°F or less. Some of these coatings have been used successfully at higher temperatures, but surface preparation (white metal) is so critical to coating performance at these higher temperatures that numerous failures have been reported. Extreme caution should be exercised when selecting a coating for greater than 150°F in immersion service, and inspection during application should be very comprehensive. The vessel portion of many heater treaters may be coated, but attempts at fire tube coating have been generally unsatisfactory.

While cathodic protection can usually prevent corrosion of holidays in immersion coatings, there is an undesirable side effect. Hydrogen is formed at the cathode in the electrolytic corrosion cell. Since the entire tank is the cathode when attached to the sacrificial or impressed current anodes of a cathodic protection system, hydrogen is formed at holidays in the coating. This hydrogen, forming at the juncture of bare metal and coating, as shown in Figure 10-10, tends to get under the coating and break the bond between substrate and coating. All coatings are far from equal in their resistance to this phenomenon, which is called cathodic disbondment. Many companies use this as one of their principal screening tests in selecting and approving coating systems for immersion service, as well as for screening coatings for external protection of pipelines. Such tests are performed in standardized salt solutions in test tanks or

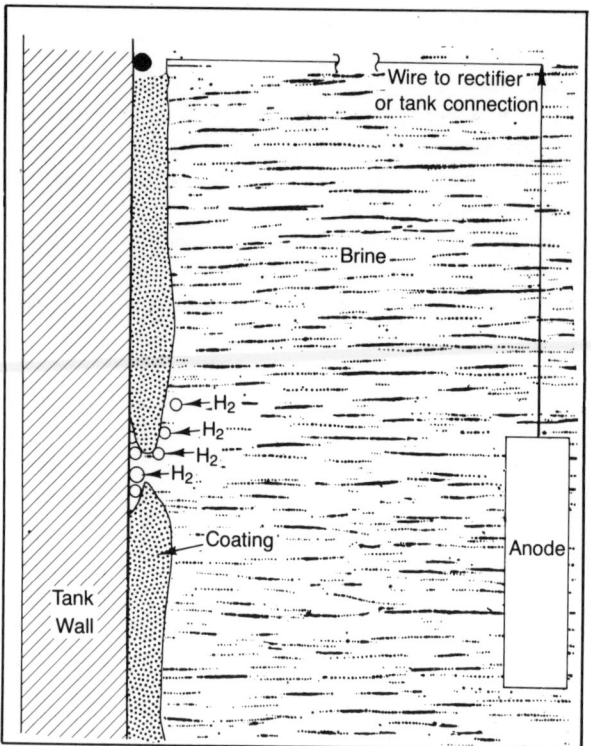

FIG. 10-10—*Cathodic disbondment.*

crocks. The test is commonly referred to as the salt crock test.

Coal-tar epoxy systems exhibit superior resistance to cathodic disbondment over a wide range of conditions. Because many contractors object so strenuously to handling and application problems associated with coal-tar, more and more operators are using the straight epoxies, epoxy-phenolics, or glass-filled polyester. Nevertheless, properly applied, coal-tar epoxy is usually an economical selection for this service.

In colder climates where it may be difficult, even with heaters, to attain the necessary minimum temperatures to properly cure the coal-tar or straight epoxies, which typically require 40° or 50°F, vinyl systems are applicable for water service.

External Pipeline Coatings—Several categories of coating are used to reduce corrosion of buried and subsea pipelines. The National Association of Pipe Coating Applicators (NAPCA) has published a manual describing the various plant-applied coatings and specifications for their application.[7]

Table 10-7 lists coatings that appear in the NAPCA manual. The manual also describes systems and pro-

TABLE 10-7
NAPCA Plant-applied Pipe Coating Systems

Coal Tar Enamel
Asphalt Enamel
Extruded Polyolefin Jacket
Extruded Polyolefin Wrap
Fusion-bonded Epoxy
Plant-applied Tape Wrap
Combination Corrosion-Weight (Mastic Type) Coating

cedures for protecting the field joint area and fittings with mastics, tapes, and heat shrink sleeves.

In addition to plant applied coating systems, over-the-ditch coatings, especially tapes, are frequently used. The selection of pipe coating and joint protection systems depends on such factors as (1) line size, (2) installation method, (3) terrain and soil conditions, (4) corrosiveness of the soil, (5) joining system, (6) operating temperature, and (7) economics and availability.

In production operations on land, coal-tar or asphalt enamels are used extensively to coat pipelines with tape to protect the joints. The less expensive over-the-ditch tapes have given adequate service over a wide range of applications. The fusion-bonded epoxy systems show much promise in testing.[2]

Offshore pipe-laying barges present a particularly severe test for external pipeline coatings. Tensioners used to prevent buckling of the pipe and straighteners used to straighten the pipe as it is uncoiled from the reel on a reel barge frequently cause mechanical damage to external coatings that must be repaired.

Cathodic protection is used frequently to protect buried and submerged pipelines at voids and holidays in the coating. Coating and cathodic protection of certain categories of pipelines and handling hazardous materials such as natural gas and natural gas liquids, are required by law in certain jurisdictions.

Although rural gas-gathering systems generally do not come under these regulations, many operators have found it economical and prudent to use external coatings and cathodic protection. Since pipelines are frequently expected to last several decades, even mildly corrosive soil conditions can cause deterioration over the long exposure time, leading to hazardous conditions, product loss, and costly repairs.

Additional information on pipeline coating systems may be found in "Control of Pipeline Corrosion" by A. W. Peabody.[9]

Paints—Paints generally serve two purposes. They improve the aesthetic appeal of the equipment, and they provide a barrier to atmospheric corrosion. Atmospheric corrosion varies greatly in severity. Corrosion rates in the continuously wetted splash zone of an offshore platform may exceed 100 mils per year. In desert regions, a bare piece of steel may last a hundred years with no significant weight loss. Selection of paint systems for these two regions would involve very different considerations.

Table 10-8 shows categories of paint systems and typical uses. Important considerations in selecting a paint system for production equipment include (1) method of surface preparation, (2) method of application, (spray or brush) (3) corrosiveness of atmosphere, (4) difficulty and cost of repainting, and (5) operating temperature.

Since a wide variation exists in the price of paints, a trade-off is frequently necessary in paint system life versus cost of repainting. Inspection during surface preparation and application is very important, since even the most carefully selected paint system can fail if applied improperly or applied to improperly cleaned

TABLE 10-8
Generic Paint Systems and Uses

System	Typical Use in Production Operations
Oil base, alkyd	House paint, rusty steel Tanks, production equipment where atmospheric corrosion is very mild
Chlorinated rubber	Marine equipment where fairly frequent maintenance is possible
Catalyzed epoxy	Offshore platforms and equipment where coating is expected to last several years
Vinyl	Offshore platforms and equipment where coating is expected to last and particularly where cold makes epoxy application difficult
Silicone	High temperature exhaust stacks on prime movers, heater-treaters, and similar applications
Urethane	Used as a thin topcoat where gloss and appearance, as well as corrosion resistance, are important
Zinc-rich	Used as a primer on offshore platforms and production equipment where atmospheric corrosion is severe

steel. Surface preparation becomes increasingly important with severity of service.

Metallic and Ceramic Coatings—Although corrosion may be a consideration where metallic or ceramic coatings are used, it is frequently secondary. Wear or erosion resistance is often the main concern where ceramic coatings are used, although, in the absence of such coatings, corrosion may combine with wear or erosion to lead to earlier failure than if wear or erosion is acting alone. When two surfaces rub together, the softer of the two will wear away. This is important to remember, whether drilling with hard-faced tool joints or pumping with sucker rods whose couplings have had a hard metal coating applied. The wear in both cases is not stopped, merely transferred to the other surface. In the first example, this means the casing contacting the tool joint, and in the second, the tubing contacting the sucker rod coupling are subject to continuing wear.

Metallic coatings have another inherent danger where the metal used in the coating is cathodic with respect to the substrate. An example is nickel or high-nickel alloys, which are sometimes used to coat downhole equipment for corrosion protection. Such coatings may be beneficial on some equipment for certain environments. However, where the equipment is immersed in a corrosive brine, such as in the case of an electric submersible pump in a high water-cut well, metallic coatings should not be used. Even though the coating may start holiday-free, it is difficult, if not impossible, to install downhole equipment without scraping off at least a little of the coating. If this happens with nickel on steel, a galvanic cell with a very unfavorable cathode/anode area ratio is created, and accelerated corrosion occurs.

Inhibition with Chemicals

Chemical inhibition is widely used to reduce corrosion. Inhibitors control corrosion in tanks, flow lines, tubing, well casing, waterflood equipment, and gas plants. There are two general types of inhibitors based on chemical composition:

1. Inorganic inhibitors, which include chromates, phosphates, nitrites, arsenic, and other chemicals.
2. Organic inhibitors, which include a wide variety of high molecular weight compounds.

Inorganic inhibitors are used in closed cooling systems, in high temperature acidizing and in the treatment of steel surfaces in preparation for painting. Sodium chromate is effective in closed fresh water systems, but it promotes pitting if used in insufficient concentrations in salt solutions. At least 5,000 ppm sodium chromate is required to prevent pitting in salt solutions.

Organic inhibitors have wide application in petroleum production. They provide an effective means for controlling corrosion in gas condensate wells and sour oil wells and in acidizing oil and gas wells. Their composition is often not available to the user. Therefore, the oil company has no reliable way to predict whether or not a particular inhibitor will be effective in a given application. It is necessary to depend on the supplier to recommend an inhibitor and then to determine, perhaps first in laboratory tests, the relative effectiveness of an inhibitor and whether it is compatible with the fluids in a specific well.

Most of the more effective chemicals used in oil and gas wells, service wells, and lease equipment are long-chain nitrogen compounds. With this type of inhibitor, a film formed on the wall of the pipe or vessel increases resistance to corrosion current. The film can be formed and maintained on metal surfaces by continuously adding the inhibitor to a flow stream, or it can be created by "slugging" the flow stream with a concentration of at least 1,000 ppm of inhibitor for one hour with frequent retreatments to maintain the film.

Film efficiency depends on inhibitor concentration and contact time with the metal surface. In slug treatments, the ability of the inhibitor to adhere to the metal determines the frequency of retreatments. Several techniques are used for application of inhibitors in oil wells and gas wells.

Gas Well Inhibition—CO_2 in the gas stream is the most usual cause of corrosion in gas wells. Corrosion usually occurs in the top portion of the tubing string and in surface flow lines where water vapor condenses on the walls of the pipe. However, if formation water is being produced, the corrosion can be scattered over the entire tubing string. Also, hydrogen sulfide can be present in gas wells. Corrosion will occur wherever water is present, either as condensation or water produced from the formation. If no water is present, no corrosion occurs.

In conjunction with an inhibition program, corrosion rates are monitored to evaluate the effectiveness of the chemical and the application method. Corrosion rate tests for monitoring include iron counts, corrosion coupons, hydrogen probes, caliper surveys, and other methods. If hydrogen sulfide is the major cause of corrosion, iron counts and caliper surveys are not suitable for rate measurements since the iron sulfide is insoluble, forms a scale on the tubing wall, and covers corrosion pits. If a tubing caliper is needed in wells with iron sulfide scale, tubing is usually acidized before making the survey.

There are four primary types of treatment for gas-well inhibition: (1) batch-down tubing, (2) tubing-displacement method, (3) continuous treatment down the casing-tubing annulus or through concentric, macaroni, or kill strings, and (4) nitrogen squeezes.

Batch-down Tubing for Gas Wells—The inhibitor is diluted with water, lease condensate, or other hydrocarbon such as diesel. The mixture is pumped into the tubing and the well shut-in long enough to allow the inhibitor to fall to bottom. One rule-of-thumb used to determine minimum shut-in time in nondeviated wells is one-half hour per thousand feet of tubing above a restriction (such as a storm choke) and one hour per thousand feet of tubing below the restriction plus one hour for each restriction. For example, the calculated minimum shut-in time for a ten-thousand foot well with a storm choke at 4800 ft would be: min shut-in time = $(4800 \div 1000) \times 1/2$ hour + $(5200 \div 1000) \times 1$ hour + 1 hour = 8.6 hours. Wells are frequently left shut-in for somewhat longer times where it is convenient or where the hole is deviated.

This treating method has the lowest cost per application, even if the carrier fluid must be purchased. It is preferred over the tubing displacement method for weaker wells because less hydrostatic pressure is imposed on the formation than when using tubing displacement, and the wells can unload the fluid more readily. Frequency of retreatment is based on corrosion monitoring or film-life instruments.

Tubing Displacement Method for Gas Wells—The tubing is filled completely with a solution of inhibitor and usually shut-in for a few hours. This is a more effective method than the batch method and provides a longer film life. It is more expensive if the carrier must be purchased and presents an unloading problem in low pressure wells. Corrosion rates and/or film life are monitored to determine the frequency of retreatment.

Continuous downhole treatments require either an open annulus, which is uncommon in gas well completions, a concentric or parallel kill string, or a macaroni string (small tubing string) to allow continuous injection of the inhibitor into the production tubing, or a side door injection valve from the casing-tubing annulus. Continuous injection, frequently at a rate of around one quart of inhibitor per million cubic feet of gas, has the advantages of (1) requiring no shut-in, (2) running no risk of killing the well, and (3) usually assuring more complete inhibition since the constant presence of inhibitor rapidly replaces any film breakdown.

One place where packer-less tubing completions are used, partly to enable continuous treatment, is in the very deep, hot, high pressure, sour gas wells of the Piney Woods fields in Mississippi. Here, a special oil/inhibitor mixture is pumped continuously into the annulus at a rate of several hundred bbl/day. Temperature, pressure, H_2S concentrations, and the extremely lean gas make conventional inhibitor treatments impossible in these wells. The oil/inhibitor mixture is separated from produced gas, processed to remove sulfur, and recycled into the annulus.

Macaroni strings may be used for continuous injection of inhibitor into the production tubing downhole. Full-size parallel kill strings, which are sometimes used to provide a means for killing a gas well in an emergency, are also a convenient means for continuous treating. The macaroni string may be almost any size. Systems are available that use continuous capillary tubing as small as .094 in. OD. This tubing, originally developed for use in monitoring bottom-hole pressures, is strapped to the outside of the production tubing and threaded into a special sub, which allows the inhibitor to enter the ID of the production tubing. The inhibitor must be well-filtered to avoid plugging of the small ID of the capillary tubing. The friction pressure drop may limit the selection of inhibitors or require dilution to reduce the viscosity of the inhibitor. Slightly larger ($1/8$–$1/4$ in OD) tubing reduces the likelihood of plugging caused by solids in the inhibitor without greatly increasing the cost of the system.

Nitrogen Squeezes for Gas Wells—Four barrels or more of a 25% solution of an oil-soluble inhibitor is atomized into a nitrogen gas stream by jet nozzles. The atomized slug of inhibitor and nitrogen gas is then displaced with nitrogen gas down the tubing and into the formation. Total tubing volume is over-displaced by about three equivalent barrels. Volume of nitrogen will depend on wellhead pressure—the higher the pressure the more nitrogen required due to compressibility of the gas.

The well is shut-in about 12 hours, and then the nitrogen gas is blown to the atmosphere prior to returning the well to production. This treatment has the greatest cost per application but may be economical for weaker wells that cannot unload small fluid columns imposed by the batch method. Corrosion rates are monitored to determine the frequency of treatment.

Inhibitor selection is very important regardless of which method of treatment is used. It must be able to withstand bottom-hole temperature, pressure, and produced fluid composition without vaporizing or partially vaporizing and gunking. And it must not be detrimental to flow properties of the rock, especially when treating is by "squeezing" into the formation. Laboratory or field tests for solubility and compatibility with well fluids, including possible emulsion formation, should always be conducted before using an inhibitor in a gas well. In very high pressure and temperature situations, inhibitor phase behavior should be investigated. Before squeezing any inhibitor into a formation, the possibility of oil-wetting or developing an emulsion in the formation should be determined through core tests. Finally, not all inhibitors exhibit the same film persistency. If other considerations are equal between various inhibitors, the inhibitor that will allow the greatest time interval between batch treatments should be used for batch treating. Laboratory procedures may be used to screen inhibitors for film persistency at bottom-hole conditions.

Oil Well Inhibition—Corrosion in rod-pumped wells is directly related to water-cut of produced fluid. Wells producing more than 30% water frequently have serious corrosion problems. If hydrogen sulfide is present, corrosion is more severe.

When used properly, corrosion inhibitors are about 90% effective in rod-pumped wells. The chart in Figure 10-11 presents a preferred approach to select the method for corrosion inhibition in oil wells.

There are six methods for oil-well inhibition:

1. *Batch flush* is applicable to low-volume low-pressure wells producing less than 100 bbl of fluid per day, and with a working fluid level of 700 ft or less above the pump. With this treatment, one to two gallons of inhibitor is dumped down the casing-tubing annulus and flushed with ½ bbl of fluid per 1,000 ft of casing.

This procedure exposes the well equipment to an inhibitor concentration of at least 1,000 ppm in the produced fluid for one hour or longer. A high degree of protection usually results.

2. The *circulation method* is used in moderate to low working fluid level wells where produced vol-

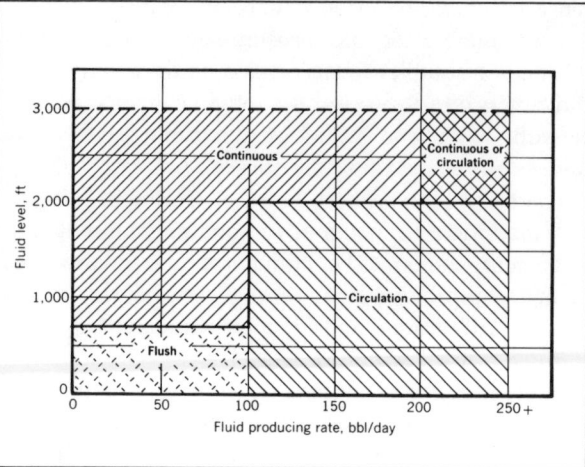

FIG. 10-11—*Selecting the method for corrosion inhibition in oil wells.*

umes are greater than 50 bbl/day. With this method a slug of three to five gallons of inhibitor is injected down the casing-tubing annulus, and then all produced fluid is diverted down the annulus for some time period. For an effective treatment, the inhibitor concentration at the wellhead should be up to 1,000 ppm for one-hour.

The required volume of inhibitor and circulation time is related to the working fluid level, casing-tubing annulus volume, and pump displacement. Larger volumes of inhibitor are required for high fluid level wells because the inhibitor slug is diluted as it falls through fluid in the annulus. The circulating time to move the inhibitor slug to the pump suction will be affected by fluid level, casing-tubing annulus volume, pump displacement, and the fall rate through the vapor zone.

3. *Continuous injection* is frequently employed in high–working fluid level wells. With this system, a small quantity of inhibitor is injected continuously into the casing-tubing annulus along with a side stream flush of one to two barrels of produced fluid per day. The produced fluid should have a concentration of 25–50 ppm of inhibitor.

4. *Squeeze injection* may be employed if production tubing is set on a packer. A quantity of inhibitor is selected to yield an average concentration of 20 ppm in the water produced during the life of the treatment, which may range from six weeks to six months. If treatment life is less than six weeks, squeeze treatment is probably not economical.

The well is shut-in for 24 hours to permit the inhibitor to adsorb on the formation rock. If the treatment is successful, the inhibitor is slowly removed from the formation rock by produced fluid and returns to the well to protect the tubing, pump, rods, and flow line for several months. A large percentage of the chemical invariably returns during the first few days. Chemicals must be selected with great care to prevent formation damage.

5. *The drop method* involves the dropping of a weighted material containing corrosion inhibitor down the well. The weighted corrosion inhibitor falls through the fluid column to the bottom, and the inhibitor is released slowly to provide some protection to the downhole equipment. The rate of release is difficult to predict and impossible to control. Weighted inhibitors are used only in very special situations where no other method of treatment is available.

6. *Automatic chemical injectors* provide automatic treatment, including proper chemical additions where the batch flush or circulating methods are used to treat pumping wells. The injector saves labor if the treatment frequency is greater than once each week and reduces the chance of missed or delayed treatment. Maintenance of this equipment is important because malfunctions may result in a well being without treatment for a considerable period of time. This could cause irreparable damage, particularly to rod strings.

Each application method has its advantages and disadvantages. The operator should consider all data on each well prior to selecting an inhibitor, the method of treatment, and the amount of inhibitor required. Safety and environmental precautions outlined in the material safety data sheet (available from the inhibitor supplier) should be followed.

Removal of Corrosive Gases

Many sources of water used in waterflooding contain up to 8 ppm oxygen. With 1 ppm oxygen, steel or iron corrodes several times as fast as in oxygen-free water. Oxygen causes growth of bacteria, which further aggravates corrosion and increases solids content of the water. Also, mixing oxygen-containing water with oil field waters containing either dissolved iron or hydrogen sulfide causes precipitation of iron oxide or iron sulfide. Oxygen is generally removed if present. If there is no oxygen present in a water system, exclusion is the best approach.

Three methods that are economically feasible for the removal of oxygen are chemical scavengers, vacuum deaeration, and counter flow gas stripping.

Chemical Scavengers—Sodium sulfite, ammonium bisulfite, or sulfur dioxide may be added to water to react and remove oxygen. Sodium sulfite (Na_2SO_3) and ammonium bisulfite (NH_4HSO_3) are used primarily in low-volume systems, and those containing a low concentration of oxygen such as the effluent from a vacuum deaeration tower. The reactions are:

$$Na_2SO_3 + \tfrac{1}{2}O_2 \rightarrow Na_2SO_4$$

$$NH_4HSO_3 + \tfrac{1}{2}O_2 \rightarrow NH_4HSO_4$$

Approximately 8 ppm of Na_2SO_3 or NH_4HSO_3 is recommended per ppm of oxygen removed. A 10% excess is also required to drive the reaction to completion, plus a catalyst such as cobalt chloride or some other metallic ion to speed the reaction. Some waters contain sufficient dissolved metal ions, so no additional catalyst is required. This should be determined by laboratory tests with the scavenger and samples of the water being treated. In any case, the catalyst should be added at the scavenger injection point to prevent reaction of the scavenger with air prior to injection. Since hydrogen sulfide precipitates all the metallic catalyst ions, an excess of catalyst may be required in sour waters.

Sulfur dioxide (SO_2) is an economical scavenger. The reaction is $SO_2 + H_2O + \tfrac{1}{2}O_2 \rightarrow H_2SO_4$ (acid) and requires approximately 4 ppm SO_2 per ppm O_2 removed. A 10% excess of SO_2 and a catalyst such as cobalt chloride are required.

Two sources of SO_2 are used: (1) bottled liquid SO_2 and (2) SO_2 gas generated by burning sulfur. Bottled liquid SO_2 is economical in treating small systems with a low concentration of oxygen. The cost of liquid SO_2 is about half that of sodium sulfite; however, more equipment is required for application of SO_2.

The sulfur burner is economical to produce SO_2 gas where larger volumes of water are treated. SO_2 gas is produced by burning sulfur. Then the gas is dissolved in a sidestream of the water to be treated and pumped through a packed column of the contact tower and back into the line. This type of plant experiences severe corrosion and operational problems and requires more labor for operation.

Possible disadvantages of the use of sulfur dioxide are:

1. SO_2 reacts with oxygen to produce acids in so-

lution. If there is insufficient bicarbonate alkalinity in the water to react and consume the acid, a corrosive condition exists. As a result of the low pH, SO_2 does not react readily with oxygen.

2. If the water contains barium, SO_2 will react to form $BaSO_4$, an insoluble scale.

3. If the water contains a high percent of $CaSO_4$, the additional sulfate formed may cause scale precipitation.

4. The SO_2 treatment may not be effective in waters containing hydrogen sulfide.

All of the disadvantages of the use of SO_2 except item 1 also apply to the other chemical scavengers.

Vacuum Deaeration—Creating a vacuum in a packed tower and passing water over the packing will reduce the oxygen content in water. The low pressure and the small amount of oxygen in vapor contacting the water causes the dissolved oxygen to bubble out of solution. The vacuum pump pulls the oxygen from the top of the tower, together with water vapor and other gases. Several passes are required to reduce oxygen to less than 0.1 ppm.

Vacuum deaeration is applicable where chemical treatment is uneconomical or where the addition of sulfur dioxide or sodium sulfite would form barium or calcium scale. CO_2 will be removed from the water if present, and the accompanying pH change may result in scale that can be controlled with scale inhibitors. CO_2 must be considered when sizing the vacuum deaeration system.

Gas Stripping—Natural gas in a counter-flow stripping column causes an oxygen-free environment, which permits dissolved oxygen to escape from water. Either a packed column or tray-type column can be used. The tray-type column is preferred where fouling with suspended matter or bacterial slime is a problem. This system is usually designed to use not more than two cubic feet of gas per barrel of water being stripped of oxygen.

Economics of the gas stripping column is almost entirely a function of the cost of the gas unless the stripping gas can be compressed and sold. Gas stripping requires a natural gas that is free of oxygen and hydrogen sulfide, containing less than 2% CO_2.

Combination Vacuum Deaeration and Gas Stripping—Frank[4] in 1972 reported an efficient removal of oxygen from 40,000 bbl of source water per day with vacuum deaeration supplemented by hydrocarbon gas stripping. Through the use of a vacuum of 1.0 in Hg absolute and 0.1 cu ft of sweet hydrocarbon gas per barrel of water, oxygen content was reduced from 4.7 ppm to 0.05 ppm. Corrosion rate on steel was reduced from 14 mpy to 1.6 mpy.

NACE has issued a recommended practice (RP-02-78) for design and operation of oxygen removal systems. Prevention of corrosion due to acid gases, carbon dioxide, and hydrogen sulfide is usually cheaper than acid gas removal.

Cathodic Protection

Corrosion damage occurs where electrical currents discharge from metal into an electrolyte, for example, from pipe into soil, tank wall into salt water, or casing wall into the formation. If an outside electrical power source (see Fig. 10-12) is used to impose a counter current with sufficient voltage to overpower the voltage of the corrosion cell, corrosion current stops flowing and corrosion stops.

This technique is called cathodic protection—all of the steel becomes a cathode. All previously anodic areas are suppressed as long as adequate current is applied.

As current flows onto steel, hydrogen is evolved on the surface and creates a counter voltage called polarization. This causes the current to spread out over the steel surface and protect it. Polarization can be measured indirectly using a copper-copper sulfate cell as a reference electrode. This is accomplished by placing the copper-copper sulfate cell in the electrolyte (water or soil) and contacting the steel surface through a voltmeter, as shown in Figure 10-13. A voltage reading between -0.85 and -1.2 usually in-

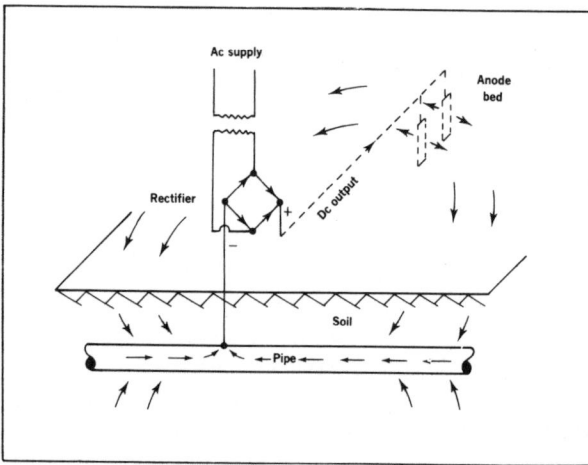

FIG. 10-12—*Cathodic protection using outside power source.*

Corrosion Control

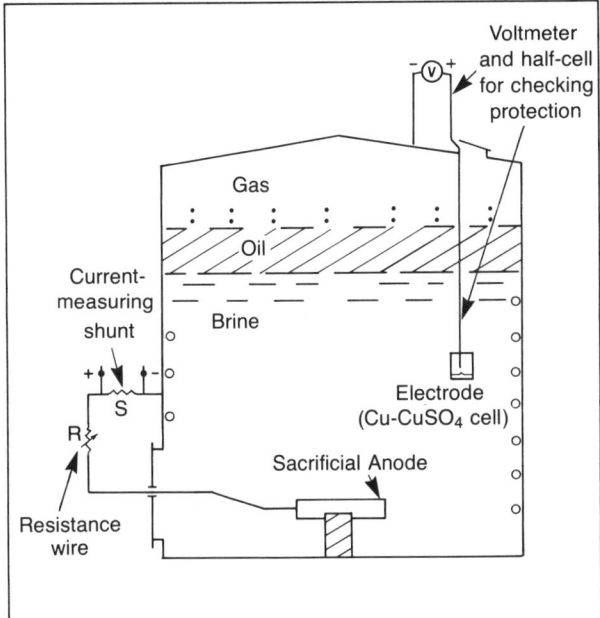

FIG. 10-13—*Cathodic protection of interior of saltwater tank with galvanic anode.*

dicates adequate protection.

Protection of the inside of a 1,000-bbl saltwater tank through the use of magnesium or aluminum anodes is illustrated in Figure 10-13. These "sacrificial" anodes require no external power source. Their life is a function of their weight and the amount of current required to protect the tank. Large aluminum anodes may last several years in a coated tank. Magnesium, while providing higher current output, may require replacement as frequently as every three months.

Design of Cathodic Protection—A cathodic protection system should be designed so that the output of the anodes will provide the minimum-required current density to all parts of the protected structure. On coated structures, the required current cannot be predicted exactly because the exposed area is not known. In these cases, the corrosion engineer must use voltage measurements to fix current requirements and the location of anodes or an estimate based on prior experience with similar coated structures.

On some structures, such as saltwater tanks, it is easy to obtain fairly uniform distribution of the current from an anode near the center of the tank. Protection of the interior of bare pipelines is practically impossible except on very large short lines, because current will extend only about three pipe diameters from the anode. Cathodic protection is frequently used to supplement coating. In the case of water handling vessels, coating is usually more expensive than cathodic protection. However, sacrificial anodes may be readily used to protect coated tanks. These anodes require almost no maintenance and provide uninterrupted protection. Thus the combination of coating and sacrificial anodes is a good choice over the more complicated rectifier systems used in bare tanks where maintenance is not always reliable.

For coated pipe, cathodic protection is needed to prevent leaks from developing at holidays formed in the coating during pipe installation. Cathodic protection will extend for great distances if the average resistance of the coating is high.

Cathodic protection requires a direct current source. This may be a rectifier that converts AC to DC, a thermoelectric generator, solar panel, wind generator, or a galvanic anode that provides its own potential and current flow with no outside current source. NACE has published a number of recommended practices that are helpful in deciding which type of cathodic protection to use. Table 10-9 lists these documents. Committee work is in progress on additional recommended practices for cathodic protection of other oil field equipment, most notably for well casings.

Cathodic protection is so universally used in production operations that many companies have cathodic protection specialists. Numerous consultants and contractors are in business strictly to design, install, and maintain cathodic protection systems for production and pipeline equipment.

External Protection of Well Casings—The first step in the external protection of well casing is cementing of centralized casing through potentially corrosive saltwater zones. This frequently requires stage cementing since the corrosive zone may be far above the productive interval and also may be a highly permeable "thief" zone. While cathodic protection has been successful with no cement behind the casing

TABLE 10-9
NACE Recommended Practices for Cathodic Protection

NACE Document	Equipment Covered
RP-01-69	Pipelines
RP-05-72	Deep Ground beds
RP-05-75	Oil Treating Vessels
RP-06-75	Offshore Steel Pipelines
RP-01-76	Offshore Platforms

across the corrosive zone, it is usually desirable to obtain the added protection of the cement sheath rather than rely on cathodic protection alone.

Measurement of the current in the flow line at the wellhead has shown that electric current is nearly always flowing into the wellhead. This current flows down the casing and must leave the casing and enter the formations at some depth. Most of the current leaves the casing in low resistivity, permeable saltwater zones.

Figure 10-14 shows a typical galvanic circuit between the casing and flow line. If this current is relatively small, 0.1 ampere or less, and if it leaves the casing over a large area, considerable time is required to cause casing leaks. However, if the current is one ampere or more and leaves the casing in a few pits, leaks will soon occur.

The current from the flow line into a well can be stopped at a small cost by placing an insulating joint in the flow line at the wellhead. As a general rule,

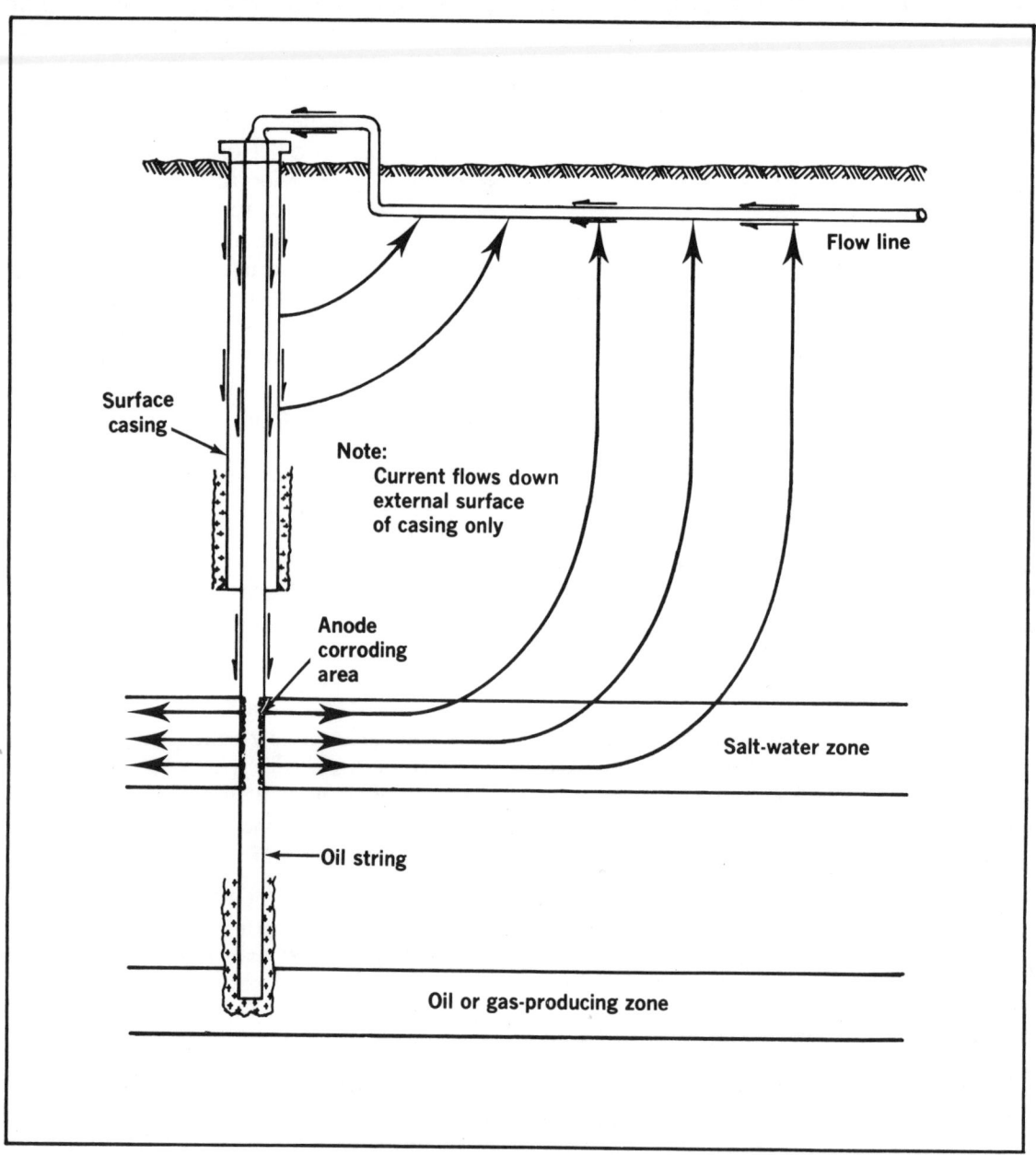

FIG. 10-14—*Current flow between flowline and well casing.*

a flow line insulator should always be installed at the wellhead.

Having taken these precautions, the next question is—should cathodic protection be applied to the wells of some particular field? The question is not easily answered, but these factors should be considered:

1. History of casing leaks in this area or similar nearby fields.

2. What happens when a leak occurs? Does water, gas, or mud flow into casing or into low pressure zones near the surface?

3. The probable life of the well, including prospects for enhanced recovery.

4. The cost of cathodic protection.

5. The probable effectiveness of cathodic protection if installed.

6. Method and cost of repairing casing leaks. The cost and effectiveness of extensive workovers with possible oil string replacement and recementing to the surface, or loss of a well if corrosion is extremely severe.

7. The existence of other cathodic protection systems in the area.

Cathodic Protection Installation—A typical cathodic protection installation for a well casing is shown in Figure 10-15, which consists of a rectifier connected to an electric power source and a ground bed. If electric power is not available at the well,

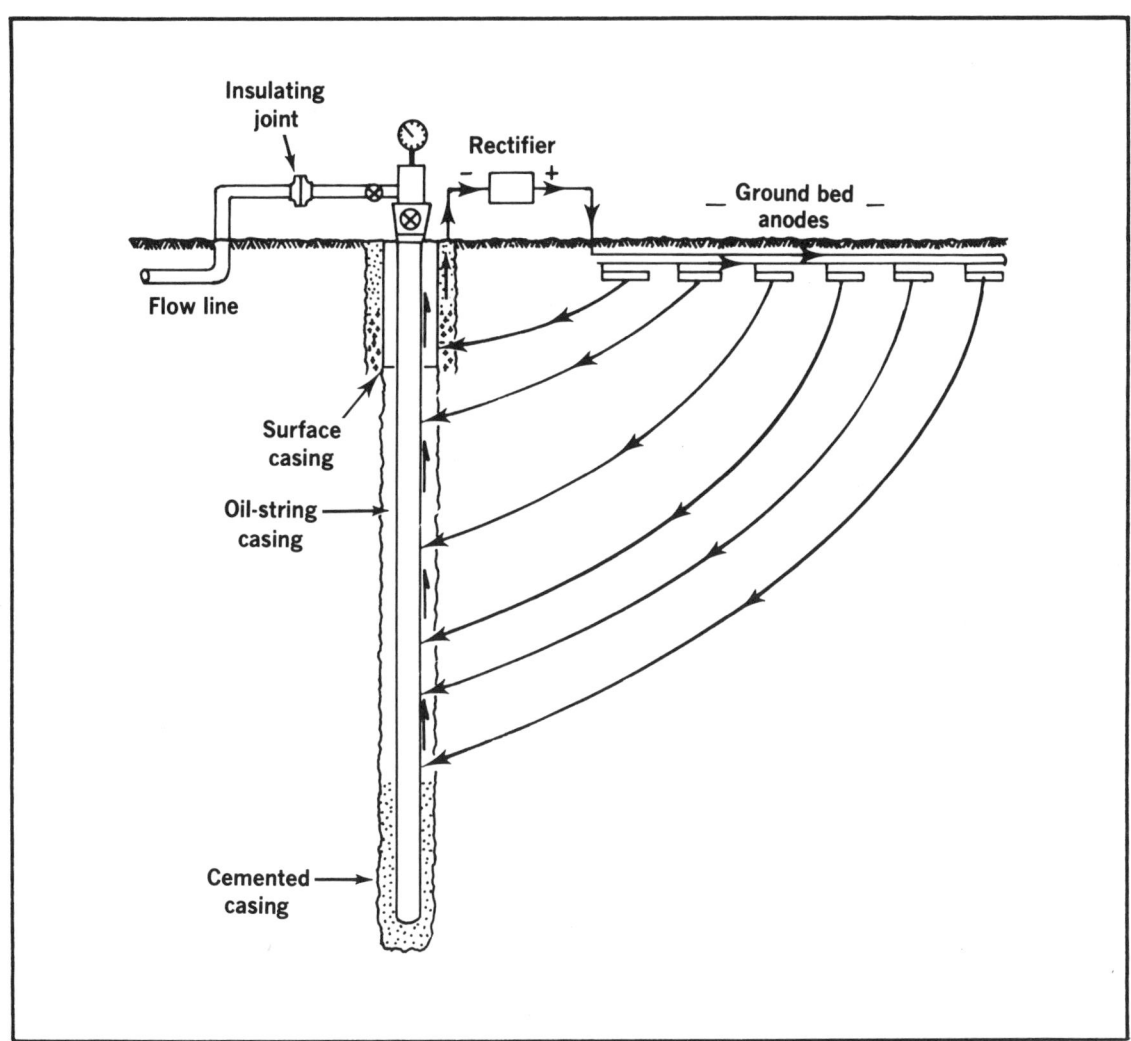

FIG. 10-15—*Typical cathodic-protection installation for a well casing.*

power distribution lines may have to be built or a small engine generator unit used as a source of electric power. Solar or thermoelectric generators may be considered in some situations.

On shallow wells with 4½ or 5½ in. production casing, it may be preferable to use sacrificial magnesium anodes and thereby eliminate the need for an external power source. Estimated actual consumption of sacrificial anodes in soil:

Zinc—26 lb/ampere-year
Magnesium—17 to 20 lb/ampere-year

If the flow-line coating has high resistance, the flow lines can be used to distribute the low–voltage direct current to the wells from a large rectifier and ground bed at the tank battery. Required current will vary from ½ amperes for 1,500 ft of 4½ in. casing to as much as 20 amperes for multicased deep wells. Surface casing should be well-cemented where cathodic protection is to be used, or it will "rob" the current intended for the production string.

The voltage of the rectifier can be adjusted as required, depending on the needed current and the resistance of the ground bed.

The ground bed should be constructed to provide the lowest possible resistance to ground and the longest possible life. It is the most important part of the installation because most of the rectifier power is used to overcome electrical resistance at the ground bed. It should be placed about 100 ft from the wellhead and removed as far as possible from the flow line and other pipe lines. If the ground bed is much nearer the well than 100 ft, more current will usually enter the casing at shallow depths. To protect the lower part of the casing, the upper part of the casing must be overprotected. This does no harm to the casing, but it wastes electric current. The resistance of the ground bed will be proportional to the resistivity of the soil in which it is buried, so a location having the lowest possible resistivity should be chosen. This is usually in an old mud pit at a depth that is always moist. As the ground bed size is increased, electrical resistance is decreased, but installation cost is increased. There is usually an economic balance between the size of

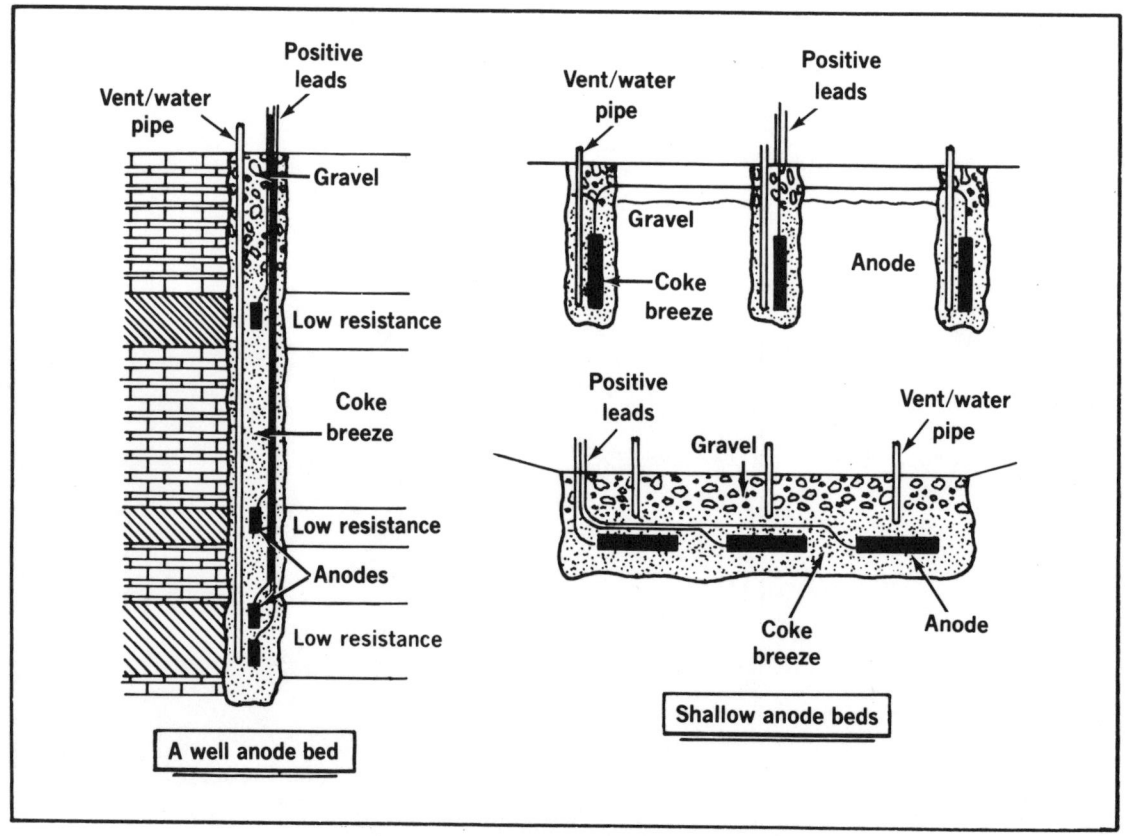

FIG. 10-16—*Ground bed for anodes.*

Corrosion Control

the ground bed and the cost of electric power.

If the ground bed is constructed of scrap steel pipe, the steel will be consumed at the rate of about 20 lb per ampere year so it would soon have to be replaced. Figure 10-16 shows a cross section of the trench of a more permanent ground bed. The anodes that are connected to the copper cable are either graphite, Duriron (14% silicon cast iron) rods about 2–3 in. in diameter and 4–6 ft long, or prepackaged platinum-plated rods. Duriron corrodes at less than one pound per ampere year, and platinum is consumed at a rate of only a few micrograms per ampere year. The anodes are imbedded in coke breeze, which makes a larger and more permanent ground bed at small additional cost. The coke is consumed rather than the graphite or Duriron. Anode selection is dependent on cost, life required, ground bed configuration, and current output.

Figure 10-16 depicts both the shallow ground bed and the deep well-type ground bed. A single deep ground bed is frequently used to protect as many as three oil, gas, or injection wells. In some very dry or desert areas, anode wells are absolutely essential. Some companies offer a deep ground bed that features a retrievable anode string, allowing the well to be used again when the anodes are depleted.

NonMetallic Materials

Where operating temperatures and pressures allow, plastic pipe and plastic tanks are frequently used to eliminate corrosion. Fiberglass reinforced plastics (FRP) are replacing steel in many environments where only steel was available to handle the stresses a few years ago. Fiberglass-reinforced polyester sucker rods are available with ultimate strength exceeding that of some steel rods, although the fiberglass must be used at a lower percentage of ultimate strength than steel.[11] Careful selection of plastics can often result in a system that costs less than steel initially and is virtually corrosion-free for its entire life.

Nonmetallic Pipe—Three major categories of nonmetallic pipe are in common use in production operations today. Table 10-10 shows these three categories and briefly outlines their limitations. API standards for the various categories of plastic pipe are helpful in designing, ordering, quality-control during manufacture, and installation of plastic pipe systems.

TABLE 10-10
Nonmetallic Pipe and Its Limitations

Thermoplastics	Maximum Temperature °F	Maximum Operating Hoop Stress (psi)	Joining Systems
PVC	150[1,2]	4,000[6,7]	SW, M, T[3]
Polyethylene	100[1]	625[6,7]	HW, M
Thermosetting			
FRP (Polyester)	180[4]	4,000–16,000[5,6]	AB, T, M
FRP (Epoxy)	180[4]	4,000–16,000[5,6]	AB, T, M
Cement-Asbestos[10] (Transite)	200	—[8]	M

[1]Maximum operating stress must be derated as temperature increases over 73 F.
[2]Chlorinated PVC (CPVC) may be used to 180 F but may not be available in large sizes.
[3]Schedule 80 and heavier may be threaded.
[4]Some manufacturers claim to have products good to higher than 180 F, but derating factors should be used at over 180 F. Some systems are good only to 150 F.
[5]Allowable stress level is dependent on the pipe grade; consult API standard 5LR for minimum grade property requirements.
[6]The hoop stress formula may be used to calculate design pressure (P) given allowable stress (S), wall thickness (t), and outside diameter of pipe (OD).

$$P = \frac{2St}{OD - t}$$

[7]Plastic pipe used in natural gas service is limited as to allowable operating pressure in some jurisdictions—notably all lines covered by Materials Transportation Bureau regulations in the United States.
[8]All sizes generally are limited to 200 psi operating pressure or less—consult manufacturer.
[9]The joint or fittings may be the limiting factor in determining design pressure—consult manufacturer for joint and fitting system.
[10]Plastic-coated transite should be used for handling water with pH less than 5.5.
Joining system code: HW, Heat weld or fusion; AB, Adhesive bond; M, Mechanical; SW, Solvent weld; T, Threaded.

TABLE 10-11
API Standards for Plastic Pipe

API Standard	Product or Subject Covered
Spec 5LE	Polyethylene Line Pipe
Spec 5LR	Fiberglass Reinforced Line Pipe
RP 5L4	Care and Use of Fiberglass Line Pipe
Spec 5LP	PVC and CPVC Line Pipe
Spec 5AR	Fiberglass Reinforced Tubing and Casing

Table 10-11 lists applicable API standards.

These API specifications and recommended practices cover many aspects of ordering, manufacturing processes, manufacturing materials, minimum physical properties, quality control tests, physical dimensions, hydrostatic design, end finish, marking, handling, installation, and leak and proof testing. They are invaluable references for any user or potential user of plastic pipe.

FRP Tanks—Fiberglass-reinforced tanks and gun barrels range in size from small chemical tanks of 500 gal or less to 500 bbl or larger. The glass/resin ratio must be controlled in the inner layers to yield a resin-rich, leak-proof layer at least 0.10 in. thick containing no more than 30% glass. A ten to one safety factor is usually specified between maximum operating hoop stress and ultimate strength of the laminate. Since hoop stress increases with diameter and height of the tank, shell thickness will depend on the tank size and tensile strength of the laminate. The most economical construction frequently includes a tapered shell, i.e., one that is increasingly thick toward the bottom. The U.S. Department of Commerce has issued a standard (PS 15-69), which may be used in qualifying manufactureres of fiberglass tanks. Hundreds of small companies have entered the fiberglass tank business in the U.S. because very little capital or space is required to make fiberglass tanks by hand lay-up procedures. Under these conditions, the quality of fiberglass tanks can be expected to vary greatly, and manufacturers should be investigated thoroughly. A reputable manufacturer will normally submit samples for qualification under PS 15-69 standards. Insurance requirements and local safety rules must be investigated before investigated before using fiberglass tanks to store or handle flammable fluids.

FRP Sucker Rods—Fiberglass sucker rods (at least the rod body) are immune to corrosion. However, their present cost and the 25–35% of the string that must be steel to allow the rods to "fall" does not justify their use where reduction of corrosion failures is their only expected benefit. FRP sucker rods may be justified by increased production resulting from lighter polished-rod loads and pump "over travel" where the well is not producing to its full capability and the load limit has been reached on the existing pumping unit. The lighter loads may also reduce lifting costs per barrel, but this is not sufficient justification for the present higher cost of fiberglass sucker rods.

Increasing steel costs or reducing cost for the fiberglass rods through economies of scale or technological breakthrough could change the outlook for fiberglass sucker rods. For the present, however, their application appears limited.

REFERENCES

1. Adams, Gene H., and Rowe, Hunter G.: "Slaughter Estate Unit CO_2 Pilot–surface and Downhole Equipment Construction and Operation in the Presence of Hydrogen Sulfide Gas," SPE 8830, Joint SPE/DOE Symposium on Enhanced Oil Recovery, Tulsa, Oklahoma, April 1980.

2. Choate, Leonard C.: "New Coating Developments, Problems, and Trends in the Pipeline Industry," Paper No. 34, presented at the NACE 1974 meeting in Chicago, Illinois.

3. Fontana, M. G., and Greene, N. D.: "Corrosion Engineering," 2nd ed., McGraw-Hill.

4. Frank, W. J.: "Efficient Removal of Oxygen in a Waterflood by Vacuum Deaeration," SPE 4064, October 1972.

5. Heinrichs, H. J., Ingram, W. O., and Schellenberger, B. G.: "Cathodic Protection Requirements for Well Casing," Petroleum Society of CIM, June 1977.

6. Keane, John D., Ed.: *Steel Structures Painting Manual*, Volume 2–Systems and Specifications, Second Edition, Steel Structures Painting Council, Pittsburgh, Pennsylvania, 1964.

7. "National Association of Pipe Coating Applicators Specifications and Plant Coating Guide," National Association of Pipe Coating Applicators, Shreveport, La., 1979.

8. Newton, L. E., Jr. and McClay, R. A.: "Corrosional and Operational Problems, CO_2 Project, Sacroc Unit," SPE 6391, Permian Basin Oil and Gas Recovery Conference of SPE of AIME, Midland, Texas, March 1977.

9. Peabody, A. W., "Control of Pipeline Corrosion," National Association of Corrosion Engineers, December 1967.

10. "Practical Tips in Trouble Shooting Pumping Wells," *Petroleum Engineer,* February 1969, pp. 60–66.

11. Saul, Harry E. and Detterick, Jerry A.: "Utilization of Fiberglass Sucker Rods, *J. Pet. Tech.,* August 1980.

12. Schremp, F. W. and Roberson, G. R.: "Effect of Supercritical Carbon Dioxide (CO_2) on Construction Materials," *Society of Petroleum Engineers Journal,* June 1975, p. 227.

13. Sum X Corporation: "Corrosion Due to Use of Carbon Dioxide for Enhanced Oil Recovery," Final Report of Work performed for the U.S. Department of Energy under

contract DE-AC21-78MCO8442, September 1979.

14. Uhlig, H. H.: *Corrosion and Corrosion Control,* 2nd ed., New York: John Wiley and Sons, 1971.

15. Wheeler, D. W., and Weinbrandt, R. W.: "Secondary and Tertiary Recovery with Sea Water and Produced Water in the Huntington Beach Field," SPE 7464, Oct. 1978.

16. Wright, C. C.: "Applying Instantaneous Corrosion Rate Measurements of Waterflood Corrosion Control," *J. Pet. Tech.,* March 1965, p. 269.

National Association of Corrosion Engineers, Recommended Practices (RP) Material Requirements (MR), and State-of-the-Art documents. Available from NACE Publications Dept., P.O. Box 218340, Houston, Texas 77218.

17. MR-01-75 (1980 Revision), "Sulfide Stress Cracking Resistant Metallic Material for Oil Field Equipment."

18. MR-01-76, "Metallic Materials for Sucker Rod Pumps for Hydrogen Sulfide Environments."

19. RP-01-69 (1976 Revision), "Control of External Corrosion on Underground or Submerged Metallic Piping Systems."

20. RP-05-72, "Design, Installation, Operation and Maintenance of Impressed Current Deep Groundbeds."

21. RP-04-75, "Selection of Metallic Materials to be Used in All Phases of Water Handling for Injection into Oil Bearing Formations."

22. RP-05-75, "Design, Installation, Operation and Maintenance of Internal Cathodic Protection Systems in Oil Treating Vessels."

23. RP-06-75, "Control of Corrosion on Offshore Steel Pipelines."

24. RP-01-76, "Control of Corrosion on Steel, Fixed Offshore Platforms Associated with Petroleum Production."

25. RP-02-78, "Design and Operation of Stripping Columns for Removal of Oxygen from Water."

26. TM-01-70, "Visual Standard for Surface of New Steel Airblast Cleaned with Sand Abrasive."

27. Technical Committee State-of-the-Art Report, "Monitoring Internal Corrosion with Hydrogen Probes," 1981.

28. T-1-3 Committee Report, "CO_2 Corrosion Literature Survey–Selected Papers, Abstracts, and References," 1982.

API Recommended Practices and Specifications, Available from American Petroleum Institute, Production Department, 211 North Ervay, Suite 1700, Dallas, Texas 75201.

29. RP 11BR, "Care and Handling of Sucker Rods," Fifth Edition, 1969 (Plus supplement 1-1973).

30. Spec SAR, "Specification for Reinforced Thermosetting Resin Casing and Tubing, (Tentative) First Edition, 1976 (Plus supplement 1-1976).

31. Spec 5LE, "Specification for Polyethylene Line Pipe, (Tentative) Second Edition, 1976.

32. Spec 5LP, "Specification for Thermoplastic Line Pipe (PVC and CPVC), Fourth Edition, 1976.

33. Spec 5LR, "Specification for Reinforced Thermosetting Resin Line Pipe (RTRP), Fourth Edition, 1976.

34. RP5L4, "Recommended Practice for Care and Use of Reinforced Thermosetting Resin Line Pipe (RTRP), Second Edition, 1976.

GLOSSARY OF CORROSION TERMS

Adsorption Adsorption occurs at the interface or boundary of a solid or liquid that is in contact with another medium. A substance has been adsorbed when its concentration at the interface or surface is higher than in the interior of the contiguous phases. (Different than absorption).

Anode The electrode of an electrolytic cell at which a net oxidation reaction occurs. In corrosion processes, the anode is the electrode that has the greater tendency to go into solution.

Bactericide An agent destructive to bacteria.

Cathode—The electrode of an electrolytic cell at which a net reduction reaction occurs. In corrosion processes, the cathode is usually the area that is not attacked.

Cathodic Protection Reduction or prevention of corrosion of a metal surface by making it cathodic, for example, by the use of sacrificial anodes or impressed currents.

Concentration Cell An electrolytic cell, the emf of which is caused by differences in the composition of the electrolyte at anode and cathode areas.

Corrosion The deterioration of a substance, usually a metal, because of a reaction with its environment.

Corrosion Fatigue Reduction of fatigue durability by a corrosive environment.

Corrosion Resistant This term apparently has no precise meaning, and some writers avoid it. We have used it to designate any metal that corrodes less than unalloyed low carbon steel.

Couple A pair of dissimilar electronic conductors in electrical contact.

Dezincification Corrosion of a zinc-containing alloy, usually brass, involving loss of zinc, and a residue or deposit in-situ of one or more less active constituents, usually copper.

Electron Negatively charged particles, always with the same mass and charge. One of the structural units of atoms.

Electrochemical Equivalent The weight of an element or group of elements oxidized or reduced at 100 percent efficiency by a unit quantity of electricity.

Electrochemistry The branch of physical chemistry that concerns itself with the interrelation of chemical phenomena and electricity.

Electrolysis Producing of chemical changes by passing an electric current through an electrolyte. Frequently used by chemical salesmen as an excuse for poor results from inhibitor treatments. Not a precise type or cause of corrosion since all corrosion is electrolytic.

Electrolyte An ionic conductor.

Erosion Deterioration by the abrasive action of fluids, usually accelerated by the presence of solid particles of matter in suspension. When deterioration is further increased by corrosion, the term erosion-corrosion is often used.

Free Energy The free energy of a substance refers to its formation from its elements, and the free energies of all elements are taken as zero in their standard states. The free energies of the more stable compounds are negative and the more negative the free energy, the more stable the compound.

Galvanic Cell A cell in which chemical change is the source of electric energy. It usually consists of two dissimilar conductors in contact with each other and with an electrolyte or two similar conductors in contact with each other and with dissimilar electrolytes.

Grounds Machine frames, meter cases, cable sheaths, etc., are connected to the ground mass if they contain conductors that may be at 150 volts or more above ground. The ground mass of a station consists of ground electrodes buried in the earth to provide earth connection to the station. The building steel is usually tied to this network.

Inhibitor A chemical substance that retards corrosion when added to an environment in small concentration.

Ion An electrically-charged atom or group of atoms. (They are formed when salts disassociate in solution.)

Intergranular Corrosion Corrosion that occurs preferentially at grain boundaries.

Polarization The displacement of electrode potential resulting from the effects of current flow measured with respect to either equilibrium (reversible) or steady state potentials.

Resistivity or Specific Resistance The reciprocal of conductivity, measured by the resistance of a body of the substance of unit cross section and unit length. Also called volume resistivity. The unit may be indicated as the ohm-cm.

Sacrificial Protection Reduction or prevention of corrosion of metal in an environment acting as an electrolyte by coupling it to another metal that is electrochemically more active in that particular electrolyte.

Stress Corrosion Cracking Spontaneous cracking produced by the combined action of corrosion and static stress (residual or applied).

Chapter 11 Workover and Completion Rigs, Workover Systems

Production rigs—utilization, selection, efficiency
Non-conventional workover systems
Concentric-tubing workovers
Conventional and hydraulic systems
Coiled-tubing system
Through-tubing operating practices
Through-flowline techniques

An important factor in overall cost and sometimes the success of a completion or workover job is the choice of the conventional rig or workover system equipment to do the job. Completion and workover techniques have advanced to the point where usually several methods could be employed to accomplish the same result.

Conventional workover rig equipment has been specialized and refined so that it is now advantageous to pick equipment to do a particular phase of the workover job, and after that phase is completed to pick another or move in auxiliary equipment—pumps, power swivel, squeeze unit, etc., to do the next phase.

In many well-treating or workover situations our primary requirement in setting up to accomplish the job is that we establish a means of circulating to the bottom of the well. In these situations small diameter Concentric Tubing units, Coiled Tubing units, or small diameter Snubbing units have application. Concentric tubing units and the second generation, Long-Stroke Hydraulic Snubbing units have the added capability of low torque drilling. With the dynadrill-type downhole motor, coiled tubing units have the capability of very low torque rotation.

Where only circulation is required, wells set up initially for the "Pump-Down" system can be worked over or serviced without the need of a rig, by pumping a work string of tools to the proper point in the well, operating the tools through application of pressure, and reverse circulating the tool string to the surface after the job is accomplished. The pump-down system should not be thought of as strictly a through-flowline system. As more experience is gained in its use, it should see increasing application in wells deviated from a central platform location onshore or offshore.

Formation damage must always be a primary concern in well workover. The ability to run tubing into a well through the christmas tree and normal producing conduit under surface pressure means that fluids can be circulated to bottom with no differential pressure into the producing zone. Thus, there is no preliminary flow of filtrate or solids into the zone to cause damage before the treating fluid arrives.

Where circulation to bottom is not required, certain jobs can be performed by use of an electric line or a wire line.

CONVENTIONAL PRODUCTION RIGS

Where producing string tubing must be moved the pipe handling capacity of a conventional Production Rig is usually required.

Rig Utilization

The following recommendations should be used to guide utilization of equipment for completion and workover operations.

Drilling Rig Versus Production Rig—On new

conventional wells, release drilling rig and complete with a workover rig or rigs selected to do the particular job required. Reasons:

1. Average day-rate cost of drilling rig may be four to six times that of a 10-hour day workover unit.
2. Drilling rig crews often lack sufficient training to perform specialized completion operations, such as running tubing and setting packers.

Shutdown at Night—Operate production rigs in daylight hours only—except:

1. In offshore operations.
2. Where procedures are sufficiently hazardous to preclude overnight shutdown—prolonged fishing jobs—high pressure situations.
3. Where time is of prime importance in returning well to production.

Use Proper Workover Unit—It is often economical to release one type of production rig and move in another more suitable to accomplish the particular job at hand. Failure to utilize the proper rig even if operations are delayed several days, results in additional rig time, special tool rental and costly mistakes. Situations to be avoided include:

1. Using heavy-duty production rigs (and certainly drilling rigs) to run small (1¼-in.) diameter tubing.
2. Using heavy-duty production rigs on shallow conventional wells.
3. Using rigs that pull only singles in lieu of a double rig on conventional wells involving workovers or jobs requiring many trips. From a depth standpoint, the breakover point between a singles and a doubles rig occurs at about 7,500–8,000 ft.

Production Rig Selection

Basically the problem of rig selection is one of picking a rig capable of doing the particular workover or service job for the lowest overall cost.

Depth or Load Capacity—Rig capacities are commonly spoken of in terms of depth rating with a particular size tubing, usually $2^7/_8$-in. Rig capacity depends upon a number of factors, principally: braking capacity, derrick capacity, and drawworks horsepower.

Braking Capacity—Braking capacity is a prime consideration in rig selection. Considerable energy is developed in lowering pipe in the hole. This energy, converted to heat by the braking system, must be effectively dissipated.

TABLE 11-1
Typical Braking System vs. Depth Rating Comparison

Nominal horsepower range	Effective brake area sq in.	Type cooling system	Auxiliary brake size[a]	Depth rating w/$2^7/_8$ in. tubing, ft
100–150	1200	Air	—	4,000
150–200	1600	Air	—	5,000
		Spray	—	7,000
200–250	2000	Air	—	6,000
		Spray	—	8,000
		Spray	15-in. SR	10,000
250–400	2400	Air	—	7,000
		Spray	—	9,000
		Spray	15-in. SR	11,000
		Spray	15-in. DR	13,000
400–600	2800	Air	—	8,000
		Spray	—	10,000
		Spray	15-in. DR	15,000
		Spray	22-in. SR	18,000

[a] Hydrotarder Braking Capacity at 1,300 rpm:
  15-inch Single Rotor—260 hp
  15-inch Double Rotor—525 hp
  22-inch Single Rotor—2500 hp

Braking effort is a function of the area of friction blocks forced against the rim; thus, effective brake area is an important comparative factor for rig selection.

Lowering a 50,000-lb tubing string in the hole at the rate of one double in 10 seconds develops about 545 hp or 4,000 Btu of heat. Thus brake cooling is a primary concern. Generally the larger the brake block area the faster the heat dissipation. Shallow depth rigs often depend on air cooling. Rigs rated below about 4,500 ft usually use water for cooling with either a spray or circulating system.

At depths below about 10,000 ft an auxiliary brake may also be required. The Parkersburg Hydrotarder is a hydrodynamic device that absorbs power by converting mechanical energy directly into heat energy within its working fluid (usually water) through pumping action. The power-absorbing capacity depends on the speed of the rotor. If the rotor speed doubles, the power-absorbing capacity increases eight times.

Table 11-1 presents accepted relationships between drawworks horsepower, effective brake area, type of cooling system, auxiliary brake system and depth rating with $2^{7}/_{8}$-in. tubing.

Derrick Capacity—Two types of masts are used with production rigs. The conventional pole-type mast is shown in Figure 11-1. Both single and double pole masts are used. With double pole type it is possible to install a racking board and hang sucker rods, and set back tubing, whereas with the single pole type, rods and tubing must be laid down. Thus, the double pole is more popular.

The self-guyed structural mast is shown in Figure 11-2. Load-bearing guys are permanently attached directly to the carrier vehicle; however, external guys are usually required under manufacturers recommendations.

The load rating and guying of new portable masts should conform to API Standard 4E (Drilling and Well Servicing Structures). Many portable masts, however, were designed based on API Standard 4D (Specifications for Portable Masts). Typical specifications of fabricated masts are shown in Table 11-2.

The foundation under the derrick legs must support the hook load and the weight of the derrick. Failure of the foundation under one leg, for example, can cause failure of the derrick at much less than rated capacity. Safebearing capacity of soils are shown in Table 11-3.

Drawworks Horsepower—Drawworks horsepower basically determines the speed at which pipe can be pulled from the hole. Loaded hook speed must be compromised with the cost and weight of the engine, transmission, and drawworks required

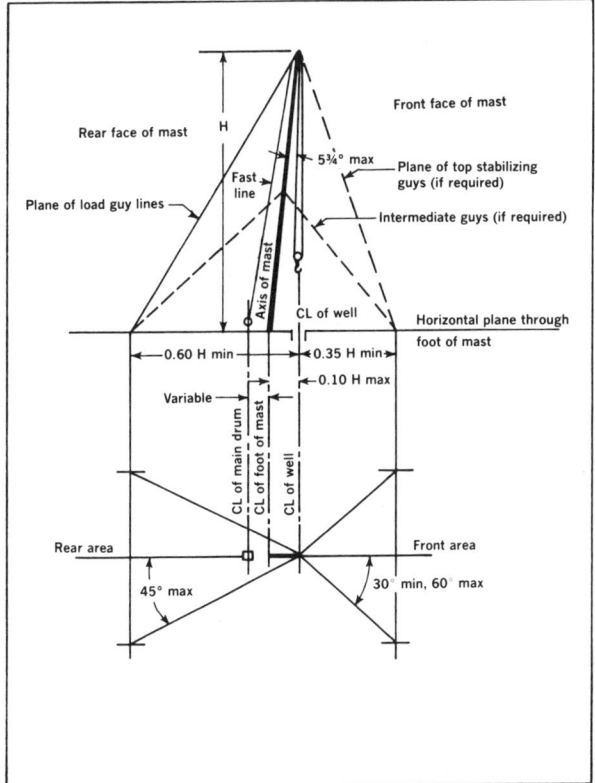

FIG. 11-1—Guying diagram for masts having guy lines anchored independently at the mast base (API Standard 4D).

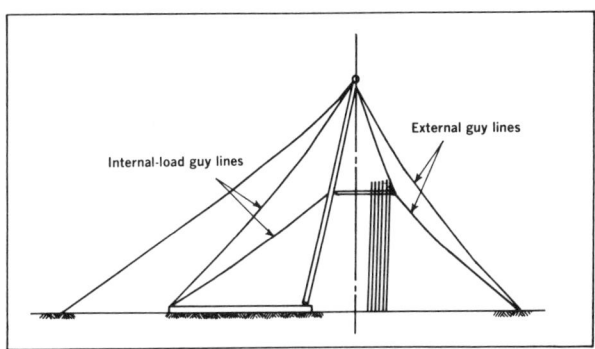

FIG. 11-2—Structural mast guying diagram, internal and external load guys (API Standard 4E).

TABLE 11-2
General Specifications of Fabricated Masts

Nominal		Hook load capacity w/6 lines, lb	Racking capacity—ft	
Height, ft	Weight, lb		2⅞ in. tubing	⅞ in. rods
69	8,000	140,000	Singles— 7,200	Doubles— 7,500
90	13,000	180,000	Doubles— 9,600	Doubles—10,500
96	15,000	215,000	Doubles—16,000	Thribbles—11,500
108	20,000	250,000	Doubles—18,000	—

TABLE 11-3
Safe Bearing Capacity of Soils
(pounds per square foot)

Solid ledge of hard rock, such as granite, trap, etc.	50,000–200,000
Sound shale and other medium rock requiring blasting for removal	20,000– 30,000
Hard pan, cemented sand and gravel difficult to remove by picking	16,000– 20,000
Soft rock, disintegrated ledge; in natural ledge, difficult to remove by picking	10,000– 20,000
Compact sand and gravel requiring picking for removal	8,000– 12,000
Hard clay requiring picking for removal	8,000– 10,000
Gravel, coarse sand, in natural thick beds	8,000– 10,000
Loose, medium, and coarse sand, fine compact sand	3,000– 8,000
Medium clay, stiff but capable of being spaded	4,000– 8,000
Fine loose sand	2,000– 4,000
Soft clay	2,000

to develop the power and translate that power into line pull and line speed.

The actual power required to pull tubing is not as great as might be supposed. Table 11-4 illustrates this for a typical rig situation.

TABLE 11-4
Power Required to Pull Tubing
10,000 ft, 2⅞ in., 6.7 lb/ft

Net hook power, hp	Off-bottom pulling speed ft/min.	Net hoisting time to pull out, minutes
150	100	60
75	50	120

Typically, 150 net hook horsepower requires about 250 engine horsepower. With 150 net hook horsepower tubing could be started off bottom at 100 ft/min, and would reach a maximum pulling speed of about 300 ft/min when two-thirds of the way out of the hole.

A diesel engine develops maximum torque within a fairly narrow range of shaft speed, as shown in Figure 11-3. Thus, there are a number of factors other than rated engine horsepower that influence the speed of pulling pipe. These are:

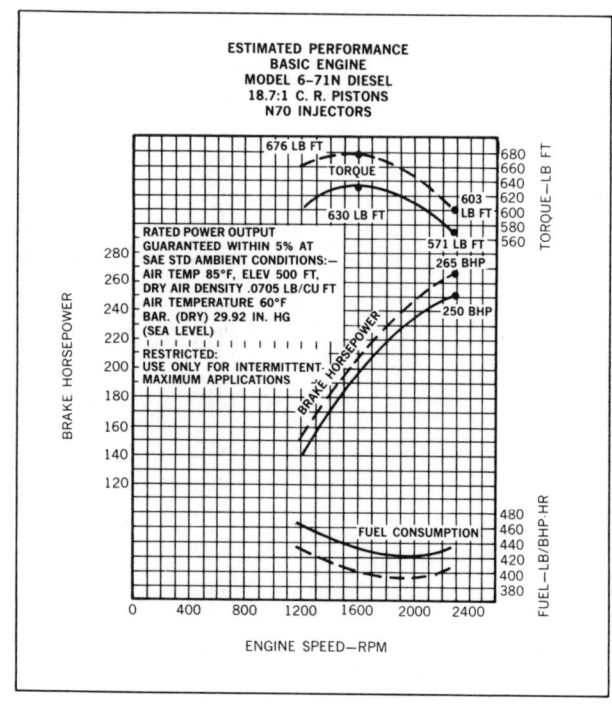

FIG. 11-3—*Engine-performance curve.*

1. Engine gear box ratios or torque converter performance.
2. Sprocket ratios between gear box propeller shaft and the drawworks drum.
3. Effective diameter of drawworks drum.
4. The number of travelling lines strung up between the crown and travelling blocks.

Figure 11-4 represents a typical power train situation for a medium depth workover rig equipped with the GM V6-71N engine of Figure 11-3, a 4-speed gear box, a two-speed drawworks, and a 4-line string-up. Assume elevation is 500 ft above sea level and air temperature is 85°F.

With this situation, maximum single line pull

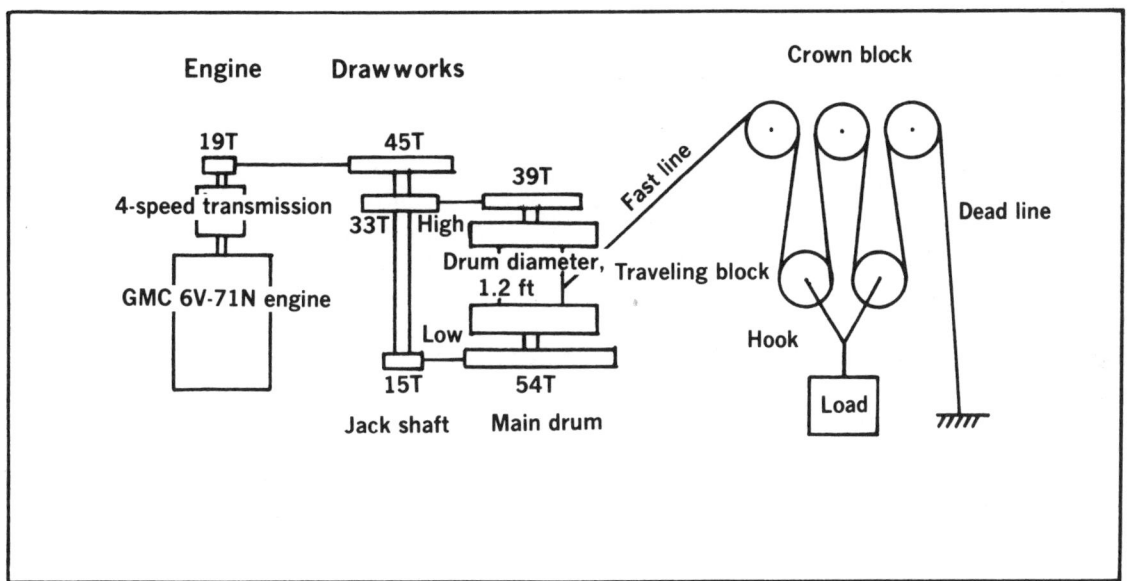

Specifications

Engine: GM V6-71N diesel
    500 ft elevation—Air temperature 85°F
    Maximum torque: 630 ft lb at 1,600 rpm
    Maximum speed: 2,000 rpm
Transmission: Low ratio: 6.5 to 1
    High ratio: 1 to 1
    Power loss through transmission—5%
Drawworks: Chain drive—5% power loss through each chain
    Gear box to jack shaft ratio: 19 to 45
    Jack shaft to high drum: 33 to 39
    Jack shaft to low drum: 15 to 54
    Drum diameter: 1.2 ft

Calculations:

1. Maximum Single Line Pull:

$$630 \text{ ft lb} \times .85 \times \frac{6.5}{1} \times \frac{45}{19} \times \frac{54}{15} \times \frac{2}{1.2 \text{ ft}}$$

$$= 49,500 \text{ lb}$$

2. Line speed at maximum load:

$$1,600 \text{ rpm} \times \frac{1}{6.5} \times \frac{19}{45} \times \frac{15}{54} \times \pi \times 1.2 \text{ ft}$$

$$= 109 \text{ ft/min}$$

3. Maximum block speed (no load):

$$2,000 \text{ rpm} \times \frac{1}{1} \times \frac{19}{45} \times \frac{33}{39} \times \pi \times 1.2 \text{ ft} \times \frac{1}{4}$$

$$= 674 \text{ ft/min}$$

FIG. 11-4—*Typical power train for medium workover rig.*

would be about 49,500 lb and fast line speed would be 109 ft/min. Pulling maximum load, block speed would be about 27 ft/min. With no load, maximum block speed would be 674 ft/min, assuming maximum engine speed of 2,000 rpm.

If any of the factors of Figure 11-4 change, then block speed and line pull will change. If drum diameter increases (due to adding another wrap) then maximum line pull will be reduced but fast line speed would be increased.

Rig designers must balance engine horsepower, gear ratios, etc. to produce a workable rig. Normally an empty block should run about 5 to 7 ft/sec. Single line pull should be about 35,000 lb with a 140,000-lb hook load capacity mast having four lines strung to the traveling block. A fully loaded block should run about 1.0 ft/sec.

Operational Efficiency

Operational efficiency refers to the efficiency or speed per unit of overall cost with which the required operations can be performed.

Obviously personnel capability, training and motivation are primary factors affecting "operational efficiency" in addition to the design and application of the mechanical equipment.

Primary rig operations include: move in, rig up, tear down; pulling and running tubing and rods; and rotating and circulating. Proper choice of accessories maximizes operational efficiency.

Move In—Rig Up—Tear Down

The efficiency of move in, rig up, and tear down depends upon several factors:

Roadability—Most modern rigs are designed to move over good roads at speeds of about 40 mph. Usually the drawworks engine powers the vehicle on the road.

Mobility on Location—A drive-in rig is somewhat easier to position than a back-in type. Positioning must be fairly precise in order that the hook is centered over the wellhead without excessive adjustment of the derrick.

Type of Mast—The hydraulically-raised self-guyed derrick has reduced time requirements for this operation almost to insignificance. For example, raising and leveling a 90-ft self-guyed derrick requiring two operations: (1) tilting mast upright and (2) extending derrick, takes about 15 minutes.

Blocks remain rigged up and leveling is done by hydraulic jacks or screws. Raising and securing a comparable pole mast with external guys might require 40 minutes.

Pulling and Running Rods and Tubing

Repetitious operations in pulling and running rods and tubing lend themselves to automation. The conventional four-man crew can be reduced to three men (eliminating one floorman) through use of air-operated slips, hydraulic tubing tongs and hydraulic tong positioners. The continuous motion block eliminates deadtime while making up or breaking out tubing joints.

Air-operated Slips—Air slips controlled by the rig operator are proved time savers. They should have button-type slip dies in order to serve as back-ups for tongs. Button slip dies minimize pipe damage.

Hydraulic Tubing Tongs—Hydraulic tubing tongs greatly reduce time to make and break tubing. The Foster-type tongs remain on the tubing at all times and are raised and lowered with a hydraulic cylinder as necessary to clear the elevators. Byron Jackson tongs do not remain on the tubing after makeup or breakout, but are pulled off the pipe with a spring.

Hydraulic tubing tongs properly equipped with a pressure gauge and bypass valve offer the advantage of controlled torque makeup.

Performance characteristics of a typical hydraulic tubing tong (Foster Model 74) are:

Torque range, high gear	100 to 800 ft lb
Maximum torque, low gear	2,500 ft lb
RPM, high gear	110
RPM, low gear	30
Hydraulic requirements	25 gpm at 2,000 psi

Hydraulic Rod Tongs—These are justified primarily on the basis of providing controlled makeup torque. With 5/8-in. and 3/4-in. rods there appears to be little speed advantage. With 7/8-in. and larger rods a slight speed advantage or operator fatigue advantage may result.

A typical hydraulic rod tong (Byron Jackson Mark IV) develops maximum torque of 1,370 ft-lb, maximum speed of 100 rpm, and has hydraulic power requirements of 12 gpm at 2,000 psi.

Continuous Motion Block—The continuous motion block (sometimes known as the split block) and transfer elevators can provide significant time saving, probably 20% on a rig pulling singles from 6,000 ft. Savings would be greater pulling doubles.

The time saving results from the fact that in pulling tubing, the block is falling while the stand of tubing in the derrick is being broken out. Going in the hole the block is run up while the stand of tubing is being made up. Transfer elevators, and a small hydraulic or air hoist operated by the derrick man, handle the single stand of tubing in the derrick.

In pulling rods, one sheave of the block can be unpinned and tied down so that the remaining single sheave and block run twice as fast.

Time Analysis—A typical time analysis is shown in Table 11-5 for a service job on a 6,000-ft well with 2⅜-in. tubing. The rig has a 66-ft derrick-type mast pulling tubing in singles and rods in doubles. It is rated for 8,000 ft with 2⅞-in. tubing. Pipe handling equipment consists of a conventional travelling block, hydraulic tubing tongs and positioner and air slips.

The operation of pulling or running an average single joint of tubing under the conditions of Table 11-5 is analyzed in Table 11-6.

TABLE 11-5
Typical Time Analysis—6000-ft Well—2⅜-in. Tubing

Operation	Time required minutes
Spot rig and set sills for derrick foundation	15
Raise and level derrick	15
Pull rods (manual rod wrench)	55
Rig block to pull tubing	10
Set BOP	20
Pull tubing	90
Run tubing	70
Flange up wellhead	15
Rig block to run rods	10
Run rods	45
Tie block back—clean tools	10
Lay down derrick	15
	370

TABLE 11-6
Analysis of Single-joint Cycle Time Average of 6000-Foot Trip

	Time required—seconds	
Operation	Pulling	Running
Loaded block motion	8	4
Empty block motion	5	5
Makeup or breakout	12	10
Operate slips	2	2
One complete cycle	27 seconds	21 seconds

Cutting one second off the time required to cycle one joint of tubing in and out would reduce trip time on a typical 6,000 ft. well by about 4%.

Rotating and Circulating

Rotating and circulating are intermittently required in workovers, during such operations as cleaning out, drilling cement or permanent packers, deepening, and perhaps washing over or milling.

Rotating—The power swivel or power sub is well adapted to workover rig rotary operations. Advantages are accurate torque control through a pressure regulator bypass, and much easier makeup on the drill string compared with a kelly. Table 11-7 shows typical capacities.

The larger capacity subs are suitable for deepening jobs in 5½-in. or 7-in. casing.

Circulating—Pumping operations involve: (1) drilling or milling where relatively high circulating

TABLE 11-7
Typical Power Sub Capacities

Swivel model	Circulating rate, gpm	Maximum pressure, psi	Horsepower	Max. rpm	Max. torque ft lb
Skytop 615	40	2000	45	100	1330
Cooper 6060	43	2000	60	113	2000
Cooper 6090	62	2000	90	190	2400

Table 11-8
Hydraulics for Conventional Circulation
150 ft/min. Annular Velocity

Nominal size, in.		Circulating rate, gpm[a]	Pressure loss in pipe and annulus per 1000 ft[b] psi	Pressure loss thru surface connections & bit or mill[c] psi	Hydraulic horsepower needed at 6,000 ft
Tubing	Casing				
1 1/4	2 1/2	20	40	50	4
1 1/4	3 1/2	40	55	80	10
2	4 1/2	60	50	150	16
2	5 1/2	115	125	275	69
2 1/2	7	190	110	400	118

[a] Circulating rate to provide 150 ft/min rising velocity in annulus.
[b] Based on 9.5 lb/gal saltwater.
[c] Higher pressure drop through bit or mill may be desirable to provide additional jetting action.

volume may be required to provide the annular velocity needed to lift cuttings to the surface, and (2) well killing where high pressures may be required to inject into the well.

Drilling or Milling—To provide effective cutting lifting with typical saltwater workover fluids annular velocities of 150 ft. per minute are needed. Table 11-8 shows circulating rates needed for a rising velocity of 150 ft/min. in the annulus. Also shown are estimated surface pressures and pump hydraulic horsepowers. Surface pressure and required hydraulic horsepower may vary somewhat, depending on pressure loss through the bit or mill.

Reverse Circulation—Where the rig pump is limited on volume, reverse circulation may provide sufficient rising velocity. This, of course, requires a suitable control head or stripper head. Shop-made stripper heads have proved satisfactory in most instances.

Hydraulics for reverse circulation
150 ft/min rising velocity

Tubing Nominal size, in.	Circulating[a] rate, gpm
1 1/4	12
2	25
2 1/2	36

[a] 150 ft/min rising velocity in tubing.

Well Killing—The major factor in efficient well killing is to maintain sufficient wellbore pressure at the formation face to prevent flow into the wellbore. If this is done, high circulating volumes are not needed. Pump pressure needed to kill a well can often be reduced by setting a wireline plug near the bottom of the tubing and bleeding the tubing pressure down before attempting to fill the tubing or circulate the well. In some areas tank trucks having a suitable plunger pump are used to kill wells before moving in a workover rig.

Pumps and Auxiliary Systems—For tubingless completions where high circulating rates are not needed to provide cutting lifting annular velocity, 75 to 100-hp pumps such as the Gardner-Denver PE-5 triplex plunger pump driven through a four-speed transmission by a GMC 4-71 (165 hp at 2,800 rpm) or GMC 4-53 (130 hp at 2,800 rpm) engine are entirely suitable.

Where higher volume capacities are necessary in conventional well pumps such as the Gardner-Denver PA-8 triplex plunger pump driven by a GMC V8-71 (350 hp at 2,300 rpm) engine has proved suitable.

NON-CONVENTIONAL WORKOVER SYSTEMS

Non-conventional workover equipment or systems can be categorized as:

Concentric tubing workover rigs are essentially smaller versions of production rigs with perhaps a higher degree of hydraulic control to assist in operating within limits of their smaller equipment.

Long-stroke hydraulic snubbing units, the most recent addition to non-conventional workover equipment, incorporate most of the advantages of a concentric tubing rig plus they have the pulling and rotating capacity of a larger workover rig, and the ability to work under surface pressure. At this point they appear to be the rig of the future both for offshore platform work and land work.

Coiled-tubing units have seen a rapid rise in application particularly for operations such as washing, spotting, or displacing fluids. Primary advantage is speed of running pipe. Primary limitation is inability to rotate effectively for drill-out operations.

Small-diameter short-stroke snubbing units have many of the same advantages as coiled tubing units (i.e., ability to work under reasonable surface pressure, and portability) and the same limitations (i.e.: inability to rotate). Another serious limitation is the time required to run tubing. This equipment has been obsoleted by the long-stroke hydraulic unit.

The pump-down system, sometimes called the through-flowline system, was originally developed for ocean floor completions, but also has significant application in deviated holes where wireline work required for normal servicing operations is made more difficult by drag or line friction.

Through-tubing electric line or wireline systems can handle such operations as plugging back, bailing or removing tubing obstructions as well as normal well maintenance functions of paraffin scraping, gas lift valve servicing, etc.

Tubing extension system has generally been made obsolete by further developments, but was the "grandfather" of techniques presently used.

CONCENTRIC TUBING WORKOVERS

The term "concentric tubing workover" refers to a system whereby a small work string of "macaroni" tubing or drill pipe is run inside the existing well tubing. This system offers a rather simple and economical means of working over a well without removing the existing production tubing. An individual completion in a multicompleted well can be worked over without disturbing other completions.

For a high pressure well, macaroni tubing can be snubbed in without killing the well. The avoidance of exposure of producing formations to drilling mud or packer fluids will result in less damage to the producing formation and higher well productivities. Lighter equipment can be employed in offshore operations or isolated areas.

Concentric tubing workover should be considered in these situations:

1. Wells with tubing set on a packer with gelled mud or other packer fluid in casing-tubing annulus.
2. Multiple completions.
3. Offshore wells.
4. Tubingless completions.

Equipment for Concentric Tubing Workovers

Primary equipment needed to conduct concentric tubing workovers includes: (1) a hoisting and rotating unit, (2) blowout preventers, (3) a macaroni workstring, (4) a high-pressure low-volume pump, and (5) small bits or mills.

Hoisting and Rotating Unit—The hoisting and rotating unit may be a concentric tubing rig, a hydraulic snubbing unit, or where rotation is not required, a coiled tubing unit.

1. Concentric-tubing-rig hoist unit. Typical characteristics of a concentric tubing land rig in the U.S. are:

—70-ft telescoping derrick with 125,000-lb hook load capacity;
—Double-drum drawworks with swab line;
—150-hp engine;
—Light substructure or working platform with power tongs and air slips;
—Hydraulically-operated power swivel;
—With a three or four man crew, rig should be able to move in, rig up, run 6,000 ft of macaroni tubing, perform squeeze job, pull workover tubing, and rig down, all in about 10 hours.

2. Long stroke hydraulic snubbing unit. Characteristics of one available unit are generally:

—18-ft hydraulic cylinder rigged up to provide a 36-ft pull or snub stroke.
—Maximum lift 120,000 lb—sufficient to pull dual strings of 2⅞-in. pipe if needed. Can run up to 4-in. tubing.
—Maximum snub against well pressure, 70,000 lb.
—Hydraulic rotary—maximum torque 1,000 ft/lb.
—Pulling or running speed 130–150 joints per hour.
—Hydraulic drive powered by 235-hp GM 671 motor.
—Weight of maximum lift in moving 9,800 lb. Total unit can be moved in 8 loads.

TABLE 11-9
Typical Concentric Tubing Workover Strings

Size Type of connection	Grade	Weight, lb/ft	Tube od, in.	Tube id, in.	Coupling od, in.	Tensile strength,* lb	Torsion yield, ft-lb
¾-in.							
NU API	J-55	1.14	1.050	0.824	1.313	8,740	—
EUE API	J-55	1.20	1.050	0.824	1.660	18,290	—
1 in.							
NU API	J-55	1.70	1.315	1.049	1.660	15,060	—
EUE API	J-55	1.80	1.315	1.049	1.900	27,160	—
EUE API	N-80	1.80	1.315	1.049	1.900	39,510	—
CS Hydril	N-80	1.80	1.315	1.049	1.552	39,510	—
CS Hydril	N-80	2.25	1.315	0.975	1.600	51,000	—
1¼ in.							
NU API	J-55	2.3	1.660	1.380	2.054	21,360	710
EUE API	J-55	2.4	1.660	1.380	2.200	36,770	1080
EUE API	N-80	2.4	1.660	1.380	2.200	53,480	1560
CS Hydril	N-80	2.4	1.660	1.380	1.883	53,480	1560
"MT" tool jt.	N-80	3.2	1.660	1.264	1.750	72,800	2290

*No safety factor.

Blowout Preventers—A twin-ram manually-operated preventer containing two sets of pipe rams is usually sufficient with the conventional hoist unit. The 5,000 psi Bowen-Itco is preferred. With the long stroke snubbing unit a hydraulically-operated dual stripper-type preventer (2,500 psi) is used above the ram-type preventers.

The stripper is essential for operations where the workover string must be run under pressure. The christmas tree master valve serves as a blank ram.

Work String—Various sizes of macaroni tubing with many different connections are available. Integral joint connections must be used where the work string is run through a stripper. Choice depends on: required clearance, tensile strength and torque limits, hydraulics, frequency of use, and retirement and salvage aspects.

Table 11-9 illustrates typical workover strings:

The most economical string for use inside 2⅞-in. tubing or casing is usually 1¼-in. J-55 with API NU or EUE connections on standard weight tubing. Continuous weld (JAL-CON-WELD 55) costs about 20% less than seamless pipe, and has proved to be quite satisfactory as a work string. Above 6,000 ft, 1¼-in. J-55 NU tubing will satisfy normal strength requirements. Below 6,000 ft, EUE tubing can be added to upper section. Below 9,000–10,000 ft N-80 tubing or drill pipe is needed in the upper section for tensile strength.

For servicing inside 2⅜-in. tubing, 1-in. non-upset tubing is employed. If more than 6,000 ft are required, external upset tubing with couplings turned down to 1.667 in. is used.

When drilling with API macaroni tubing, maximum torque is usually limited to about 400 ft-lb. Rotary speed is usually 50 to 80 rpm and bit weight in the 500 to 1,500-lb range.

Pump—High friction losses and small fluid volumes require a pump such as the Gardner-Denver PE-5 Triplex plunger pump which is rated at ½ bbl/min. at 5,000 psi, or 2 bbl/min. at 3,000 psi.

Pressure required to circulate saltwater under various conditions are shown in Table 11-10.

Low circulating rates can be tolerated due to the small capacity of the pipe and annulus. For example, cement can be circulated off bottom from 10,000 ft in 20 min when circulating at ½ bbl/min through 1-in. tubing.

Bits and Mills—Drag bits, rock bits and mills are readily available. A 1²⁵⁄₃₂-inch bit or mill is run inside 2⅜-in. tubing, and 2¼-in. tool is run inside 2⅞-in. tubing. A scraper is normally run above the bit.

Coiled Tubing System

The coiled tubing workover system primarily supplies a means of running small-diameter tubing

TABLE 11-10
Pressure Required to Circulate Through Small Diameter Pipe

Tubing	Depth, ft	Circulating rate bbl/min	Circulating pressure, psi
¾-in. EUE	4,000	½	1,000
	8,400	½	1,700
	12,600	⅓	2,500
1-in. Integral joint	7,000	½	1,200
	11,900	¾	1,500
1¼-in. NU	1,800	⅔	250
	3,000	⅔	400
1¼-in. Drill pipe	5,000	¾	1,100
	10,000	½	1,600

through the usual well tubing, thus providing a circulating path to the bottom of the well. The following advantages are apparent:

1. Coiled tubing can be run in the well against reasonable surface pressure, thus the well can be controlled with a low density clean fluid. For many applications the high pressure well does not have to be killed with a damaging fluid.

2. Weight and size of the various components are such that transportation to an offshore platform is facilitated. Two systems are commercially available: Brown Equipment and Service, and NOWSCO (Bowen). Both are similar with the primary exception of the tubing injector unit.

3. Tubing can be run at relatively high speeds—150–200 feet per minute. Circulation is possible while running and pulling. There are no collared connections in the string.

Primary components are the injector-hoist unit, continuous tubing, storage reel, and the blowout preventer stack.

Injector Hoist Unit—This consists of friction gripper blocks mounted on an endless chain driven by a hydraulic motor system. These blocks grip the small tubing to run in or pull out. The unit also serves to straighten the coiled tubing as it is run into the well and to yield it again as it is pulled out.

Continuous Tubing—This is ¾-in. or 1-in. od electric weld steel pipe. It is manufactured in 1,000–3,500 ft lengths and must be welded to form longer lengths. Welding requires special equipment and techniques and these facilities must be available in the area where these units operate.

Pipe life is a function of work performed. It can be quite short, varying from ten trips where high pulls and high pressures are involved, to 40 trips with low-tensile, low-burst conditions. Pipe closest to the reel core sees most severe conditions and must be replaced first. One-inch tubing welded and ready to use costs about $0.45 per ft. In operations, cost per ft run is about $0.02.

Coiled-tubing specifications are shown in Table 11-11.

TABLE 11-11
Specifications—Republic Steel Electric Weld Pipe
(Min. Yield 60,000 psi; Tensile 75,000 psi)
(Elongation 28–30%)

	1-in. od	¾-in. od
Actual od, in.	1.00	0.75
Nominal id, in.	0.870	0.652
Weight/ft, lb	0.649	0.367
Wall area, sq in.	0.1909	0.1080
Operating ultimate, lb.	95–11,000	55–6000
Tensile, operating max., lb	9000	5000
Test pressure, psi	5500	5500
Burst, operating max., psi	4500	4500
Wellhead pressure max., psi	3000	3000
Capacity, bbl/1,000 ft	0.74	0.41

Useful depth limit is about 12,000 ft; performance is better at 10,000 ft. Accurate measurement of pipe strain is important and under high tensile loads internal pressure must be reduced accordingly.

Friction pressure loss in 1-in. tubing is a real

problem; 250–300 psi/1000 ft with water, or about 30% less with a guar gum or polymer fluid. With foam, pressure loss is about 60 psi/1000 ft. With 3/4-in. tubing, loss is much more severe.

The tubing reel, normally 10 ft in diameter, is powered by hydraulic drive to maintain tension on the tubing, and uses a "level wind" mechanism to coil the tubing uniformly.

The BOP stack contains a stripper assembly, two pipe rams, one blind ram, hydraulic slips and a hydraulic pipe cutter.

Concentric Tubing Operating Practices

Operations performed with concentric tubing techniques include squeeze cementing, washing out sand, plugging back, plastic consolidation, gravel packing, and deepening. Where rotation is required the concentric tubing rig, or long stroke snubbing unit, must be used. Where only circulation is required the coiled tubing unit can be used.

Squeeze Cementing—Squeeze cementing can be done between packers on a dual completion or below the bottom packer. See Figure 11-5. Suggested procedure for a typical job:

1. Run the "macaroni" work string to the lowest perforation to be squeezed.
2. Batch mix low fluid loss cement with retarder to provide more than adequate pumpability time.

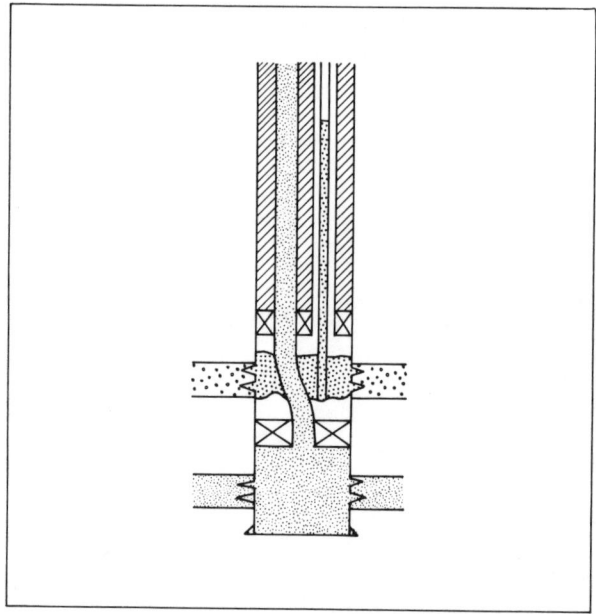

FIG. 11-5—*Concentric-tubing squeeze cementing.*

3. Pump in about 4 bbl of fresh water, required cement—usually about 2 bbl for short zones, and 2 bbl of fresh water, followed by appropriate displacing fluid—usually saltwater.

4. For short zones, pump cement to bottom of "macaroni" string, and circulate cement around bottom until about 1/2 bbl of cement has been circulated above uppermost perforation. The objective is to spot about 50 ft of cement in the casing opposite and above perforations.

5. Pull macaroni string until it is about 15 ft above top perforations, with the hydraulically-operated stripper-type blowout preventers closed on pipe.

6. Close BOPs and perform a low pressure squeeze job with maximum squeeze pressure, usually 300 to 500 psi above anticipated reversing pressure.

7. If a cement plug is desired opposite perforations, reverse out excess cement, raise pipe about 300 ft, keeping hydraulically-operated stripper closed, and wait on cement to set. For tubingless completions, it may be desirable to drill out cement when recompleting lower.

8. To remove all cement opposite squeezed perforations leaving only dehydrated nodes of cement inside perforations, start reversing out cement with macaroni tubing above perforations. After cement returns are obtained, continue reverse circulation while slowly lowering the tubing through the perforated interval.

9. For conventional completions, equipped with an inside flow string set above completion zones, it is usually preferable to circulate out cement to bottom because of difficulties in drilling cement from casing with a much smaller diameter bit.

Dump bailers are available for cement plugbacks, but are not too satisfactory for wells with appreciable reservoir pressure. Primary difficulty—feed in and mixing of fluids with cementing material.

Perforating—Through-tubing type guns can be used to perforate either conventional or tubingless completions. See Figure 11-6.

Conventional multiple completions can be perforated between packers by use of a perforator equipped with a special kickover arm, surface indication of direction of adjacent tubing strings, and gun rotating device. Figures 11-7 and 11-8 show this equipment and perforating arrangement.

Sand Washing—To minimize plugging on bottom,

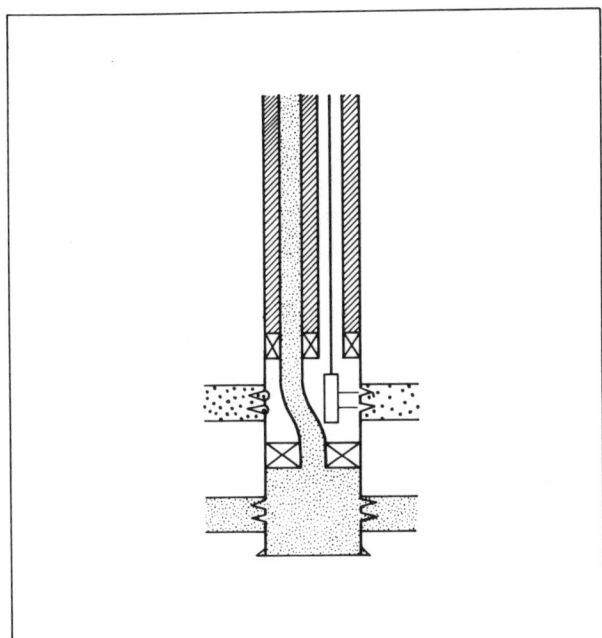

FIG. 11-6—Through-tubing perforating between packers.

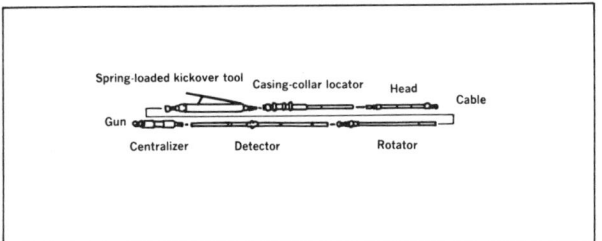

FIG. 11-7—Perforator-orienting tool.

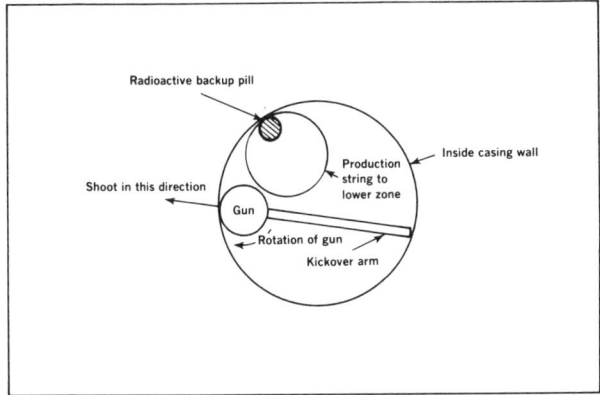

FIG. 11-8—Orienting device for between-packer perforating.

bottom end of the tubing string is cut off diagonally, forming a "muleshoe." Saltwater or oil can be used as the circulating fluid. See Figure 11-9.

Normal practice is to jet about 60 ft into the sand, then reverse circulate the sand out to avoid accumulating too much sand in the annulus. A mill is sometimes used if sand is extremely hard.

Sand Control—Concentric tubing can be used to place plastic for sand consolidation. This system has been very satisfactory in offshore operations. Small diameter tubing is advantageous in sand consolidation because of improved control of small volumes of chemicals used.

Small diameter equipment is available for gravel packing to control sand; however, the small diameter screen required appreciably limits fluid volumes.

Acidizing—It is frequently advantageous to acidize long zones, using concentric tubing so that acid can be circulated adjacent all zones.

In high pressure wells it may be desirable to snub in regular concentric tubing or coiled tubing to avoid damage to formation during stimulation operations.

Deepening—Concentric tubing techniques can be used to drill in with a non-damaging fluid when small diameter casing has been set and cemented above a productive zone. This system is also useful to deepen through zones previously plugged back with cement.

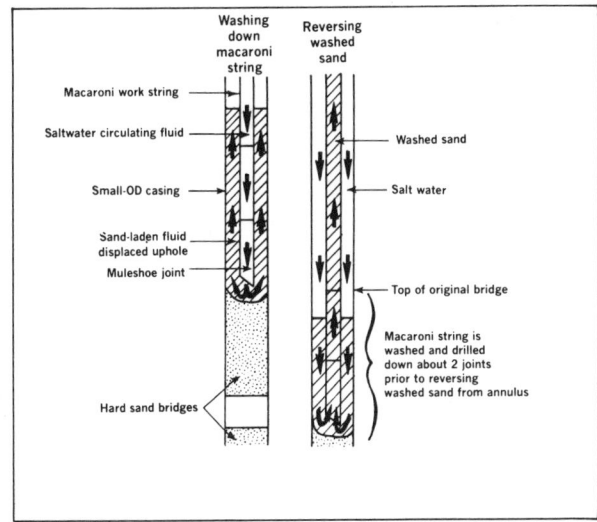

FIG. 11-9—Washing sand bridges with concentric tubing.

Through Flow Line Maintenance and Workover Techniques

To perform well servicing operations, a system has been developed by Otis Engineering to convert "wireline" techniques to "pump-down" techniques. The system requires two flowlines and at least two tubing strings to permit circulation of fluids from a central location to the bottom of the well, as shown in Figure 11-10. It has application in highly deviated holes, as well as in ocean floor completions.

Locomotive—Heart of the system is the piston unit or locomotive—(Figure 11-11) which is used to install or retrieve through flowline tools. Each piston unit is provided with a bypass restriction which absorbs energy in the form of differential pressure, but does not prevent circulation completely. By sharing loads between pistons, considerable thrust can be generated. Six pistons would develop a pull (or push) of 4,500 lb with 1,500 psi differential across the string of tools inside 2⅜-in. tubing.

Auxiliary Tools and Maintenance Operations—Other tools, such as running and pulling tools,

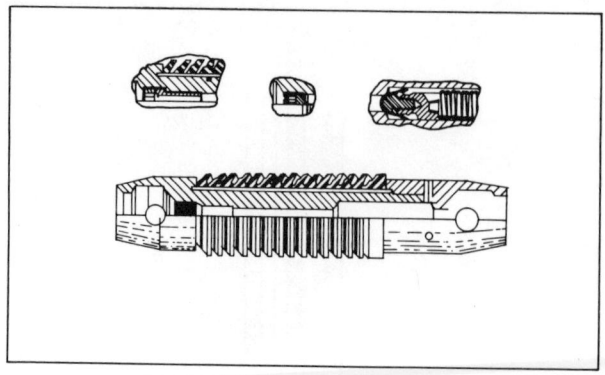

FIG. 11-11—*Piston unit or locomotive.*[5] *Permission to publish by The Society of Petroleum Engineers.*

hydraulic, or mechanical jars, stem, and paraffin cutters, are available for use with the pump-down system. Thus, all the "conventional" wireline operations can be performed by a properly set up pump-down system including running or pulling safety valves, operating sliding sleeves, running bottomhole instrumentation, or fishing.

Paraffin cutting using the pump-down system is actually simpler than with wireline tools. The cutter with several locomotives above is pumped through the flowline and down the well tubing. Some fluid bypasses the tools washing cut paraffin ahead. After the tools have been run to below the level of deepest

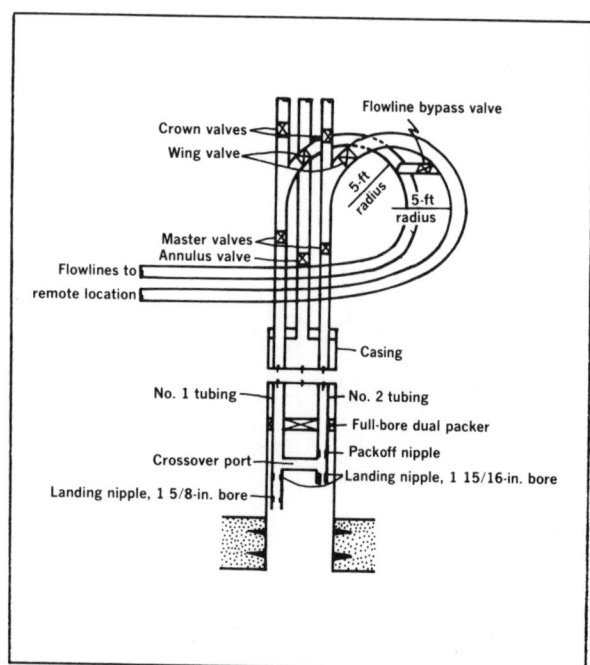

FIG. 11-10—*Flow-line and tubular hookup for pump-down systems.*[5] *Permission to publish by The Society of Petroleum Engineers.*

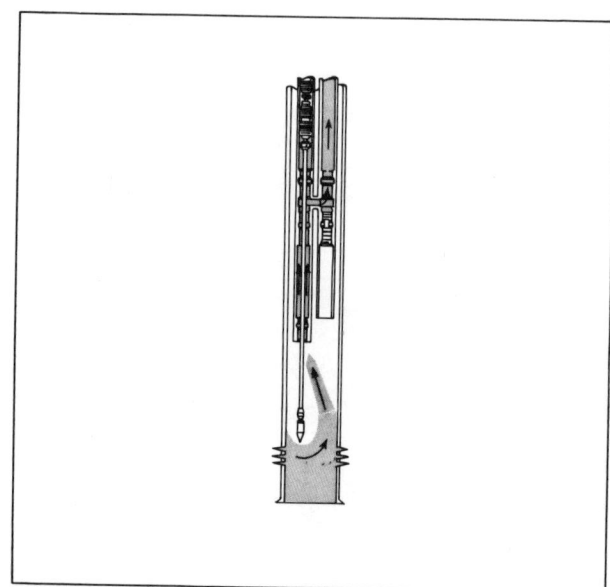

FIG. 11-12—*Sand-washing operation.*[5] *Permission to publish by The Society of Petroleum Engineers.*

paraffin accumulation, the well is opened up and paraffin and tools are returned to the remote production platform.

Workover Operations—Tubing extension (flexible ¾-in. tubing of any reasonable length) can be run where necessary to perform operations such as sand washing (Figure 11-12), squeeze cementing, plasticizing or acidizing. Perforating can be done (oriented perforating if necessary) by "locating" the perforator with respect to a landing nipple, the position of which has previously been established with respect to the formation.

Surface-Control Equipment—Surface equipment located on a remote production platform consists of: a pump (about 150 hydraulic hp); a horizontal lubricator to permit inserting tools against well pressure; and a manifold and instrument skid for controlling flow, regulating back pressure, measuring rates and volumes, and recording pressures.

The location of tools in the flowline and tubing is determined primarily by the recorded pressure "signature" of the particular well configuration as the tools pass through the christmas tree and the various other flow restrictions.

REFERENCES

1. Goeken, R. J.: "Report on Methods of Concentric Tubing Workovers," API Paper No. 926-3-B, February, 1958.

2. Corley, C. B., Jr., and Rike, J. L.: "Tubingless Completions," API Paper 926-4-6, March, 1959.

3. Rike, J. L.: "A Small Coiled Tubing Workover Rig," API Paper No. 926-12-I, March, 1967.

4. Frank, W. J., Jr.: "Improved Concentric Workover Techniques Offshore," J. Pet. Tech., April 1969, p. 401.

5. Raulins, G. M.: "Well Servicing by Pumpdown Techniques," J. Pet. Tech., February 1970, p. 161.

6. API Spec 4 E Drilling and Well Servicing Structures, March, 1974—Supplement 1, February, 1976.

Chapter 12 Workover Planning

Problem-well analysis
Alleviation of formation damage
Permeability and reservoir pressure problems
Reduction of extraneous water or gas production
Sand control and mechanical problems
Adjustment of completion intervals
Workover economics

Maintenance of wells in optimum producing condition is a primary objective of workover operations. In addition to solving specific well and reservoir control problems, workovers can provide a check on lateral and vertical movement, and current location of oil, water, and gas in specific zones and reservoirs. Judicious use of workovers often can appreciably increase economically recoverable oil and gas reserves.

Workovers, especially when well stimulation is involved, may be an alternative to the drilling of additional wells to provide required production of oil and gas. Because there is a "before and after" producing record on worked over wells, many improved well completion, workover, and well stimulation techniques have been proved through workovers.

Workovers are historically a most profitable business if optimum creative skill and imagination are exercised in (1) fact-finding, (2) analysis, (3) experimentation, (4) job execution, and (5) job evaluation. However, both the well and reservoir must be considered in problem diagnosis to intelligently plan action that may be either remedial type of work or well stimulation.

Reasons For Workovers

1. Remedial action on problem wells is usually planned to increase productivity, to eliminate excessive water or gas production, or to repair mechanical failure.
2. Work on nonproblem wells may be done to increase production through well stimulation, recompletion or multiple completion to evaluate potentially productive zones, or to provide service wells.
3. If large increases in profit are the objective, well stimulation of high productivity wells frequently offers the best profit opportunities.
4. In stratified reservoirs, frequent workovers may be necessary to maintain dynamic control of oil, water, and gas in various zones or layers in each well.

Problem-well analysis—A complete analysis should be made on all problem wells. This analysis may be a well, area, or reservoir study. The "Problem Well Analysis" chapter covers details on well analysis, including the problem well analysis check list. Use of this check list should increase the probability that all data are considered relative to each problem.

Usual reasons for low producing rate are:

—formation damage,
—low permeability,
—pressure depletion,
—excess water production,
—excess gas production in oil wells,
—sand, mud, or other debris in well,
—mechanical failure,
—high-viscosity fluid, or
—ineffective artificial lift.

Workover Planning to Alleviate Formation Damage

All oil and gas wells have some impairment to production. Therefore, bypassing or removing blockage from the tubing, wellbore, perforations, formation matrix, and formation fractures, if present, should be considered on all workovers. A usual first approach is to check bottom with a wire line or with tubing to detect any fill in the casing or open hole. Cleanout, reperforating, chemical treating, acidizing, fracturing, or a combination of these techniques may be used to reduce damage.

Plugged perforations—For relatively unconsolidated sandstone wells, wash perforations with double-cup perforation washer having 3 in. to 1-ft spacing between wash cups. A surge tool may be employed on some sandstone wells to be consolidated with plastic.

Perforations may be broken down with ball sealers. Employ water or oil for breakdown of sandstone perforations and oil, water, or acid for carbonate perforations.

Reperforate when necessary.

Scale-Damaged Wells—For scaled tubing, acidize, chemically remove, or ream out the tubing as required.

For scale in casing perforations, reperforate and, if necessary, remove remaining scale by chemical treatment or acidizing.

Scale in Open Hole—Chemically treat or acidize to remove soluble scale; drillout or underream to remove soluble or insoluble scale.

Open hole may be perforated with standard or open hole shaped charges prior to chemically treating or acidizing.

If the well has been previously fractured and propped, or fractured and etched with acid, it may be necessary to repeat these operations after removing scale from the casing or open hole.

Paraffin or Asphalt Plugging—Wax in the tubing, casing, or wellbore may be reamed out, removed by steam, hot oil, or hot water, or dissolved with a solvent.

The preferred method of removing paraffin or asphalt from the wellbore or formation is to clean out with solvents. Employ a very low-rate low-pressure solvent-surfactant squeeze into formation followed by a 24 to 72-hour soaking.

Normally, hot oil or hot water should not be used to remove wax from perforations, wellbore, or formation because a portion of the melted wax will cool, solidify, and remain in the formation.

Emulsion or Water Blocks—Emulsion or water block damage may be alleviated with surfactants. Most water blocks are self correcting over a period of weeks or months.

For sandstone wells, an HF-HCl acid-surfactant treatment may be preferred to remove emulsion damage.

For matrix damage in carbonates, the usual approach is to bypass damage with acid.

Emulsions formed during fracture acidizing of carbonates may be broken by pumping a 2 or 3% surfactant solution into the fracture.

Clay or Silt Damage to Formation Matrix—In sandstone wells, remove clay or silt damage with HF-HCl acid treatment or bypass damage by fracturing and propping.

For carbonate wells, dissolve carbonates near the wellbore with HCl or acetic acid, or bypass damage through fracture etching or hydraulic fracturing and propping.

Loss of Mud in Fractures—Mud should never be pumped into propped fractures. However, if this is done, it may be possible to partially remove mud by pumping, at very high rates, a large volume of viscous gel or water with a dispersant into the fracture to drive the mud and proppants far into the formation; then, the well can be refractured and propped if desired. If this approach is not successful, it may be necessary to plugback, sidetrack, drill a new hole, and then fracture the formation.

Workover Planning for Low Permeability Well

An effective artificial lift system is usually required for any low permeability oil well. This may defer or eliminate the need for workover of some wells. Limited increases in production may be obtained from some oil or gas wells by removing damage or enlarging pores in the formation near the wellbore with acid. Hydraulic fracturing to develop a linear flow system is often the most effective approach to appreciably increase production from very low permeability wells.

Hydraulic fracturing and propping is applicable in both sandstone and carbonate reservoirs. Conductive fractures, extending a great distance from a wellbore are usually desirable.

Fracture acidizing is an alternative to hydraulic fracturing and propping in carbonate reservoirs. To be successful, the fracture must be properly etched to provide a high conductivity linear flow system.

Workover Planning of Wells in Partially Pressure-Depleted Reservoirs

Prior to considering workover of wells in a partially pressure-depleted oil reservoir, an effective artificial lift system should be planned. Pressure maintenance or improved recovery is usually the best long range approach to increase production rate and recovery from partially pressure-depleted reservoirs.

If a workover is required, give first consideration to removing or bypassing any formation damage. Formation damage is very serious in pressure-depleted reservoirs because less pressure is available to remove damaging solids and fluids including emulsions from formation pores or fractures.

Hydraulic fracturing and propping applies to all types of oil and gas reservoirs. Fracture acidizing is an alternative to hydraulic fracturing and propping for stimulation of carbonate reservoirs.

Matrix acidizing may offer limited stimulation in either sandstone or carbonate reservoirs. Atomized acid may be beneficial in low-pressure reservoirs. Jet acidizing of the formation matrix is especially applicable to wells that cannot be circulated because of low reservoir pressure.

Workover to Reduce Water Production in Oil and Gas Wells

If production logs indicate flow behind the casing from a water zone, this communication can usually be stopped with a low-pressure low-fluid loss cement squeeze. The well can then be reperforated in the desired interval.

Fingering of Water in Stratified or Layered Reservoirs—The most reliable approach to locate water-producing zones in flowing oil or gas wells is to run through-tubing production logs to determine watercut and volume of fluid flow from each porous interval. For perforated casing completions, squeeze water-producing zones with a low-pressure low-fluid loss cement squeeze.

For open hole completions where the lowest zones are producing water, a plugback may be satisfactory. However, if the remaining oil-producing or gas-producing zones are lower in the well than the water-producing zones, it is usually necessary to cement a liner in open hole to shut off water and then recomplete in the desired oil-producing or gas-producing interval.

Water Coning—Water coning may occur in reservoirs that have appreciable vertical permeability—either matrix or fracture permeability.

The usual procedure to minimize water coning is to plugback and recomplete as high above the oil-water or gas-water contact as practicable. If there are at least partial barriers to vertical flow, plugback can be quite effective.

Alternatives to Workover—An initial approach to reduce coning is to shut in the well for 1 to 3 months to allow the water cone to recede. When the well is opened, it should be initially produced at a rate much lower than the previous rate. Then the producing rate can be increased on a step-wise basis over a period of days or weeks. This approach is frequently impractical because producing rates to prevent coning may be too low for profitable oil or gas production.

If the reservoir can be produced as a unit by a single operator, it is usually more profitable to produce only wells high on the structure to avoid water coning until most of the reservoir oil or gas has been recovered.

Control of water in high permeability reservoirs where all zones appear to be producing with a high water cut and squeeze cementing is ineffective—If water cannot be shut off satisfactorily by squeeze cementing, it may be practical to produce the well by high volume pumping or gas lift with all zones open to the wellbore. For this system to be applicable, permeability feet open to the wellbore must be sufficiently low for high volume pumps or gas lift to provide appreciable pressure drawdown at the wellbore.

A second approach is to squeeze all zones with one of the available polymer sealants. This tends to selectively enter and seal off the more permeable zones, which are probably producing the highest volume of water. Because this method may squeeze off appreciable oil or gas, it is usually not employed unless all other methods fail.

Workover to Reduce Gas Production in Oil Wells

In stratified or layered reservoirs, gas-producing zones can usually be squeezed off. The well can then be recompleted in zones with lower gas-oil ratios.

If gas flow is due to an unsatisfactory cement job, this may be effectively squeezed off.

Alternatives to Workover—If increased gas production is due to pressure depletion in a dissolved gas drive reservoir, workover is not required unless a secondary gas cap has been formed and is in communication with the wellbore.

Where moderately high gas-oil ratio is due to gas fingering in stratified reservoirs, it may be more practical to produce with high ratios and return the gas to wells higher on the structure. With this system, the reservoir is produced on the basis of "net gas-oil ratio," as defined here:

Net GOR =

$$\frac{\text{Total gas prod.} - \text{Gas returned to reservoir}}{\text{Oil produced}}$$

If a reservoir has a primary gas cap, one of the purposes of gas return is to prevent shrinkage of the gas cap and movement of oil into gas zones. If oil invades a dry gas zone, a portion of the invading oil will be unrecoverable.

If a reservoir has a large gas cap, it may be practical to initially produce only those wells located low on the structure in order to reduce gas production.

If gas production from an oil reservoir does not cause loss of oil recovery, the objective should then be to make optimum economic use of the excessive gas produced. If the gas cannot be economically used, it may be practical to store the gas in the same or another reservoir for future use. A good approach is to use some excess gas for gas lift where applicable.

If the problem is gas coning and there are no effective barriers to vertical permeability, the best approach is to recomplete the well as low on the structure as practicable. If this does not alleviate the problem, the usual procedure is to reduce producing rates; the well is shut in for several weeks or months to give the gas cone time to retract. Then the well may be produced at low rates, gradually increasing rates to determine the maximum producing rate without coning.

If the problem is gas cap expansion into the oil zone due to a reduction in reservoir pressure, water or other liquids injected lower on the structure may stop gas expansion and reduce the need for workovers to shut off gas. However, in some steeply dipping reservoirs, or reservoirs with a high relative permeability to water, it may be desirable to use the expanding gas cap as the primary recovery mechanism rather than injecting water down dip in the reservoir.

Workover for Sand Control

Gravel packing is the best approach to control sand in long-zone, single completions. It may also provide higher productivity in short zones.

Clean fluids are the key—perforations should be clean. Size gravel properly in relation to formation sand. Squeeze gravel into perforation tunnel. Water-wet gravel if oil is the placement fluid used.

Put well on production immediately after gravel packing, but start with a low rate and increase gradually to higher rate.

Sand consolidation with plastic is a clear choice only in perforated short zone multiple completions. Clean perforations, clean fluids, and uniform injection of plastic throughout zones are the keys.

If sands contain considerable clays either natural or induced, pre-treat sands using the HF-HCl acidizing technique described in the sandstone acidizing chapter. Spent acid should be removed from the well prior to performing the consolidation job.

Injection of plastic-coated sand using either a normal viscosity or a high viscosity placement fluid has application in medium length zones where sand has been produced or a cavity has been created.

Milling a window in the casing, underreaming open hole, and then performing an open-hole gravel pack job has been very successful where high rates of production are required.

Alternatives to Workover for Sand Control—Plugged perforations and formation damage will cause abnormally high flow velocities through a few perforations, and cause sand production. Thoroughly clean completions with no perforation plugging or formation damage may eliminate the need for any special sand control measures in many reservoirs.

When perforations are plugged, flow velocity and resulting sand problems can be reduced by (1) reperforating, (2) perforating additional section, (3) unplugging perforations through selective breakdown with ball sealers using oil or water, or (4) by cleaning perforations with a surge tool or perforation washer.

Workover to Repair Mechanical Failure

Mechanical failure includes (1) primary cement failures, (2) casing, tubing and packer leaks, (3) wellbore communication in multiple completions, and (4) other downhole failures.

Other Considerations—Prior to working over a well for mechanical repair, consideration should be given to alleviating at the same time any other problems existing in the well, especially all reservoir problems.

Consideration should always be given to making any desired changes in the mechanical arrangement of completion equipment to optimize future operation, workover and servicing. This includes locating the bottom of the tubing above all current or future production zones. Tubing should be open-ended with an inside bevel or with an inside beveled collar on bottom to facilitate running through-tubing tools out of the bottom of the tubing.

Other factors should be considered relating to future through-tubing workovers such as making certain that tubing is not bowed or corkscrewed by excessive weight on the packer. Landing nipples and "no-go" nipples in the tubing string should have the largest possible inside diameter to allow the largest possible perforating guns, flowmeters, and other tools to be run through these nipples.

Workover to Change Zones or Reservoirs

The usual workover procedure for changing zones or reservoirs is to squeeze cement and reperforate in new zone or reservoir in perforated casing completions. In open hole, it may be necessary to deepen or plug back, or to cement a liner and selectively perforate a specific zone or reservoir.

Because of the many types of well problems, all possible reservoir or well problems should be considered prior to reaching the decision to change zones or reservoirs.

Workover to Multicomplete or Change to Single Completion

Multiple completions in different oil or gas reservoirs or zones are relatively simple operations if perforation and recompletion can be carried out through-tubing. If it is necessary to perforate in mud or dirty water with conventional perforators in the process of making multiple completions many perforations will be totally plugged. Plugged perforations cause many problems, including sand problems, loss of production, and premature well abandonment.

For oil wells that will ultimately require gas lift, gas lift mandrels should be installed in the tubing so that gas lift valves can be run when needed with a minimum of cost.

Alternatives to Multiple Completions—If reservoirs can be operated as a unit by one operator, it may be more practical to operate all wells as high-rate single completions. When individual zones are depleted in one well, completion can then be made in other zones or reservoirs as required, frequently on a through-tubing basis.

If a number of reservoirs penetrated by a given well have the same type of drive, similar oil, and are near the same depth, it may be more profitable to commingle production from different reservoirs in some or all wells.

Increasing Production in High-Viscosity Wells

Thermal stimulation is usually applicable in high viscosity oil reservoirs.

If a well is plugged near the wellbore with asphaltenes, paraffin, or heavy hydrocarbons, steam stimulation will remove these plugging agents.

If it is neither convenient nor desirable to remove waxes with steam, a solvent soak treatment should be employed. The first step is to select the best and most economical solvent by chemical tests in the laboratory of various available solvents. A surfactant is normally used with the solvent.

Inject solvent and surfactant into formation at a slow rate and let it soak for 24 to 72 hours. Swab, pump, or flow back the dissolved wax, silt, and other debris from the well.

Fracturing and propping with high conductivity proppants will increase well productivity in limestone, dolomite or sandstone wells. This applies to relatively high permeability as well as low permeability reservoirs.

Wells in limestone and dolomite reservoirs may be fracture-acidized to provide linear flow channels to the wellbore.

If high viscosity is due to an emulsion, a surfactant injected uniformly into all zones may be helpful in breaking the emulsion.

Alternatives to Workover—Production may be increased by decreasing viscosity of heavy oil in the tubing and flowline through the use of heat. This may be accomplished by installation of a bottomhole heater or by circulating hot oil or hot water down the casing-tubing annulus and up a separate return tube.

If a high viscosity water-in-oil emulsion is produced, viscosity can be reduced and production increased by continuous breaking or inverting the emulsion by injecting a surfactant continuously down hole. When produced fluid is less than about 10% water, sufficient water should be added to the surfactant to bring the water percentage to approximately 10% by volume of produced fluid.

Continuous circulation of light hydrocarbons down the casing-tubing annulus and mixing with the heavy oil is a practical but somewhat costly approach to increasing production from wells producing very high viscosity oil.

Workover Economics

Consider Workovers on Program Basis—Because of the somewhat experimental nature of individual workovers, it is preferable to plan workover expenditures and profits for all but the simplest type of workover as part of a reservoir, field, or area program rather than on an individual well basis.

Due to the large number of probable causes of well problems and an even larger number of possible solutions, conventional investment-type economic analysis does not always apply to workovers on a single-well basis. When workovers can be planned on a multi-well program basis, workover economics are more comparable to investment in low-risk potentially high-profit exploratory drilling programs.

Properly planned and executed workover programs in the U.S. should pay out in 9 to 12 months or less, with a 70% or more success rate for workovers. In areas outside the U.S., payout is often noted within days or weeks.

Success Directly Related to Preplanning—Because of the large number of unknowns, initial workovers in an area program usually should be carried out as a carefully controlled experiment or on a "Researched Engineering" basis.

Due to a very high rate of return from properly planned and executed workover programs, many oil companies could appreciably increase their cash flow, profits, and recoverable reserves by increasing skilled engineering and geologic study time on workover planning and execution.

Expenditures for workover studies should be increased until profits and cash flow begin to decrease as a result of excessive technical study costs on workovers. Ten percent or more of total workover costs can probably be justified for prework study, planning, and evaluation of workovers.

As a rule, these studies should be carried out by one or more engineers and geologists assigned full-time to the study of workover projects in one or more areas. It is preferable to have completed reservoir studies as a background for any well workover study.

Economic Considerations—Cash flow rate of return, workover payout, and various other economic yardsticks are used in calculating profitability of workover programs.

It should be ascertained whether workover or well stimulation will actually increase ultimate profits and recoverable reserves or whether the project will merely accelerate income.

Some oil companies in the U.S. calculate payout of workovers on a net profit per barrel or MCF basis after considering royalties, taxes, and lifting costs per barrel. However, in cases where lease lifting cost less workover cost will increase very little, it is often more realistic to consider payout of workovers after deducting royalty and taxes only, rather than using the lease lifting cost per barrel of oil or MCF of gas produced.

For oil wells, sale of produced gas is added to the value of oil produced to determine gross income. For operations outside the U.S., the different profit incentives in each area must be considered in justifying workover operations.

Risk Analysis on Payout of Individual Workovers—Although consideration of workovers should be made on a program basis where possible, the formulation of a program depends on the analysis of individual wells. Some combination of risk of failure and potential gain from a successful workover should be considered for each well in a program. This will tend to encourage relatively high-risk workovers where potential profit is high and discourage low-risk, low-potential profit workovers.

A comparative analysis of various programs is presented in Table 12-1 to illustrate the value of analysis on a program basis. The selection of wells for each program is based on estimated net income on a risk basis. A number of companies have developed computer programs for economic planning of workovers. These programs incorporate risk and the various economic yardsticks being employed by an individual company. A cash flow analysis on a program basis may be required.

Low-risk, high-potential profit workover justification may be made on an individual basis rather than a program basis. However, as fields become older, there may be an accumulation of relatively high-risk workovers where management has been reluctant to workover these wells. The following approach is suggested where individual wells cannot be justified on a single workover basis.

For the recommended four-well program in Table 12-1, it is assumed that some of the workovers may fail to pay out and that estimated income may be too high or too low. However, as the number of wells in a program increases and as experience is gained with this type of approach, errors tend to balance out on an overall program basis.

In determining economic success or failure of high-risk workover programs, only the overall program

TABLE 12-1
Selection of Wells for a Workover Program

Well no.	Estimated cost	Estimated net income to workover cost ratio	Estimated risk or success factor	Estimated net income on a risk basis
Recommended 4-well program:				
1	$ 30,000	4	0.5	$ 60,000
2	21,000	10	0.40	84,000
3	60,000	2.5	0.75	112,500
4	30,000	1.5	0.9	40,500
Program cost	$141,000		Estimated total net income of program	$297,000
			Estimated net profit of program after payout	$156,000
Proposals rejected for program:				
5	$ 30,000	1.5	0.80	$ 36,000
6	60,000	4	.25	60,000
7	15,000	1.1	1.00 (no risk)	16,500
Total	$105,000		Estimated total net income of program	$112,500
			Estimated net profit after payout	$ 7,500

should be considered. For example, on a ten-well program of high-risk workovers, 60% may fail to pay out. However, if the overall programs pay out with a good profit, the program is well justified, assuming workovers in the program could not be justified on a single well basis using economic yardsticks of the particular oil company.

Summary

To simplify the discussion of workover planning, each type of well problem has been considered separately. In actual practice the final workover plan should represent an integrated solution to all or a majority of indicated problems. The aim of the workover plan should be to maximize current profits, and/or future net income and recovery of oil and gas as opposed to minimizing cost of workover.

REFERENCES

1. Weaver, J. D.: "A New Water-Oil Ratio Improvement Material." SPE 7574, Oct. 1978.
2. Weeks, S. G.,: "Through-Tubing Plug-Back Without Depth, Temperature, or Pressure Limitations," SPE 12104, Oct., 1983.

Symbols and Abbreviations, Vol. 2

Fracturing

σ = stress, psi
σ_v = total vertical stress, psi
σ_h = total horizontal stress, psi
$\tilde{\sigma}_v$ = effective vertical rock matrix stress, psi
$\tilde{\sigma}_h$ = effective horizontal rock matrix stress, psi
ρ = rock density, lb/ft^3
D = depth, ft
p_r = formation pore pressure, psi
ν = Poisson's ratio
$(P_i)_h$ = borehole pressure to initiate horizontal fracture, psi
$(P_i)_v$ = borehole pressure to initiate vertical fracture, psi
S_v = vertical tensile strength of rock, psi
S_h = horizontal tensile strength of rock, psi
k = permeability, md
k_f = proppant permeability, md
L = fracture length from wellbore, ft
h = fracture height, ft
w = fracture width, in
r_e = drainage radius, ft
r_w = wellbore radius, ft
A = well spacing, acres
r_c = radius of closed fracture near wellbore, ft
n' = flow behavior index, log slope of shear rate-shear stress plot, dimensionless
K' = consistancy index, lbf · sec^n/ft^2
μ_a = apparent viscosity, cp
V = flow velocity, ft/sec
D = conduit diameter, in
Q = flow rate, bbl/min
E = Young's modulus, psi
ε = strain, in/in
C_o = ultimate compressive strength of rock, psi
G = shear modulus, psi/radian
F = force, psi
A = cross sectional area, in^2
θ = angle of deformation, radian
K = bulk modulus, psi
v = rock volume, in^3
V_p = compressional wave velocity
V_s = shear wave velocity
Δt_s = shear wave travel time, microsec/ft
Δt_c = compressional travel time, microsec/ft

English/Metric Units
Standards for Metric Conversion Factors*

The following conversion factors are those published by the American Society for Testing and Materials (ASTM) in E380-76. These same units may be found in literature published by all U.S. Technical Societies, i.e., API Bulletin 2563, American National Standards Institute ANSIZ 210.1, Society of Petroleum Engineers, The Canadian Petroleum Association (CPA) and others.

The metric units and conversion factors adopted by the ASTM are based on the "International System of Units" (designated SI for Systeme International d'Unites), fixed by the International Committee for Weights and Measures. This system has been adopted by the International Organization for Standardization in ISO Recommendation R-31.

Conversion factors herein are written as a number equal to or greater than one and less than ten with six or less decimal places. This number is followed by the letter E (for exponent), a plus or minus symbol, and two digits which indicate the power of 10 by which the number must be multiplied to obtain the correct value. For example:

(1) 3.523 907 E−02 is $3.523\,907 \times 10^{-2}$
or
0.035 239 07

(2) 3.386 389 E+03 is $3.386\,389 \times 10^{3}$
or
3 386.389

(3) Further examples of conversion are:

To convert from:	To	Multiply by:		
pound-force per square foot	Pa	4.788 026 E+01	means-	1 lbf/ft^2 = 47.880 26 Pa
inch	m	2.540 000 E−02		1 inch = 0.0254 m (exactly)

To convert from	To	Multiply by
	ANGLE	
degree (angle)	radian (rad)	1.745 329 E−02
minute (angle)	radian (rad)	2.908 882 E−04
second (angle)	radian (rad)	4.848 137 E−06
	AREA	
acre (U.S. survey)	meter^2 (m^2)	4.046 873 E+03
ft^2	meter^2 (m^2)	9.290 304 E−02
hectar	meter^2 (m^2)	1.000 000 E+04
in^2	meter^2 (m^2)	6.451 600 E−04
mi^2 (U.S. survey)	meter^2 (m^2)	2.589 988 E+06
yd^2	meter^2 (m^2)	8.361 274 E−01

CAPACITY
(See Volume)

DENSITY
(See Mass Per Unit Volume)

*The following conversion factors have been selected from Halliburton Services Technical Data Handbook Section 240

To convert from	To	Multiply by
ELECTRICITY AND MAGNETISM		
abampere	ampere (A)	1.000 000 E+01
abohm	ohm (Ω)	1.000 000 E−09
abvolt	volt (V)	1.000 000 E−08
ampere hour	coulomb (C)	3.600 000 E+03
ohm centimeter	ohm meter (Ωm)	1.000 000 E−02
statampere	ampere (A)	3.335 640 E−10
statohm	ohm (Ω)	8.987 554 E+11
statvolt	volt (V)	2.997 925 E+02
ENERGY (includes Work)		
British thermal unit (International Table)	joule (J)	1.055 056 E+03
British thermal unit (mean)	joule (J)	1.055 87 E+03
British thermal unit (thermochemical)	joule (J)	1.054 350 E+03
British thermal unit (39°F)	joule (J)	1.059 67 E+03
British thermal unit (59°F)	joule (J)	1.054 80 E+03
British thermal unit (60°F)	joule (J)	1.054 68 E+03
calorie (International Table)	joule (J)	4.186 800 E+00
calorie (mean)	joule (J)	4.190 02 E+00
calorie (thermochemical)	joule (J)	4.184 000 E+00
calorie (15°C)	joule (J)	4.185 80 E+00
calorie (20°C)	joule (J)	4.181 90 E+00
calorie (kilogram, International Table)	joule (J)	4.186 800 E+03
calorie (kilogram, mean)	joule (J)	4.190 02 E+03
calorie (kilogram, thermochemical)	joule (J)	4.184 000 E+03
erg	joule (J)	1.000 000 E−07
ft·lbf	joule (J)	1.355 818 E+00
ft·poundal	joule (J)	4.214 011 E−02
kilocalorie (International Table)	joule (J)	4.186 800 E+03
kilocalorie (mean)	joule (J)	4.190 02 E+03
kilocalorie (thermochemical)	joule (J)	4.184 000 E+03
kW·h	joule (J)	3.600 000 E+06
therm	joule (J)	1.055 056 E+08
ENERGY PER UNIT AREA TIME		
Btu (thermochemical)/ft^2·s	watt per meter^2 (W/m^2)	1.134 893 E+04
Btu (thermochemical)/ft^2·min	watt per meter^2 (W/m^2)	1.891 489 E+02
Btu (thermochemical)/ft^2·h	watt per meter^2 (W/m^2)	3.152 481 E+00
Btu (thermochemical)/in^2·s	watt per meter^2 (W/m^2)	1.634 246 E+06
cal (thermochemical)/cm^2·min	watt per meter^2 (W/m^2)	6.973 333 E+02
FLOW (See Mass Per Unit Time or Volume Per Unit Time)		
FORCE		
dyne	newton (N)	1.000 000 E−05
kilogram-force	newton (N)	9.806 650 E+00
ounce-force	newton (N)	2.780 139 E−01
pound-force (lbf)	newton (N)	4.488 222 E+00
poundal	newton (N)	1.382 550 E−01

To convert from	To	Multiply by

FORCE PER UNIT AREA
(See Pressure)

HEAT

Btu (International Table)·ft/h·ft^2·°F (k, thermal conductivity)	watt per meter kelvin (W/m·K)	1.730 735 E+00
Btu (International Table)/ft^2	joule per meter^2 (J/m^2)	1.135 653 E+04
cal (thermochemical)/cm·s·°C	watt per meter kelvin (W/m·K)	4.184 000 E+02
cal (thermochemical)/cm^2	joule per meter^2 (J/m^2)	4.184 000 E+04

LENGTH

angstrom	meter (m)	1.000 000 E−10
foot	meter (m)	3.048 000 E−01
foot (U.S. survey)	meter (m)	3.048 006 E−01
inch	meter (m)	2.540 000 E−02
micron	meter (m)	1.000 000 E−06
mil	meter (m)	2.540 000 E−05
mile (international nautical)	meter (m)	1.852 000 E+03
mile (U.K. nautical)	meter (m)	1.853 184 E+03
mile (U.S. nautical)	meter (m)	1.852 000 E+03
mile (international)	meter (m)	1.609 344 E+03
mile (statute)	meter (m)	1.609 3 E+03
mile (U.S. survey)	meter (m)	1.609 347 E+03
parsec	meter (m)	3.085 678 E+16
yard	meter (m)	9.144 000 E+01

MASS

grain	kilogram (kg)	6.479 891 E−05
gram	kilogram (kg)	1.000 000 E−03
hundred weight (long)	kilogram (kg)	5.080 235 E+01
hundredweight (short)	kilogram (kg)	4.535 924 E+01
ounce (avoirdupois)	kilogram (kg)	2.834 952 E−02
ounce (troy or apothecary)	kilogram (kg)	3.110 348 E−02
pennyweight	kilogram (kg)	1.555 174 E−03
pound (lb avoirdupois)	kilogram (kg)	4.535 924 E−01
pound (troy or apothecary)	kilogram (kg)	3.732 417 E−01
slug	kilogram (kg)	1.459 390 E+01
ton (assay)	kilogram (kg)	2.916 667 E−02
ton (long, 2240 lb)	kilogram (kg)	1.016 047 E+03
ton (metric)	kilogram (kg)	1.000 000 E+03
ton (short, 2000 lb)	kilogram (kg)	9.071 847 E+02

MASS PER UNIT AREA

oz/ft^2	kilogram per meter^2 (kg/m^2)	3.051 517 E−01
lb/ft^2	kilogram per meter^2 (kg/m^2)	4.882 428 E+00

MASS PER UNIT CAPACITY
(See Mass Per Unit Volume)

MASS PER UNIT TIME
(Includes Flow)

lb/h	kilogram per second (kg/s)	1.259 979 E−04

To convert from	To	Multiply by
lb/min	kilogram per second (kg/s)	7.559 873 E−03
lb/s	kilogram per second (kg/s)	4.535 924 E−01

MASS PER UNIT VOLUME
(Includes Density and Mass Capacity)

To convert from	To	Multiply by
grain (lb avoirdupois/7000)/gal (U.S. liquid)	kilogram per meter³ (kg/m³)	1.711 806 E−02
g/cm³	kilogram per meter³ (kg/m³)	1.000 000 E+03
oz (avoirdupois)/gal (U.K. liquid)	kilogram per meter³ (kg/m³)	6.236 021 E+00
oz (avoirdupois)/gal (U.S. liquid)	kilogram per meter³ (kg/m³)	7.489 152 E+00
oz (avoirdupois)/in³	kilogram per meter³ (kg/m³)	1.729 994 E+03
lb/ft³	kilogram per meter³ (kg/m³)	1.601 846 E+01
lb/in³	kilogram per meter³ (kg/m³)	2.767 990 E+04
lb/gal (U.K. liquid)	kilogram per meter³ (kg/m³)	9.977 633 E+01
lb/gal (U.S. liquid)	kilogram per meter³ (kg/m³)	1.198 264 E+02
lb/yd³	kilogram per meter³ (kg/m³)	5.932 764 E−01
slug/ft³	kilogram per meter³ (kg/m³)	5.153 788 E+02

PERMEABILITY

To convert from	To	Multiply by
darcy	μm^2	9.869 233 E−01
millidarcy	μm^2	9.869 233 E−04

POWER

To convert from	To	Multiply by
Btu (International Table)/h	watt (W)	2.930 711 E−01
Btu (International Table)/s	watt (W)	1.055 056 E+03
Btu (thermochemical)/h	watt (W)	2.928 751 E−01
Btu (thermochemical)/min	watt (W)	1.757 250 E+01
Btu (thermochemical)/s	watt (W)	1.054 350 E+03
cal (thermochemical)/min	watt (W)	6.973 333 E−02
cal (thermochemical)/s	watt (W)	4.184 000 E+00
erg/s	watt (W)	1.000 000 E−07
ft·lbf/h	watt (W)	3.766 161 E−04
ft·lbf/min	watt (W)	2.259 697 E−02
ft·lbf/s	watt (W)	1.355 818 E+00
horsepower (550 ft·lbf/s)	watt (W)	7.456 999 E+02
horsepower (boiler)	watt (W)	9.809 50 E+03
horsepower (electric)	watt (W)	7.460 000 E+02
horsepower (metric)	watt (W)	7.354 99 E+02
horsepower (water)	watt (W)	7.460 43 E+02
horsepower (U.K.)	watt (W)	7.457 0 E+02
kilocalorie (thermochemical)/min	watt (W)	6.973 333 E+01
kilocalorie (thermochemical)/s	watt (W)	4.184 000 E+03

PRESSURE OR STRESS
(Force Per Unit Area)

To convert from	To	Multiply by
atmosphere (standard)	pascal (Pa)	1.013 250 E+05
atmosphere (technical = 1 kgf/cm²)	pascal (Pa)	9.806 650 E+04
bar	pascal (Pa)	1.000 000 E+05
centimeter of mercury (0°C)	pascal (Pa)	1.333 22 E+03
centimeter of water (4°C)	pascal (Pa)	9.806 38 E+01
dyne/cm²	pascal (Pa)	1.000 000 E−01
foot of water (39.2°F)	pascal (Pa)	2.988 98 E+03

To convert from	To	Multiply by
gram-force/cm²	pascal (Pa)	9.806 650 E+01
inch of mercury (32°F)	pascal (Pa)	3.386 38 E+03
inch of mercury (60°F)	pascal (Pa)	3.376 85 E+03
inch of water (39.2°F)	pascal (Pa)	2.490 82 E+02
inch of water (60°F)	pascal (Pa)	2.488 4 E+02
millibar	pascal (Pa)	1.000 000 E+02
millimeter of mercury (0°C)	pascal (Pa)	1.333 22 E+02
poundal/ft²	pascal (Pa)	1.488 164 E+00
lbf/ft²	pascal (Pa)	4.788 026 E+01
lbf/in² (psi)	pascal (Pa)	6.894 757 E+03
psi	pascal (Pa)	6.894 757 E+03

STRESS
(See Pressure)

TEMPERATURE

To convert from	To	Multiply by
degree Celsius	kelvin (K)	$t_{°K} = t_{°C} + 273.15$
degree Fahrenheit	degree Celsius	$t_{°C} = (t_{°F} - 32)/1.8$
degree Fahrenheit	kelvin (K)	$t_{°K} = (t_{°F} + 459.67)/1.8$
degree Rankine	kelvin (K)	$t_{°K} = t_{°R}/1.8$
kelvin	degree Celsius	$t_{°C} = t_{°K} - 273.15$

VISCOSITY

To convert from	To	Multiply by
Centipoise	pascal second (Pa·s)	1.000 000 E−03
centistokes	meter² per second (m²/s)	1.000 000 E−06
ft²/s	meter² per second (m²/s)	9.290 304 E−02
poise	pascal second (Pa·s)	1.000 000 E−01
poundal·s/ft²	pascal second (Pa·s)	1.488 164 E+00
stokes	meter² per second (m²/s)	1.000 000 E−04

VOLUME
(Includes Capacity)

To convert from	To	Multiply by
acre-foot (U.S. survey)	meter³ (m³)	1.233 489 E+03
barrel (oil, 42 gal)	meter³ (m³)	1.589 873 E−01
fluid ounce (U.S.)	meter³ (m³)	2.957 353 E−05
ft³	meter³ (m³)	2.831 685 E−02
gallon (Canadian liquid)	meter³ (m³)	4.546 090 E−03
gallon (U.K. liquid)	meter³ (m³)	4.546 092 E−03
gallon (U.S. dry)	meter³ (m³)	4.404 884 E−03
gallon (U.S. liquid)	meter³ (m³)	3.785 412 E−03
in³	meter³ (m³)	1.638 706 E−05
liter	meter³ (m³)	1.000 000 E−03
ounce (U.K. fluid)	meter³ (m³)	2.841 307 E−05
ounce (U.S. fluid)	meter³ (m³)	2.957 353 E−05
pint (U.S. dry)	meter³ (m³)	5.506 105 E−04
pint (U.S. liquid)	meter³ (m³)	4.731 765 E−04
quart (U.S. dry)	meter³ (m³)	1.101 221 E−03
quart (U.S. liquid)	meter³ (m³)	9.463 529 E−04
ton (register)	meter³ (m³)	2.831 685 E+00
yd³	meter³ (m³)	7.645 549 E−01

To convert from	To	Multiply by
VOLUME PER UNIT TIME (Includes Flow)		
ft^3/min	meter^3 per second (m^3/s)	4.719 474 E−04
gallon (U.S. liquid)/hp·h (SFC, specific fuel consumption)	meter^3 per joule (m^3/J)	1.410 089 E−09
in^3/min	meter^3 per second (m^3/s)	2.731 177 E−07
yd^3/min	meter^3 per second (m^3/s)	1.274 258 E−02
gallon (U.S. liquid) per day	meter^3 per second (m^3/s)	4.381 264 E−08
gallon (U.S. liquid) per minute	meter^3 per second (m^3/s)	6.309 020 E−05

WORK
(See Energy)